Nature Classics

A CATALOGUE OF THE E. A. McILHENNY
NATURAL HISTORY COLLECTION
AT LOUISIANA STATE UNIVERSITY

Nature Classics

A CATALOGUE OF THE E.A. McILHENNY NATURAL HISTORY COLLECTION AT LOUISIANA STATE UNIVERSITY

COMPILED AND EDITED BY
ANNA H. PERRAULT

WITH A PREFACE BY KATHRYN MORGAN

AND AN INTRODUCTORY ESSAY
ON NATURAL HISTORY ILLUSTRATION
BY DAVID M. LANK, F.R.S.A.

EDITORIAL ASSISTANT
SANDRA M. McGUIRE

PUBLISHED BY THE FRIENDS OF THE LSU LIBRARY
BATON ROUGE, LOUISIANA
1987

Manufactured in the United States of America

DESIGNER: Joanna V. Hill
TYPEFACE: Palatino
TYPESETTER: Printing Office, LSU
PRINTER: Thomson-Shore

LIBRARY OF CONGRESS CATALOGING IN PUBLICATION DATA
Louisiana State University, Baton Rouge.
Nature classics.

Bibliography: p.
Includes index.
1. Natural history—Bibliography—Catalogs.
2. Louisiana State University, Baton Rouge—Catalogs.
3. McIlhenny, Edward Avery, 1872-1949—Library—Catalogs.
4. Natural history illustration—History. I. Perrault, Anna H., 1944- . II. Title.
Z7409.L88 1987 [QH45] 016.508 86-15399
ISBN 0-8071-1363-8

Contents

Illustrations

Editor's Foreword

The idea that the Friends of the LSU Library publish a catalogue of the E.A. McIlhenny Collection originated with the members of the Friends of the Library Development Committee: LSU Chancellor Emeritus Cecil Taylor, Dean Emeritus of the Graduate School Max Goodrich, and Friends' Secretary Mrs. Mary Jane Kahao. To implement the idea an editorial committee was formed consisting of these three people plus the Louisiana-Rare Book Room Librarian, Miss Evangeline Lynch, and Humanities Bibliographer Anna Perrault.

The editorial committee's original concept was to use an existing computerized insurance list as the basis for a bibliography which could then be computer printed. After information in the insurance list proved too scanty and inaccurate, this idea was abandoned, in spite of the fact that much time had been spent trying to correct and expand the data. It was also impossible at that time to get an end product with machine readable data that wasn't an obvious computer printout.

At this point the committee decided to begin anew and Anna Perrault was designated as editor of the publication. Shortly thereafter, when Kathryn Morgan became the Librarian for the McIlhenny Collection she was added to the editorial committee.

This catalogue was first prepared by typing the entries using an IBM typewriter with an optical character reader ball. Then the typed pages were scanned into a Lanier word processor. During these early stages of preparation the capacity to transfer data from the word processor to the LSU mainframe computer was acquired. This catalogue was one of the first large data sets transferred to the mainframe. From that point on, all of the editing, coding, sorting, and indexing was done through the LSU Division of Research using the LSU mainframe computer. During the years since the editorial committee had first envisioned a machine readable data-base which would be printed via computer, many services had become available at LSU which were not in place at the beginning of the project. As the catalogue was nearing the end of the indexing phase, LSU Graphic Services acquired the capacity to typeset from magnetic tape or disc. The codes for typesetting instructions were inserted, some manually, some by programming using the already existing codes. The catalogue and index went straight from the mainframe and into type somewhat as the committee had originally intended.

As is the case with any publication, there are numerous people who deserve thanks. Former LSU Library director George J. Guidry, Jr., and current LSU Director of Libraries Sharon A. Hogan have provided time, understanding, and staff help in the compilation of the catalogue and the editorial assistance necessary to complete the project. The former Dean of Arts and Sciences and former Friends' President, Henry A. Snyder, provided access to the services of the Arts and Sciences Division of Research. When the Division of Research became a campus-wide service, its Directors continued to support the project. Most deserving of extra-special thanks is Randy Hebert, computer analyst in the Division of Research who wrote the programs that sorted, alphabetized, and indexed the catalogue. Randy also assisted us and LSU Graphic Services in coding the catalogue and index for computer typesetting.

One other person deserving a special commendation is Sandra McGuire who as editorial assistant performed the major tasks of inserting all the coding necessary for the sorting, indexing, and print-command programs to function.

The editorial committee was a source of support and advice, especially Kathryn Morgan. She enthusiastically cooperated, making the collection and her expertise available to me and serving as a sounding board throughout the entire project.

Special thanks also go to David Lank who graciously lent his knowledge of animal art and history of natural history illustration to write an introduction for the catalogue.

Others to be acknowledged are Robert S. Martin, who supported us after he became Assistant Director of Libraries for Special Collections; Gail Chance for her botanical knowledge; Don Morrison for photographing the illustrations; Billy Coble, Joe Engler and others at LSU Graphic Services; Joanna Hill for her classic design and publication advice; the Director and staff of the LSU Press; and numerous catalogers and staff of the LSU Library. My family—spouse Joe, and children Jean-Paul and André—gave moral support and suffered along with me. All are collectively thanked!

The Board and officers of the Friends of the LSU Library are thanked for their moral and ultimately financial support in the publication of the catalogue.

When the idea of publishing a catalogue of the E.A. McIllhenny Collection was first conceived, the title *Nature Classics* quickly came to mind. From the classical, fanciful beasts to the now 'classic' Snowy Egret saved by modern conservation efforts, the McIlhenny Collection reflects the development of the natural sciences. It contains editions of significant early scientific works—Aldrovandi, Dioscorides,

L'Écluse, Linnaeus, Ray. Major scientific voyages and expeditions are represented—Cook, Beebe, Sloane, Wilkes, the *Beagle*, the *Challenger*. Scientific classics of the modern era—Darwin, Elliot, Lea, Reichenbach—are included. The monumental achievements in natural history illustration are also present in works by Audubon, Catesby, Gould, and Redouté. There are numerous recent acquisitions of modern limited editions, such as the *Banks' Florilegium*, which are destined to become classics. The title *Nature Classics* seemed to just 'naturally' fit this catalogue. Our hope is that *Nature Classics* will itself become a classic.

ANNA H. PERRAULT

EDWARD AVERY MC ILHENNY (1872-1949)

Preface

The E.A. McIlhenny Natural History Collection first opened its doors in 1971 at Louisiana State University's Troy H. Middleton Library. It was then a small collection of rare natural history books, particularly emphasizing ornithology and botany, including many finely illustrated volumes, as well as the more strictly scientific works. The collection was formed as the result of a gift from certain descendents of the late E.A. McIlhenny of Avery Island, Louisiana. Initially it consisted of the natural history portion of E.A. McIlhenny's library as well as those materials which had been added by family members after his death in 1949.

The McIlhennys—long a prominent Louisiana family—and their forebears, the Marsh and Avery families, have owned and inhabited Avery Island since 1813, with the exception of a brief period between 1863 and 1865, during which Union forces occupied their property. Avery Island, a salt-dome of approximately 2,000 acres, protruding from the marshlands near New Iberia, Louisiana, has been, since 1868, the site of the company that concocts the piquant-tasting hot pepper sauce known around the world as Tabasco. Other island resources include several producing oil wells and one of the largest operating salt mines in the United States, from which is extracted 99.98% pure salt.

Another facet of Avery Island, however, reflects the interests of one of the more colorful and notable family members: Edward Avery McIlhenny (1872-1949), for whom the E.A. McIlhenny Natural History Collection is named. It was he who developed the famed Avery Island bird sanctuary, "Bird City," as well as the exotic "Jungle Gardens," whose beauties obscure the landscape of the island's industries. Although both the gardens and the bird refuge are well-known tourist attractions, they are also excellent examples of McIlhenny's early concern for the conservation and preservation of the earth's natural environment and resources.

Although his duties as President of the family business required much of his attention, E.A. McIlhenny pursued, in a variety of ways, his great interest in nature and its proper management by man. The greatest energies of his life were, in fact, directed to promoting wildlife management; his many contributions range from early Arctic exploration—first with Captain Frederick A. Cook, and again in 1897, when he led his own expedition to Point Barrow, Alaska—to the tamer pursuit of writing books and articles on wildlife and botanical subjects for both scientific journals and popular magazines.

Aside from a deep-rooted personal love of nature, fostered by McIlhenny's childhood amidst the vast marshes surrounding Avery Island, he was almost certainly influenced in his concern for conservation by his brother John's friend, Theodore Roosevelt. In 1898, Roosevelt and John Avery McIlhenny were soldiers in the First Volunteer Cavalry of the United States Army, later known as the "Rough Riders." Having fought together in the battle of San Juan Hill, the two men remained friends, and of their association there exists at least one well-documented account of a bear hunt in Louisiana.

On a national level, Roosevelt, as President, was one of the few men of his era to plan for the protection and preservation of America's natural resources and wildlife. During his administration, state agencies for conservation were conceived and improved, and in time, became politically powerful units interacting with other government forces on all levels.

In Louisiana, McIlhenny's similar interests led him to be the key figure in acquiring lands around the state to be used for wildlife refuges. He was very adept—and successful—at obtaining donations for this purpose, with the result that this significant undertaking marked the first time in the world wherein private lands were donated for public use as wildlife refuges. As a result, 164,664 acres of Louisiana marshland were dedicated as refuges, providing a wintering ground for millions of migratory birds which use the Mississippi flyway for their annual flight.

In his own backyard, on Avery Island, McIlhenny gave his attention to solving a growing wildlife problem—the decimation and threatened extinction of the lovely, white-plumed Snowy Egret. During the 1890's, the birds' showy plumage was much in demand to adorn ladies' hats, and the birds were wantonly killed by the thousands.

To provide a safe haven for the egrets, McIlhenny set up the sanctuary now known as "Bird City." Initially there were only a few egret families which he placed in nests there, but gradually the numbers increased, indicating the success of his relocation project. Now the birds that return annually to nest at Bird City number in the tens of thousands. As a result of McIlhenny's work, conservation agencies have long credited him with being singularly responsible for saving the Snowy Egret from extinction.

In the years following McIlhenny's death in 1949, the family businesses continued to flourish under the guidance of remaining members of the family. Of greater significance to the Collection, however, is the enthusiastic interest in the natural sciences which has been handed down to the surviving generations of McIlhennys. Since the early 19th century Avery Island has welcomed natural scientists of all varieties and interests: geologists, ornithologists, botanists, biologists, archaeologists and paleontologists; through their research there have been major contributions to the academic community concerning this unique and fascinating environment. Between 1949-1971, enduring family interest in natural history resulted in the purchase of many new titles which were added to the body of works that would one day become the E.A. McIlhenny Natural History Collection.

In 1971, when the collection was given to the LSU library, its continued support was guaranteed by a generous, but anonymous donor. At that time, the collection was significantly augmented by the addition of important titles gleaned from Middleton Library's Rare Book Collection and Science collections. Among these were the four great elephant folio volumes of Audubon's *Birds of America*, which had been purchased for the Library in 1963 with a grant from Crown Zellerbach. Over the years, gifts to the collection have been gratefully accepted from distinguished faculty members and other generous philanthropists as well.

With the passage of time, the McIlhenny Collection has grown to a total of over 4,000 volumes, representing five hundred years of printing history. Eighteenth and nineteenth century works predominate; however the collection also strives to acquire the best of modern natural history publications as well. Among the recent modern acquisitions are such works as: *Banks' Florilegium*, published by Alecto Historical Editions in association with the British Museum (Natural History); *Parrots and Cockatoos* and *Amazon Parrots* both by Rosemary Low, with illustrations by Elizabeth Butterworth; and most recently, *Six plates from John James Audubon's Birds of America*, published by Alecto Historical Editions in association with the American Museum of Natural History.

Not included in this catalogue are the numerous works of art that are a part of the collection. Among these are many fine porcelain sculptures by such well-known firms as Boehm, Royal Worcester and Limoge. In addition, there are many original oils and watercolors by notable natural history artists, both early and contemporary. In speaking of modern works, particular mention should be made concerning a unique ten-year project commissioned by LSU of the noted botanical artist, Margaret Stones. In 1976, Miss Stones began the first of what will ultimately total 200 watercolor drawings of Louisiana flora. These works, too, are under the curatorship of the McIlhenny Collection.

In its fifteen years at LSU, the McIlhenny Collection has expanded greatly in size and has undergone a series of significant changes, the most recent of which was a move from Middleton Library to Hill

Memorial Library—a building renovated specifically to house the University's special collections. In spite of consistent growth and change, the purpose of the collection clearly remains the same—to provide and conserve an ongoing resource of distinctive materials for international researchers and scholars. It is hoped that *Nature Classics* will help fulfill this goal by facilitating and encouraging the use of our materials.

KATHRYN MORGAN

Purple-breasted Carib from John Gould's *Humming birds*.

Richea Curtisiae from Margaret Stones' *Flora of Tasmania.* (Reproduced courtesy of the artist and the Queen Victoria Museum and Art Gallery, Tasmania, Australia.)

Crocus Sativus by Pierre J. Redouté from *La Botanique de J. J. Rousseau*, 1805.

(a) Insects drawn by M. Latreille from Cuvier's *The Animal Kingdom*.

(b) Hand-coloured illustration from Edward Donovan's *Natural History of British Insects*.

(a) From *The Naturalist's Library*, vol. V, "Gallinaceous Birds."

(b) From *The Naturalist's Library*. vol. I, "Hummingbirds."

(c) The Wild Duck from *British Ornithology* by George Graves.

(d) Peregrine Falcon from *British Ornithology* by George Graves.

The Blue Wampum Snake with red lily from Mark Catesby's *Natural History of Carolina, Florida, and the Bahama Islands.*

John James Audubon's Louisiana Heron from the elephant folio *Birds of America.*

Scene from *Zoological Sketches* by Joseph Wolf.

"Shooters Coming by Surprise on a Tiger", hand-coloured engraving from Thomas Williamson's *Oriental Field Sports*.

(a) "Thé" from *Flore Médicale* v. 6 by François Pierre Chaumeton, 1815-1820.

(b) "Poke" from *American Medical Botany* by Jacob Bigelow, 1817-1820.

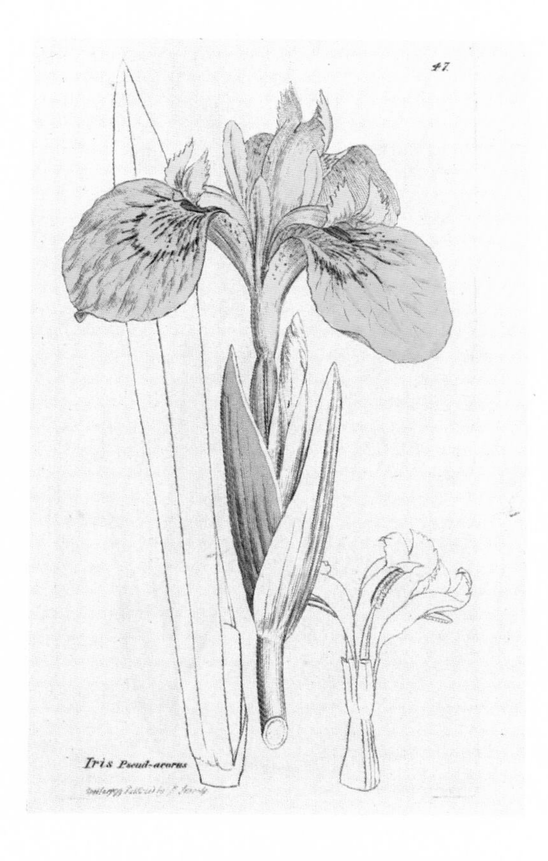

(a) Engraved hand-coloured iris from James Sowerby's *English Botany*, 2nd ed., v. 1.

(b) Gladiolus Cardinalis or Corn Flag from *Curtis's Botanical Magazine*, v. 4, 1790.

Le Saï from Jean Baptiste Audebert's *Histoire Naturelle des Singes,* Paris, 1797.

(a) Hand-coloured engraving from James Sowerby's *Coloured Figures of English Fungi or Mushrooms.*

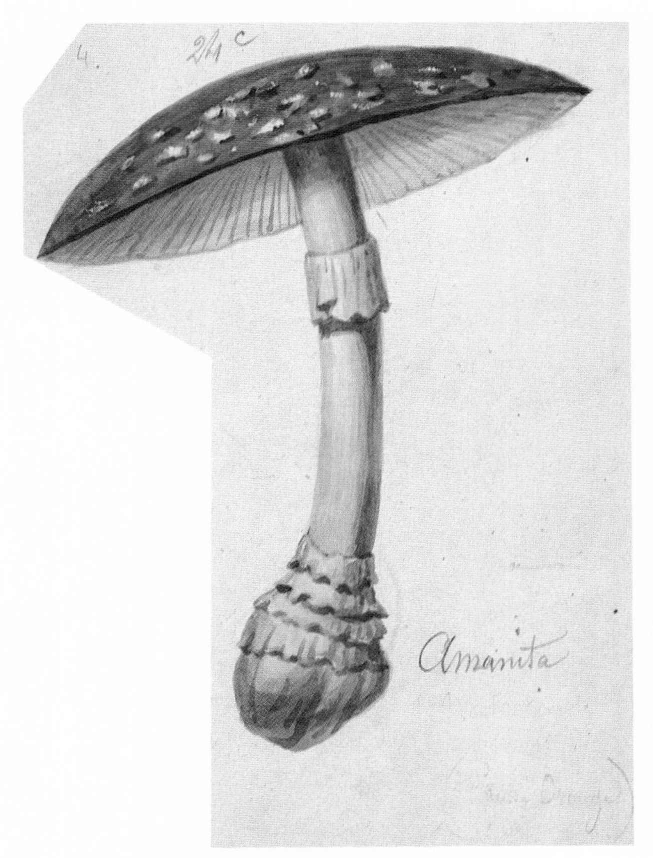

(b) Amanita Muscaria, original drawing from Lucien Quélet's *Champignons Trouvés dans le Forêt de Fontainebleau.*

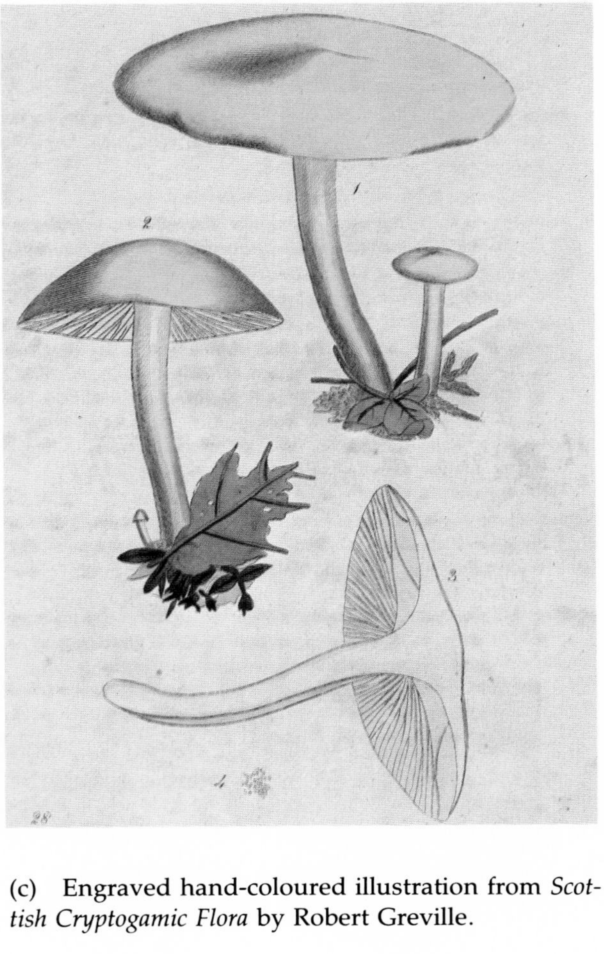

(c) Engraved hand-coloured illustration from *Scottish Cryptogamic Flora* by Robert Greville.

(a) Pink, or Roseate Imperial Sun Trochus from Edward Donovan's *Naturalist's Repository*.

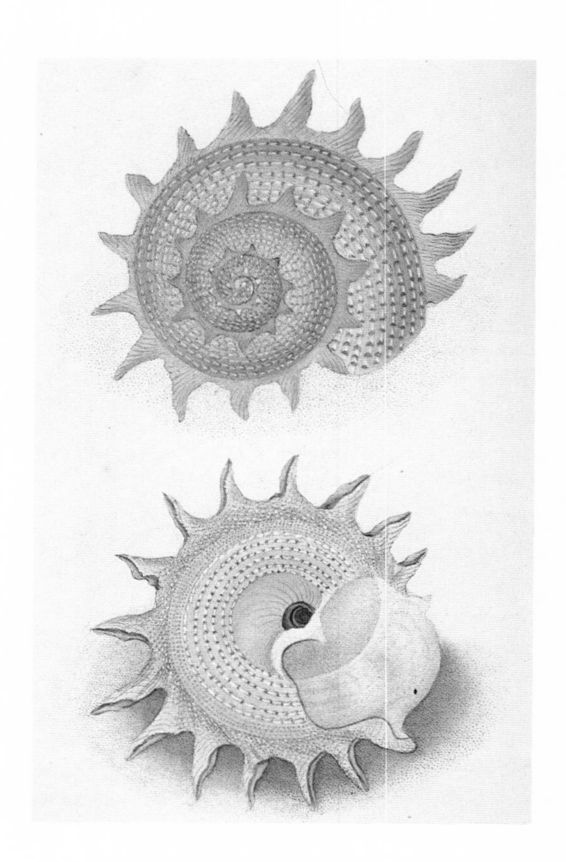

(b) Hand-coloured engraving from *American Conchology* by Thomas Say.

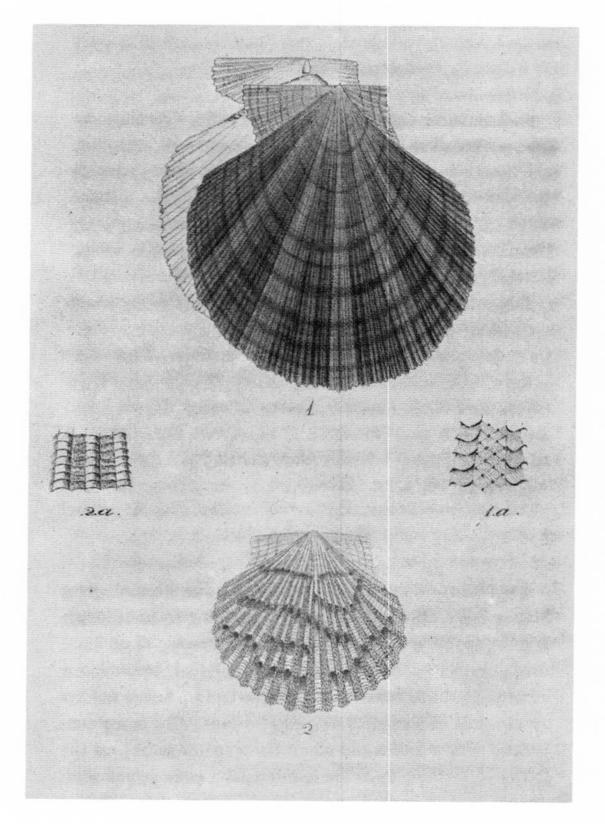

(a) Aquilegia Vulgaris from Karl Levin's *Illustrations of the Flora of St. Petersburg.*

(b) Illustration from *Deutsche Kakteen Gesellshaft.*

From *Banks' Florilegium,* drawn by Sidney Parkinson, Bustard Bay, Australia, May 1770. (Reproduced by permission of the British Museum (Natural History) and Alecto Historical Editions.)

Nature Classics

A CATALOGUE OF THE E. A. McILHENNY
NATURAL HISTORY COLLECTION
AT LOUISIANA STATE UNIVERSITY

Natural History Illustration: An Introductory Essay

BY DAVID M. LANK, F.R.S.A.

Man's emerging awareness of the world around him is chronicled in the thousands of volumes and illustrations that over the centuries form the collective heritage of our civilization. As the repositories of this cumulative knowledge, the major libraries of the world play a vital role, historically, scientifically, and artistically.

With the McIlhenny Collection, Louisiana State University has joined the ranks of the leading institutional concentrations of books and art dealing specifically with the natural world. The major emphases—botany and ornithology—reflect not only the interests of the late E.A. McIlhenny, but also share a tradition older than the printed word.

The development of science can be effectively traced through the tens of thousands of illustrations that have appeared in the volumes of man's emerging knowledge. In the classical era science first began to emerge as a subject for the study and understanding of the natural world. What is today called botany was one of the major divisions of classical science. The study of plants was so preeminent because medicine was essentially treatment with natural substances. The further back the historian probes the more botany and medicine intertwine; the broad study of zoology and its vast array of subgroups can thus be traced to botanical roots. Early botanists had an absolute obsession to identify and illustrate the medieval plants described in classical Greece by Dioscorides (364)*. As human knowledge grew through exploration and experimentation, the pharmacy of Dioscorides grew apace, and was catalogued in an outpouring of expanding medieval herbals. Subsequently, however, generations of monks copied and recopied the medieval manuscripts and the illustrations often rendered the plants unrecognizable, and, therefore, potentially lethal. Accuracy

*Works in the McIlhenny Collection are indicated by the *Catalogue* number. The initials "RBC" refer to works in the Rare Book Collection, "LA COLL" to the Louisiana Collection.

demanded an artistic discipline that was unprecedented, but which, by the late 15th century was producing astonishingly good illustrations.

The early link between botany, medicine and natural history is perhaps most clearly demonstrated by the Society of Apothecaries, one of the oldest guilds, founded in the 16th century primarily to supervise the practice of what today we would recognize as the related professions of medicine and pharmacy. The apprentices were expected to accompany their masters on their rounds, and to learn by example and osmosis the very demanding standards of their profession. An objectively defensible degree of competence included the ability to identify what were known as "the simples," that is, the drug plants that formed the basis of their pharmacopoeia. As these plants needed to be known in their fresh state, the apprentices had to make frequent visits to the botanical gardens of their guild as well as forays into the countryside. These outings were known as "herbalizings," field trips designed specifically to acquaint the apprentice with the living raw materials. The first description of such an excursion is found in the Society's records of May 1620, the very month in which the Mayflower set sail for America.

The need to record in books what had been seen in the field became increasingly important, especially as urban sprawl (not a modern phenomenon) blighted the formerly accessible "botannick gardens." One consequence of this need was the formation of scientific societies for the study and dissemination of knowledge. The first independent natural history society in the English-speaking world was the Temple Coffee House Botanic Club, formed in or around 1689; next was a small club founded around 1706 in connection with the Physics Garden at Dundee. By contrast, the earliest known organization to specialize in a specific branch of zoology was the Society of Aurelians, as lepidopterists of the era were called. The exact date of the founding is unknown, but was probably in the first decade of the 18th century.

The Royal Society of London, which was founded in 1662, included in its official title its aim: "for Improving Natural Knowledge." In exchange for royal subsidies, new knowledge would be shared with the government through its *Proceedings* and *Philosophical Transactions* (258). The Society considered it one of their prime objectives and triumphs to revise John Gerard's *The Herball or Generall Historie of Plantes* (390) which had first appeared in 1597. Early members also conducted experiments on sections of "unicorn horn."

Just as there can be no doubt that the field trips of the Society of Apothecaries were the major seminal influence in the establishment of the great field traditions that form the core of modern natural history, so too it can be stated that the illustrations in the early botanical

"The rose without prickles," woodcut from *The Herball or Generall Historie of Plantes* by John Gerard. London, 1597.

works led directly to the scientific and artistic traditions in the natural history books that were to follow. In fact, botany and zoology were frequently so wonderfully combined that it requires an arbitrary decision as to where a given work should be classified. Mark Catesby was trained as a botanist, and included extraordinary floral settings for his birds, mammals and reptiles in his 1731 *The Natural History of Carolina, Florida and the Bahama Islands* (192). The only plate not by Catesby, pl.61 of the *Magnolia grandiflora,* was contributed by the great Georg Dionysius Ehret (1708-1770), who, along with Pierre-Joseph Redouté (1759-1840), Pierre Jean François Turpin (1775-1840) and the Bauer brothers, Franz (1758-1840) and Ferdinand (1760-1826) rank supreme among floral painters.

It is not generally known that the floral settings in John James Audubon's double elephant folio *Birds of America* (892), the first plates from which appeared in 1828, were not all drawn by Audubon. It does not diminish Audubon's greatness that many were supplied by assistants such as Maria Martin, George Lehman and especially the young Swiss, Joseph Mason, whose clump of grasses in the engraving

Copper engraving from *Historie Général des Insectes de Surinam* by Maria Sibylla Merian. Paris, 1771.

of the meadowlark would have done justice to Albrecht Dürer.

John Gould's sumptuous bird folios contain thousands of flora studies, and the flowers in Edward Donovan's *The Natural History of British Insects* (694) published from 1793-1802, are frequently as important, scientifically and artistically, as the insects themselves. Nearer to our own time, Rex Brasher's four volume masterpiece published in 1932 was entitled *Birds and Trees of North America* (952), as it underlined the interrelations of plant and animal life. Surely one of the finest flower books of all time, despite its name, was Maria Sybilla Merian's 1705 and 1719 *Insects of Surinam* (784) with its magnificent hand-coloured copper engravings. The blend of botany and the sister sciences is of long standing.

Sadly, scholarship was not always sustained after a promising start. John Martyn, a passionately dedicated botanist who was also a very successful physician in London, published his grand *Historia Plantarum Rariorum* in 1728. The large full-page mezzotints* by Elisha Kirkall were beautifully printed in colour (sometimes in several colours on a single plate) and touched up by hand. So magnificent was the work that Martyn was rewarded, academically at least, by being named Professor of Botany at Cambridge in 1733. Not wishing to jeopardize his medical practice by extended or even temporary absences, Martyn did not give a single lecture after 1735, but retained his title and post of professor for another twenty-seven years when he passed them on to his son. The father justified his inactivity by the fact that the lack of a botanical garden at Cambridge made his task unduly difficult.[1]

Over at Oxford, the situation was the same. Humphrey Sibthorp dozed in the Chair of Botany for thirty-six years, from 1747 to 1783, during which period he did not publish a single scientific work, and, as far as is known, delivered but a single lecture. This lackadaisical attitude permeated the other natural history faculties as well. In Oxford, in 1730, Huddesford, the President of Trinity College, was elected Keeper of the Ashmolean Museum for £50 "whether he do anything or not."[2]

Despite such flagrant abuses, there had been a continuous increase in botanical knowledge. Much of the progress, however, was dissipated by the proliferation of local names for the same plants. Confusion reigned supreme despite some attempts at standardization, for example John Ray's comprehensive three-volume study of the world's flora, *Historia Plantarum Generalis*, 1696-1704 (526) and his more portable *Synopsis*. Botanical classifications, whether alphabetical, or based on colour or smell, were somewhat subjective. Seemingly, as many proposals were put forward as there were authors. The problem posed by national languages was partially overcome by the universal use of Latin, but this international language had the effect of perpetuating the gulf between layman and expert. Adding to an already confusing situation was the wealth of new, hitherto undescribed material resulting from the ocean navigation under Henry the Navigator, Magellan, Columbus and their fellow explorers. Standardization of plant names, at least among the learned world, was brought to some degree of order in 1623 with Kaspar Bauhin's *Pinax Theatri Botanici* (297). This work constituted a listing (pinax register) of the 6000 or so plant species known at that time as well as any synonyms applied by pre-

*A detailed description of the various printing methods must lie outside the scope of this catalogue, but, as will be seen in the text, references to the technology will be made as they relate to artistic development.

vious workers. Bauhin recognized the importance of the rank of genus and species in taxonomic schemes and assigned names to plants based on a combination of binomial (two-word name) and polynomial (multi-word name) systems. Each name was composed of a generic name, a trivial name or specific epithet and, in some cases, the name was extended to include five or so more words which were descriptive in nature. Although Bauhin was the first to employ a binomial system of nomenclature, it was Sweden's Carl Linnaeus who is now often hailed as designer and father to the system.

Linnaeus produced a monumental comprehensive classification system for all known animals, minerals and plants in his *Systema Naturae* (232-233), first published in 1735. For botany his two most acclaimed works are *Genera Plantarum*, which lists and briefly describes 1105 plant genera, and its companion *Species Plantarum* (468) which lists and describes approximately 7700 plant species. The influence of Bauhin's work is clearly seen in Linnaeus' *Species Plantarum*. Not only are many of Bauhin's binomials repeated in this work, but many of his polynomials are incorporated as brief descriptions for the species. Furthermore, Linnaeus often cites Bauhin as author to many of the names included. So, although Linnaeus cannot in truth be credited as first to conceive the system of binomial nomenclature, it was his consistent use of binomials in *Species Plantarum* which led to the present day mandatory use of such a system for naming plants.

Coming as it did during a time of unprecedented exposure to new species from around the world, the Linnean system required the rethinking of the nomenclature and classification of all known plants and incorporating them into the on-going classification process of the new but unrecorded material. This system has supplanted all others to such a degree that no botanical names dating from before 1753 are recognized as being legitimate. This, of course, had less impact on the untutored who, for instance, applied the name "batchelor's button" to more than twenty different species of flower in England.

Pre-Linnaean books, therefore, have to be placed in their historical context before harsh judgements are made. The sheer quantities of new species far outstripped the ability of the contemporary scientific mind to categorize, let alone understand, the mass of material shipped back to Europe by the irrepressible collectors. Inaccuracies concerning characteristics and distribution of the new discoveries were inevitable. Nevertheless, while much is made by historians about the voyages of discovery from the points of view of conquest, commerce, politics and economics, too little is stressed concerning the true revolutionary impact on world society by the concomitant introduction from abroad of plants and animals into Europe. Some introductions were inadvertent, others calculated; some were beneficial, others catastrophic. All that can be said with certainty is that the age

Zebra from *Relatione del Reame di Congo*, 1591.

of exploration ushered in an age of biological upheaval for the biota of the world, the ramifications of which could never have been anticipated.

With the era of exploration came the beginning of a vast body of books dealing with local, state, regional, national and continental surveys of natural history. In 1591 Filippo Pigafetta gave a detailed account of the Congo, its geography, customs and animals in *Relatione del reame di Congo* (138) in which, interestingly, is found the first known engraving of a zebra. Prospero Alpinus' *De Plantis Aegypt*, published in 1592 was one of the earliest books dealing exclusively with the flora of a non-European country. The name "De Plantis..." clearly differentiates the intent of this modest tome from his 1591 work, *De Medicina Aegyptiorum*, in which, the coffee plant was described for the first time in European literature. Similarly, books dealing with exploration down through the nineteenth century contained very significant sections devoted to botany. In 1854, Sir Joseph Dalton Hooker published his very detailed *Himalayan Journals* (133) with an emphasis on the flora that would be expected from the son of the future head of the Royal Botanic Gardens at Kew. Livingstone's *Last Journals* are of profound importance to the botanical and natural histo-

Copper engraving from *Specula Physico-mathematico-historica notabilium* . . . of Johann Zahn, 1696.

rian. And of course the *Voyages* of Captain Cook are filled with important botanic information.

Cook's voyages are in one way only now nearing completion. A passenger on one of Cook's voyages was Sir Joseph Banks, the wealthy gentleman naturalist who made Cook's life miserable by insisting on botanizing while good seamanship indicated the need to put to sea. But under Banks' supervision, 738 plates were made of the drawings by Sidney Parkinson and others to augment the catalogues of the plants based on manuscripts of Banks and David Solander. The plates lay dormant in the British Museum for almost 200 years before they were printed "à la poupée" (the method used for some of Redouté's finest books) starting in 1980. The McIlhenny Collection is one of a small number of American subscribers to *Bank's Florilegium* (290).

In the same vein, another modern publication is *Forty Drawings of Fishes* (711) from drawings by the artists who accompanied Capt. Cook on his three voyages to the Pacific in 1768-71, 1772-75 and 1776-80. These were finally printed in a large folio by the Trustees of the British Museum (Natural History) in 1968 with thirty-six coloured collotype plates.

The McIlhenny Collection contains an especially strong representation of works concentrating on the botany and wildlife of tropical areas written from the 18th to 20th centuries. These begin with *A Voyage to the Islands Madera, Barbados, Nieves, S. Christophers and Jamaica* published with copperplate engravings by Sir Hans Sloane, 1707-25 (143). This was followed by Hughes Griffith's *The Natural History of Barbados* in 1750 (220); Edward Bancroft's *An Essay on the Natural History of Guiana, in South America* in 1769 (169); and *Histoire des plantes de la Guiane Françoise* of Fusée Aublet in 1775 (287).

Geographically limited books could also specialize on specific families, as was the case with the sumptuous classics such as Bateman's *Orchids of Mexico and Guatamala* (294) and Hooker's 1853 folio of the *Rhododendrons of Sikkim-Himalaya* (RBC), the latter two titles with excellent hand-coloured lithographs. This tradition of geographically limited works extends to the present with such contributions as *The Endemic Flora of Tasmania* (571). This is of particular interest not only because of the superior botanical illustrations of Margaret Stones, but also because of the revived tradition of patronage—Lord Talbot de Malahide and his sister, The Honourable Rose Talbot, financed the publication just as Samuel Pepys and his fellow members of the Royal Society once did in the 1660's for the copper engravings in the quartos of Willughby and Ray.

The interest in local floras and faunas was paralleled by attempts to integrate all aspects of nature into some cosmic pattern. Johann Zahn's *Specula physico-mathemtico-historica notabilium* (280) from 1695 wanders over all the known natural world, including the night skies. Dozens of animal, fish, and bird species are crowded together in Noah's-ark-like scenes on the large fold-out copper engravings in which science and art are less important than comprehensiveness.

Obviously, time, technology and taste all played roles in the evolution of natural history books. It is important, therefore, to bear in mind the historical context of the seminal works. Leonhart Fuch's *New Kreuterbuch...* from 1543 (384, 1938 fac.) typifies the approach found in many of the 16th century books in which the wood cuts are executed in basic outline with no attempt at shading. Such starkness was a conscious aid for the colourists who were expected to paint each individual plate before, or as was more usually the case, after binding. Not surprisingly, virtually no two coloured copies of a given book are the same nor of equal quality. The two-stage production also explains why both coloured and uncoloured states exist for many of the early works.

Woodcuts were unable to achieve subtle shading and curved forms. The parallel grain of even the finest wood restricted the freedom of line, as all curves or gouges made across the grain tended to be awkward. Cross-hatching, the favoured method of shading, also was not

Mushrooms, woodcut from *Rariorvm Plantarvm Historia* by Charles de L'Écluse. Antwerp, 1601.

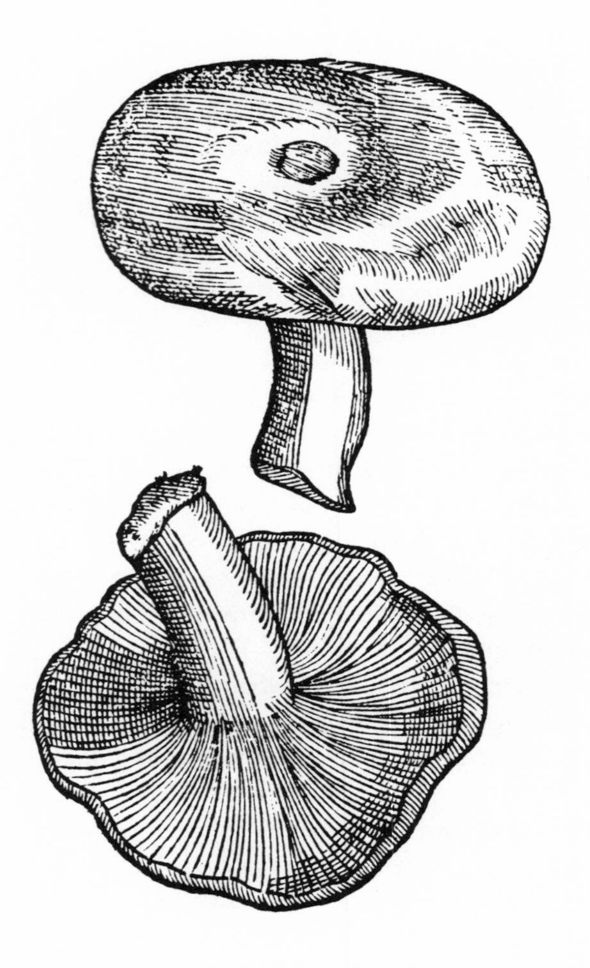

feasible as the diamond-shaped interstices tended to defy even the most skilled carver. (The cross-hatching on the 16th century woodcuts by anonymous craftsmen after the drawings of Albrecht Dürer was achieved by over-printing. A second woodblock had been engraved with parallel lines to be pressed onto the original printing at an angle to form the cross-hatch.) To contrast negatively the austere woodcuts—which were designed to be hand-coloured—with later illustrative processes is unfair given the limitation of the woodcut.

Far greater freedom of line, detail and shading could be achieved with copper rather than with wood. The contrast between the two methods is clear in the copper engraved title page of Charles de l'Écluse's *Rariorvm Plantarvm Historia* (455) from 1601, and the woodcut illustrations in the text. In 1608, P. Vallet exploited the etched copperplate in his *Le Jardin du Tres Chréstien Henri IV* for one of the first, if not *the* first time for an actual floral illustration. Three years later, Johann Theodor de Bry, of Frankfurt, was the first to use copper engravings in his *Florilegium Novum*, a new form of flower book which was the direct outgrowth of the technology involved in the

(right)Engraved title page from *Rariorvm Plantarvm Historia*, 1601.

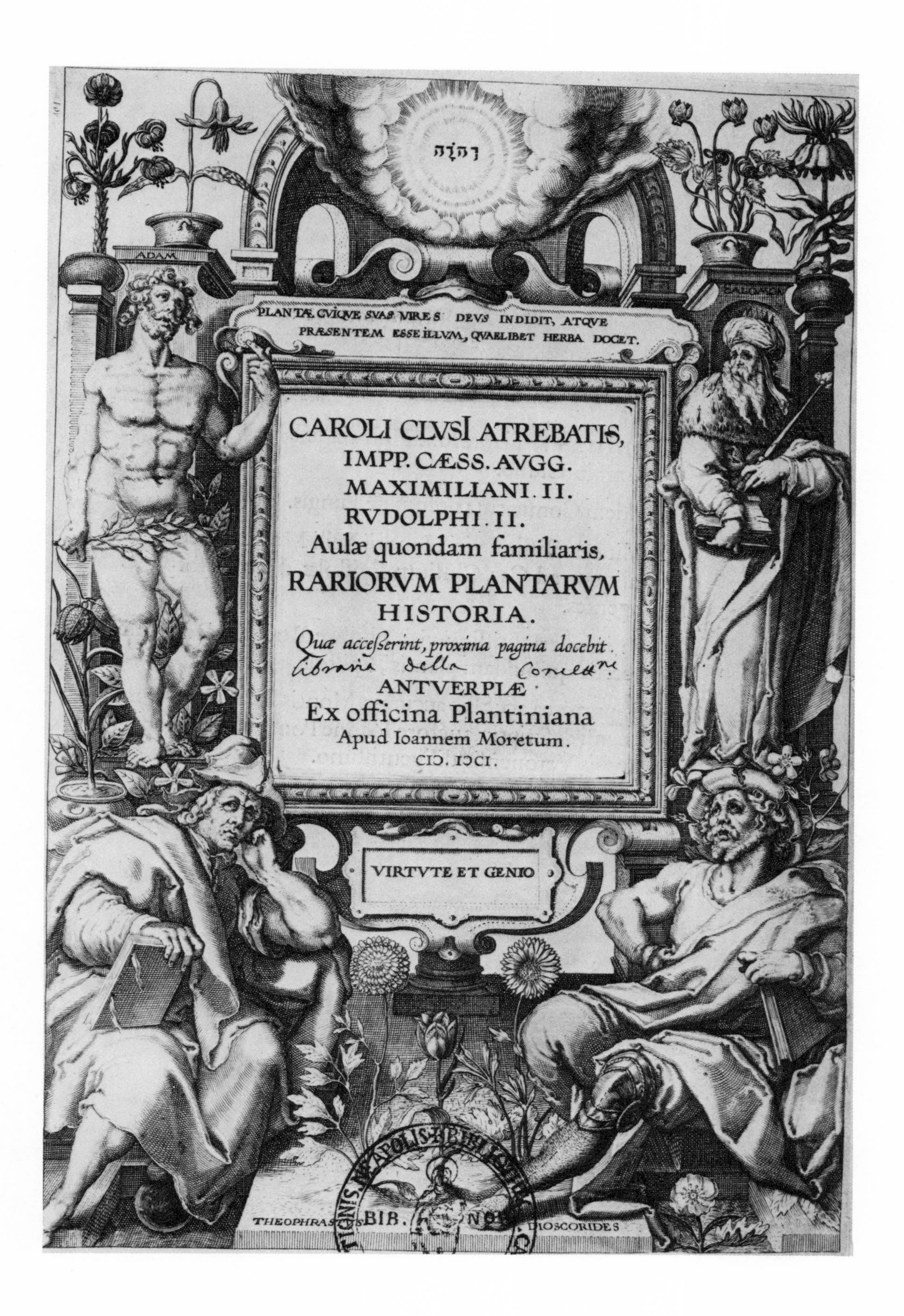

יהוה
ADAM
PLANTÆ CUIQUE SUAS VIRES DEUS INDIDIT, ATQUE
PRÆSENTEM ESSE ILLUM, QUALIBET HERBA DOCET.
CAROLI CLUSI ATREBATIS,
IMPP. CÆSS. AUGG.
MAXIMILIANI. II.
RUDOLPHI. II.
Aulæ quondam familiaris,
RARIORUM PLANTARUM
HISTORIA.
Quæ accesserint, proxima pagina docebit.
ANTUERPIÆ
Ex officina Plantiniana
Apud Ioannem Moretum.
CIↃ. IↃCI.
VIRTUTE ET GENIO
THEOPHRASTUS
DIOSCORIDES

Stipple engraving by P. J. Redouté from Charles Louis L'Héritier de Brutelle's *Stirpes, Novae* . . . Paris, 1784-85.

printing process in which the text was basically omitted, or restricted to a few words engraved directly onto the copper sheet. In 1611, therefore, a new word entered the botanical lexicon, the "florilegium," a collection of flower pictures without accompanying text. These books were frequently devoid of scientific content, being primarily aimed at gardening enthusiasts, or conceived of as objects of beauty in their own right.

Within a relatively short time, modifications and improvements were made in the copper engraving method. One such example was stipple engraving, a modified form of engraving on copper which made use of patterned surface textures with endless possibilities for tonal intensity. Above all, stipple engraving was ideally suited to floral portraiture, and it ushered in the greatest age of botanical books.

In 1788, Charles-Louis L'Heritier de Brutelle, a Frenchman visiting Sir Joseph Banks' private herbarium in London, used the stipple engraving process in his *Sertum Anglicum* (460), which contained a few of the earliest plates from his Belgian protégé, Pierre Joseph Redouté, the genius destined to become one of the most illustrious flower painters of all time. Redouté's *Roses* (530) and *Les Liliacées* have been so thoroughly dealt with in special studies and historical reviews, and have had their plates reproduced so many times, that a detailed description or discussion here would be superfluous. Their very magnitude was impressive. The consistently high level of excellence is astonishing, not only botanically speaking, but also from an artistic and graphics point of view. Suffice it to say that Redouté's work was a standard against which all subsequent floral books have been judged. The artist's influence was so preponderant that he gave his name to the entire "Age of Redouté," that golden era of the late 18th and early 19th centuries which represents the absolute pinnacle of excellence in botanical works.

It is interesting to note that Redouté's flowers and Audubon's birds were being published at the same time. The two artists met in Paris and were profoundly impressed with each other.[3]

But Redouté was not alone. Franz-Andreas Bauer supplied the incredible engravings for the *Delineations of Exotick Plants* in 1796. They were less decorative, and aesthetically less striking than those of Redouté, but in terms of botanical accuracy their superiority becomes evident when plates of the same plants by the two great artists are compared; for instance, Franz Bauer's drawing of *Erica grandiflora* in the *Delineations...*, and that by Redouté for Aimé Bonpland's *Description des Plantes Rares* from 1813.

Another great name from the same period was that of Jean-Louis Prévost, whose *Collection des Fleurs et des Fruits* from 1805 ranks as one of the most beautiful books of all time. As Alice Coats so delightfully put it in her *Book of Flowers*: "Jean-Louis Prévost belonged to the dew-drop-on-leaf school." Despite this rather cute affectation which became his trademark, his floral portraits are of transcendent beauty, exploiting to the fullest the range of richness possible with hand-finished colour stipple engraving. His blossoms and leaves may lack the incredible "bloom" and luster of Redouté, but so powerfully beautiful and accurate are his renderings that, again as Coats says, "Prévost has almost managed to convey the perfume" of his lilacs.

While unquestionably the French were preeminent in the large flower plate books, the last decades of the 18th century also witnessed the rise of British flower books. Eminent among them was

William Curtis' *Flora Londinensis* (347) whose elegant folio-sized copper engravings combined a sure sense of layout with botanical accuracy and excellent hand-colouring.

Another English claimant for inclusion in the top ranks, is Dr. Robert John Thornton. Just as John Gould would later have great influence on bird books despite the fact that he seldom completed a painting, so too did Thornton enjoy great fame even though he himself contributed only one plate, that of roses, to his masterpiece, known popularly as *The Temple of Flora* (576, 1951 repr.) Published from 1799 to 1807, the complete title revealed its true intent: *New Illustrations of the Sexual System of Carolus von Linnaeus*. It has achieved such repute that it is almost a sacrilege to refer to it in other than hushed tones. To dismiss it as being oppressively melodramatic would be as tempting as it would be unfair. As a period piece it offers an intimate glimpse into the psyche of England during the dawn of the Industrial Revolution. The romantic backgrounds within which he placed the flowers were a nostalgic if distorted view of a gentler time when the virtues associated, again mistakenly, with the countryside were still uninfected by the realities of urban society. For most of the plates, Thornton employed the best artists and craftsmen he could find, including the stipple-engraver par excellence, Bartolozzi. By today's standards, and to be completely fair, even by those of Thornton's day for natural history illustration, Thornton's nightingales, their nest and eggs, in the theatrical smorgasbord of roses on plate 7 are hard to explain away. They are truly inferior. Nevertheless, placing birds in an ecological setting was quite a revolutionary concept, and even though in this case the birds were incidental to the main theme, it still represents a foretaste of what was to become the norm after the appearance on the European scene of Audubon's *Birds of America* in the late 1820's.

The beginning of the nineteenth century ushered in new technologies for colour illustration in printed works. In 1811-1818, J.N. Mayrhoffer and F.P. von Schrank published their four-volume *Flora Monacensis* which dealt with the flowers found in the vicinity of Munich. This was the first botanical book to be illustrated by the still relatively new medium of lithography—literally, writing on stone—which had been accidentally discovered by Alois Senefelder in 1796 in Austria. One of the earliest English botanical books to use lithography was Bateman's *Orchidaceae of Mexico and Guatemala* (294). Lithography became very widespread in Europe, competing successfully in hundreds of titles offering "Views" of the countries on the Grand Tour of 19th century tourists and armchair travellers.

Another printing process was "nature printing" which became something of a craze in the second half of the 18th century, lasting intermittently well past the 1850's. It was a marriage of nature and

Copper engraving from *Flora Londinensis* by William Curtis. London, 1777-98.

technology; the actual plant was inked and used as the printing plate. It was fascinating for its ability to capture absolute fidelity of the faintest patterns in leaves and blossoms, but doomed to relative insignificance for large scale works or large press runs by its inherent weaknesses. The plants themselves were not well-suited to the inking and pressure needed to transfer their images onto paper. The specimens quickly proved fragile and brittle, soon wore out, and had to be replaced. As the plants became deformed, the images were not satisfactory, so that uniformity and consistency were impossible to achieve. No two "nature printed" books are alike. As actual specimens were used, almost microscopic detail could be achieved, but subsequent hand-colouring effectively obliterated much of the botanical imprint, thereby negating just about the only arguable advantage for nature printing over conventional engravings. They did, however, preserve "the ghosts of flowers that once bloomed." D.H. Hoppe's

1789-1793 *Ectypa Plantarum Ratisbonensium* was an innovative example, whereas Moore's *Octavo Nature-Printed British Ferns* (495) from 1859-1860 represented no significant artistic advance. Some "nature-printed" titles conjure up strange visions: *The Nature-Printed British Sea-Weeds* by W.G. Johnstone and A. Croall from 1859-60 must have posed some damp problems for Henry Bradbury, by far the most famous nature-printer of the mid-19th century.*

During the Victorian Era a fad developed for books designed specifically to be "illustrated" with actual flowers. *Wild Flowers and Their Teachings* (597) published in London in 1845 (presumably not the first edition, since the preface refers to an earlier one) is a fine example. Ruari McLean in *Victorian Book Design* says that "probably not more than five hundred, possibly one thousand copies were printed, but the collecting, preserving and mounting of the specimens (including daffodils, wild violets and some now rare grasses) must have presented just the sort of problems that never dismayed the Victorians." Also by the same publisher, Binns and Goodwin of Bath was *Ocean Flowers and Their Teachings*, illustrated profusely with actual seaweeds mounted on the pages.

So great was the interest in botany that by the end of the 18th century many small format periodicals had been developed for various audiences. An outstanding example of this type was *Curtis's Botanical Magazine* (348). From 1787 when it first appeared until 1948, the illustrations were hand-coloured, in all over 9600 plates. *Curtis's* continues today as *Kew Magazine*. Remarkably this publication survived in the great tradition of hand-coloured illustrations during a time when techniques for colour printing were developed for the popular market. Gone for the most part were the great royal and aristocratic flower gardens that had been designed by Capability Brown and Humphry Repton and illustrated in the great flower books published under the patronage of the wealthy. These were supplanted by a new type of publication mainly targeted at those people for whom a small plot of land with a few flowers was a welcome relief from the pervasive greyness of industrial England.

*Sherman Foote Denton's *As nature shows them: Moths and Butterflies of the United States, east of the Rocky Mountains* (689) was published in Boston in 1900. From the Preface we learn: "The coloured plates, or Nature Prints, used in the work, are direct transfers from the insects themselves, that is to say, the scales of the wings of the insects are transferred to the paper, while the bodies are printed from engravings (that's a relief!) and afterwards colored by hand." Denton, not only handled the printing, he also personally collected more than half the specimens. The time requirements alone limited the number of copies that could be produced. After 500 Denton called it quits, and—perhaps apocryphally—he lost most of his eyesight. The results were not satisfactory on the longer term, as the colours have lost their vibrancy in the intervening years. Fascinating though it was, nature printing was a dead end, and was by its very nature the antithesis of art, and is therefore peripheral to the history of illustration.

Women played a far greater role in botanical works than in all other branches of natural history combined. In 1807 Anne Rudge was responsible for the plates in *Plantarum Guianae Rariorum*. Mrs. Augusta Withers was "Flower Painter" to Queen Adelaide, and contributed some beautiful plates to the *Transactions of the Horticultural Society* (1352) in the 1830's; Miss Daly was the very talented artist who drew the plates for John Lindley's *Sertum Orchidaceum* (465, 1974 repr.) in 1838; Clara Maria Pope was responsible for many aquatints in Samuel Curtis' *A Monograph of the Genus Camellia* in 1819. Of special interest was Mrs. Edward Bury's *Hexandrian Plants* in 1831-1834, the plates of which were aquatinted by Robert Havell in London, the engraver whose fame would be secured by his work on Audubon's monumental *Birds of America*. In 1863, Mme. Berthe Hoola van Nooten illustrated *Fleurs, fruits et feuillages choisis de la flore et de la pomone de l'ile de Java* (502). The 20th century saw Ellen Willmott's *The Genus Rosa* (598) in 1914, and the five volumes of the *North American Wild Flowers* (590) by Mary Vaux Walcott in 1925.

The list could be expanded tremendously, due in part to the peculiarities of the educational system of the English upper classes which considered flower painting an acceptable outlet and pastime for gentlewomen. It seems that the libraries of every stately home contain original florilegia by female ancestors who had the time and the talent to indulge in the production of beauty. Thousands of such one-of-a-kind books exist, most of them anonymous. Apparently little or no thought was given to having them published. *The Diary of an Edwardian Lady* was brought to light in facsimile only in the 1980's, seventy-five years after its author/illustrator put her pen and brush to paper.

Botanical books had had as their intent everything from medicine to seed catalogues, textbooks and gardeners' aids. There were also books, including some sumptuous volumes, specifically designed to supply patterns for the textile, embroidery, tapestry, metal-working and ceramic trades. The first botanical textbook in America was published in 1803, *Elements of Botany*, (292) by Benjamin Smith Barton, physician and botanist (among his many other accomplishments) in Philadelphia. Within fourteen years, in 1817, another textbook was published, by Dr. Jacob Bigelow, *American Medical Botany* (305), the first book printed in colour in the United States as opposed to black and white plates designed for subsequent hand-colouring.

By the 19th century, natural history had begun to divide into a number of distinct scientific fields, although there continued to be large numbers of lay enthusiasts in the natural sciences. Many scientists also continued to have broad interests. Historically, one of the foremost scientists of the age was, of course, Charles Darwin whose published works include both botanical and zoological treatises.Today, however, he is best remembered for his theories on evolution.

Hand-coloured lithograph from *A History of British Butterflies* by Frances Orpen Morris. London, 1857.

When Darwin published his *On the Origin of Species* (200, 5th edition) in 1859, the controversy unleashed by his theories, which had also independently and simultaneously been arrived at by Alfred Russel Wallace, prompted many other authors and their champions to claim prior discovery. In fact, the fundamental premise of the special creationists—that species were immutable—had long been suspect. Discussion among the common man had been going on for years.

Some powerful voices, however, held out—shouted out—against Darwin, or more accurately against their interpretation of Darwin, as indeed they still do today. An extraordinary blast against Darwin's *Origin of Species* was levelled by someone who should have known better, the Rev. Francis Orpen Morris, famous as the author of *A History of British Birds* (1155-1157) which was first published from 1851 to 1857 and went through many subsequent revised editions. In a pinch,

Morris was more a Reverend than a naturalist. His rare *All the Articles of the Darwin Faith* was published in 1876 as yet another of the endless Morris pamphlets. His arguments were as vitriolic as they were absurd, and in the end he had to fall back on the quotations from Scripture, relying on their Divine Inspiration as proof in itself of Darwin's heresy. His pamphlet was "Dedicated to the Right Honorable The Common Sense of the People of England." His blinding fundamentalism is astonishing for someone whose books on birds required such exposure to nature, equally so, perhaps, in his other two popular successes, *A History of British Moths* (796, 5th edition) and *A History of British Butterflies* originally published in 1857 (794-795).

That Morris was the author of books dealing with butterflies was something of an irony. Entomology was by his own admission, "more captivatingly interesting" for him than any other branch of natural history, and gave him the "feelings of the holiest adoration and most humble worship." And yet it was evidence found in butterflies that became one of the major underpinnings of Darwin's theories of evolution. Henry Walter Bates (1825-1892) was the indefatigable collector who, during eleven years in South America, amassed 14,712 species, mostly of insects, of which no less than 8,000 were new to science. In his 1863 classic, *The Naturalist on the River Amazons* (114) (note the plural form, a far more appropriate name for that incredible braided network of waterways than the English singular in use today), there is a hand-coloured plate of a selection of thirteen butterflies from poisonous (*ithomids* and *heliconids*) and nonpoisonous (*pierids*) families which clearly demonstrate the principle of Batesian mimicry. Bates wrote, of the wings of his beloved butterflies: "on these expanded membranes Nature writes as on a tablet, the story of the modification of species." This, obviously, was grist for Darwin's mill, and an abomination for Rev. Morris. Despite the acclaim and financial success of the book, Bates later remarked that he would gladly spend another eleven years in the Amazon rather than to have to write another book.

Darwin received much support from a variety of sources. Alfred Russel Wallace—his co-author of the evolutionary theories—dedicated his 1869 *Malay Archipelago* "To Charles Darwin... not only as a token of personal esteem and friendship but also to express my deep admiration for his genius and his works." Such openness, plus Darwin's repeated tributes to Wallace, would indicate a lack of the bitterness between them that modern authors such as Brackman in his 1980 *A Delicate Arrangement* have attempted to impute.

Joseph Dalton Hooker supplied vast amounts of botanical data from his own researches as well as from Kew Gardens, to demonstrate variability in plants. And, of course, Thomas Huxley, "Darwin's Bulldog," championed the cause against such formidable opponents as Bishop Wilberforce who sarcastically queried whether Darwin traced his ancestry to the apes through his father's side or his mother's. With Huxley leading the defense, Darwin was free to continue compiling the massive empirical data on which his numerous books, on subjects ranging from orchids to earthworms, were based.

Darwin's career as a published naturalist started modestly, with some coleoptera "captured by C. Darwin, Esq." in Stephen's *Illustrations of British Insects*. True fame and recognition was immediate upon the publication of the *Journal of Researches into the Geology and Natural History of the Various Countries visited by H.M.S. Beagle* (199) in 1839. Other books soon followed, and Darwin's reputation as a scientist became firmly established. It was his very fame that caused the uproar over the *Origins*; had he been a scientific nonentity, his arguments would not have carried such weight. Today, not enough emphasis is placed on the scientific contributions of Darwin in the field of botany. These titles are still considered classics in their field.

Popular acclaim for Darwin was reinforced by the appearance in 1890 of an edition of the *Voyage of the Beagle*, illustrated by Pritchett, placing in the hands of the masses the visual record of the famous trip.

The original edition of the *Voyage* was not illustrated, but Darwin engaged John and Elizabeth Gould to supply the hand-coloured lithographs for an accompanying folio, *The Zoology of the Voyage of H.M.S. Beagle* (686) in 1841, Part III, Birds. The plates were accurate and elegantly rendered, not surprisingly in view of the fact that they were from the hands of the future "Bird Man" of England and his very talented wife. To put the Goulds into their proper perspective, however, we must retrace our steps to examine the roots of wildlife art.

A strong case can be made for surprise that there exists as much wildlife art as there is today. A specialized study can trace the development from the paleolithic cave paintings of Lascaux and Altamira, the early Mediterranean civilizations through classical Greece and Rome, through the Middle Ages, the Italian and Northern Renaissances, right down to the present with our explosion of artists who serve the art market created by sportsmen, conservationists and collectors. But this tenuous thread, though unbroken, represents a minor yet important element in the whole fabric of art history.

The main repository of animal art over the last four centuries has been in the illustrations in books, but here too the number of natural history volumes, numerous though they be, is miniscule compared to the huge body of printing since the time of Gutenberg and commercial exploitation of moveable type. What books there are, however, present a fascinating view of artistic evolution, and the McIlhenny Collection offers an excellent cross-section for study in the United States.

As John Ruskin once said, one of the three legacies of a civilization

Hand-coloured lithograph from *The Zoology of the Voyage of the H.M.S. Beagle.*

is the corpus of art it leaves. The relative scarcity of animal art—and animal illustration—can largely be traced to the perception of the status of animals held by our civilization in the "Divine Scheme of Things," and that status was pretty low. Genesis I: 28 gave man "dominion" over every creature of sea, air and land. The theologians, moralists and philosophers all found justification, through biased interpretation of this or that Biblical passage, to state that all animals were for the use of man, whether for food, labour, clothing, transport, or diversion as pets, prey or sport. Animals in art were, with incredibly few exceptions until the last century and a half, depicted in one or more of these contexts.

Animals in their relationship to man were fit for study; animals in their own right were not. This attitude was paralleled in art. Many were the Italian Renaissance masters who boasted of never having studied animal anatomy, and it shows vividly in their work. Leonardo da Vinci was, as such, an exception in Italy, as were Albrecht Dürer and his followers in the North.

This prejudice against animals was widespread, and persisted for centuries. In 1583, a British Act of Parliament required all parishes to equip themselves with nets in which to catch rooks, choughs and crows. In 1566 another Act authorized church wardens to raise funds to pay so much a head to all those who brought in the corpses of fifteen species of supposedly imperious vermin. The Act Against Cruelty to Animals was passed only in 1835. Seabirds gained legislative protection in 1869 only because, it was argued, they were necessary to guide sailors, and to show the fishermen where the shoals of herring were. Cock-fighting was officially prohibited in 1849, and as late as 1699 tigers were baited with dogs in London, a pastime so popular that the House of Commons once was unable to raise a quorum when a bloody spectacle competed with the vote on an important bill. Pepy's *Diary* showed that true blood sports in the 17th century were almost as well attended as public hangings.[4]

The attitudes in America were modified only by the presence of the wilderness. Wild animals were basically a source of food at best, or, at worst a threat to domestic herds and flocks. The Colonists in Virginia began the task of converting the Indians by offering them a cow for every eight wolves they could kill. Tentative steps towards conservation were taken quite early: *Acts and Laws Passed by the Great and General Court of Assembly of Her Majesties Province of the Massachusetts-Bay in New England* on May 31, 1710 included "An Act for the better Regulation of Fowling" which established seasons, and the kinds of camouflaged boats that could not be used by sportsmen and market hunters. Game Laws had been promulgated from very early times, but their intent was not exactly conservation. In 1733, E. and R. Nutt published a compendium of *The Laws Relating to the Game* containing "All the Laws and Statutes for preserving thereof, to the Gentry to whom it belongs..." Protection of wildlife was the offshoot of protection of privilege. Sporting literature—of which the McIlhenny Collection has a significant cross-section—contained much of the first accurate field information on hundreds of species. More was learned about migration through the sights of a gun than through the sights of a telescope.

All animals were fit to be shot, but not all were considered fit to be eaten. The fact that the French and Italians felt no revulsion to what Topsell in his 1607 *The History of Four-footed Beasts and Serpents and Insects* (856) (1967 repr. of 1658) called this "wanton eating of frogs" was an enduring subject of contemptuous comment from the English. Not surprisingly, the first decent artistic portrait of a frog was as late as 1800 in Shaw and Nodder's *A Naturalist's Miscellany*.

Man's changing view of nature is brilliantly documented by Keith Thomas, in his 1983 *Man and the Natural World - A History of the Modern Sensibility* (265). Two other essential books for those interested in philosophical context of man and nature are David Elliston Allen's 1976 *The Naturalist in Britain* (156) and Lynn Barber's 1980 *The Heyday of Natural History, 1820-1870* (170). They form part of the extensive group of reference works in the collection.

Despite an oppressively anti-animal sentiment, and lacking the medical incentives enjoyed by botany, it is suprising how many animal books did exist. The traditions of animal art in books are as old as books themselves. Embellishments on illuminated manuscripts date back to the 9th century. Animals, or at least creatures resembling them, figured prominently in medieval bestiaries. In the manuscripts the animals played primarily a decorative role, whereas in the bestiary they were objects through which the church could teach moral or allegorical lessons. In both cases animals played a subordinate role.

In fact, animals in their own right had received scant attention since the days of Pliny. For fifteen hundred years he was to be the acknowledged authority on animals. It was a brave man who questioned his descriptions in *The Historie of the World, commonly called The Naturall Historie of C. Plinius Secundus* (254). Whales were described as "600 feet long and only three feet wide which took ...up in length as much as foure acres or arpens of land." Much of Pliny's writings were original, but much was copied from Aristotle and even earlier sources.

It was not until the 1550's that animals began to be examined in even a quasi-scientific way. The Frenchman, Pierre Belon (924), and the Swiss, Conrad Gesner (719, 924, 1047) borrowed heavily from Pliny and Aristotle, but their large folios contained enough new material to mark the transition between the medieval bestiary and the modern scientific treatise. It is true that Gesner included descriptions of assorted sea monsters purported to have been seen by Olaus Magnus, the Archbishop of Uppsala, who was prone to meeting sea-satyrs and marine horrors dressed in episcopal habits, but Gesner did include the first accurate, although limited, descriptions of actual birds and beasts.

In Gesner we find some of the earliest true animal art. His folios were illustrated with woodcuts of remarkable quality. It is good to remember that the artists worked from dried and very deformed skins. Nevertheless we can recognize the European goldfinch or the Lapwing at a glance. The fine detail and fierce mien of the large heraldic eagle is so impressive that it is reproduced today after the passage of half a millenium. The artist was, in all probability, Lucas Schan of Strasbourg, and the engraver was probably Franz Oberreiter of the same city. Despite the lack of color, and despite the same technical problems imposed by grain of wood that we saw in the early botanical works, Schan and Oberreiter created figures that certainly qualify as serious, and successful, scientific draftsmanship.

True, some of the ornithological features are, by modern standards, a little weak, but let us pause here and put these woodcuts into their

Woodcut from *L'Histoire de la Nature des Oyseaux* of Pierre Belon. Paris, 1555.

proper and perhaps surprising historical perspective. Raphael, who lived from 1483-1520, just before Gesner, was responsible for a portrayal of St. George and the Dragon, with the horse anatomically ridiculous. Bruegels, the greatest master of the early Northern Renaissance, painted two deformed and arthritic monkeys. Piero di Cosimo painted the Forest Fire half a century before Gesner's time and filled the air with "birds," feathered bombs hurled by an unseen hand on trajectories over which they had no control, and which spelled certain doom. Titian, the great genius of the Italian Renaissance, was Gesner's contemporary. One of his most famous pictures was of the Emperor Charles V at Mühlenberg, astride a horse that is truly awful. Paolo Veronese, who lived from 1528 to 1588 painted Venus and Mars being interrupted by an anthropomorphically maudlin horse.

Pietro Longhi painted a rhino in 1751, a slight improvement over the woodcut by Albrecht Dürer of 1514, showing the poor beast covered in steel plate, complete with nuts and bolts. At least Dürer—a great animal painter—can be excused, as his rendering was based on a written description of an animal that he had never seen. Many consider his "Attack of the Marmalukes" to be one of Francesco Goya's greatest masterpieces, and yet despite the fact that he was born in 1746, he showed an obvious ignorance of muscle and skeleton in the impossible horses. The French are fond of referring to Géricault as the greatest painter of horses of all time (a surprise to those who appreciate Rosa Bonheur and George Stubbs), but he substituted drama for accuracy in such monumental extravaganzas as his 1817 "Horses Held By Slaves." And finally, Eugene Delacroix knew or cared nothing about the form or function of the plastic tiger and unarticulated horse in his famous "Tiger Hunt" of 1854.

The list could be expanded to even more extreme lengths to demonstrate that "bad" animal art was hardly limited to books, nor to the 16th century, the age when natural history was just starting to come into its own.

The study of natural history crossed the Rubicon into the modern era with the works of Ulisse Aldrovandi, the scholar/nobleman from Bologna, the first person ever to hold the position of Professor of Natural History. During his lifetime (1527-1605) he published four folios and posthumously nine more were edited by his followers from his voluminous notes and manuscripts (282-283). Much of his material and his general approach was borrowed from earlier authorities, including Gesner, but Aldrovandi introduced a concept which was unprecedented in its scope. For the first time we find vast amounts of knowledge gained through first-hand field observation. The "herbalizings" of the botanists had come to the zoologists.

Aldrovandi had scribes and artists accompany him on his endless treks around Europe. The illustrations were done by, among others, Jacob Ligotius, who had worked with the great Duke of Urbino. The original paintings still exist among the treasures of Bologna. They

Woodcut from *Historiae Animalium Liber II* by Konrad Gesner. Frankfurt, 1586.

were first transcribed in outline onto pear wood and then engraved by Christopher Cariolanus and his nephew, so expertly, according to Aldrovandi, "...that they seemed not to be done on wood but on copper." They were originally cut in the mid-1550's. Remember that Copernicus was still alive, and the earth, not the sun, was still center of the official universe. Kepler had not yet explained the laws of planetary motion; Newton had not yet discovered gravity; Galileo had not yet been forced to recant; and when the third volume of Aldrovandi appeared in 1637, René Descartes was still thinking, and about to publish his *Discourse on Method*. The illustrations by Ligotius are primitive, but in addition to being more than four hundred years old, they also are a direct link with the dawning of intellectual discovery of our modern times. These illustrations, with all their faults—but with all their force—come from a world as different from ours as the planets themselves. Their historical importance cannot be overstated, for they, along with those in Gesner, mark the major departure point of animal art.

The next major milestones in wildlife illustration came in the latter part of the 17th century with the works of Willughby and Ray. In 1678 Ray had edited *The Ornithology of Francis Willughby* (1280)—with 78 copperplate engravings, the descriptions, according to the title page, "Illustrated by most Elegant Figures, nearly resembling the Live Birds." The Golden Eagle is a reduced and reversed version of the great woodcut from Gesner, and many of the other birds are so stylized that without their labels there would be no way of identifying them. Others, however, represent a quantum leap beyond anything that had come before. The artist was F.H. van Houe who, it appears also engraved most if not all of the plates. He was particularly successful with those birds—such as the pigeons—with which he was personally familiar. Copper engraving, as was the case in the parallel development of botanical works, was to become the favoured medium for illustration for the next century and a quarter.

The introduction of colour was the next step. Granted, as had been the case in botanical works for several hundred years, illustrations printed in black and white had been coloured by hand after the book had been published. Not until 1728-1731, however, did a bird book appear with hand-coloured plates as an integral part of the publishing venture. Eleazar Albin produced *A Natural History of Birds* in three quarto volumes "with a Hundred and one Copper Plates, Curiously Engraven from Life. Published by the Author...and carefully colour'd by his Daughter and Self, from the Originals, drawn from the live Birds." This represented two radical departures: colour, and "drawn from live birds." In at least one case we know this was not true, as, thanks to George Edwards' attack on Albin "who would have the world believe his draught was taken from nature," Albin's "Indian

Woodcut from *De Qvadrupedibvs* Liber I of Ulisse Aldrovandi, 1648.

Pye" (actually, the Pita) was copied from an existing drawing. Edwards himself, it might be noted, was not above pilfering Mark Catesby's blue jay nor Collins' famous portrait of the dodo for his own use.

Despite his claim to the contrary, Albin did not possess a great degree of talent, his birds being stiff and awkward. There were, however, some lovely plates which, upon closer examination, were the work of his daughter Elizabeth, whose sense of colour and layout far exceeded that of her father.*

*Important women artists in birds could be counted on the fingers of one hand. Elizabeth Albin, Elizabeth Gould, Mary-Ann Meyer, for instance, are basically unknown or forgotten, eclipsed by the reputations enjoyed by their father (Eleazar Albin) or their husbands (John Gould and Herbert Leonard Meyer). (If Audubon is to be believed, his wife Lucy was responsible for the plate of the swamp sparrow, but this seems highly unlikely.) Their hand was important in the books that made their men famous. Elizabeth Albin was so much better than her father that he rewarded her in later editions of his *History of Birds*, the first book with coloured bird plates, by removing all reference to her. Mary-Ann Meyer supplied the convincing smaller passerines and the wonderful floral settings of Meyer's *British Birds*, which, incidentally were coloured by her unnamed daughters in the 1840's. Elizabeth Gould converted her husband's sketches into finished lithographs for "his" *Zoology of the 'Beagle'* and the *Birds of Australia*. To Gould's credit, he did acknowledge his wife's contribution, although, through the passage of time, this fact has been largely forgotten.

Copperplate engraving from *The Ornithology of Francis Willughby*. London, 1678.

Albin included a plate of a bird of paradise with no feet. Two hundred years after Magellan's sailors had given evidence to the contrary, some still believed that these feet-less birds, being unable to land, were born in a cloud, predestined to a life of eternal flying, truly, therefore, birds of paradise. Down to 1760, when Linnaeus named the largest species, *Paradisea apoda* (the footless bird of paradise), no perfect specimen had been seen in Europe. Albin also included "...the batt or fluttermouse...a creature between bird and beast," as he said, just to be sure. The inaccuracies were obvious, and as more up-to-date works vied for attention, the first colourplate bird book struggled through a couple more editions before succumbing to total eclipse. Because of the scientific weakness of the text, these plates—and those from similarly out-dated books in future years—have been condemned to an unfortunate obscurity, and their artistic merit unjustifiably overlooked. Had there been "avilegia" similar to the florilegia—plate books without text—some of these early bird artists would be far better known today.

Albin was more successful in his little volumes on the natural his-

Copperplate engraving from *The Ornithology of Francis Willughby*. London, 1678.

tory of "singing birds," both English and foreign, which appeared in several editions from 1731 through the 1770's (878). He and his daughter also combined for an exceedingly rare series of uncoloured copper engravings of fish which, in addition to swimming in stylized water, represented a considerable improvement in accuracy and sense of life over the Willughby plates by van Somer. But the preeminence of the Albins was short-lived.

In 1731 the first edition of *The Natural History of Carolina, Florida and the Bahama Islands* (192, 1771 ed.) was published by Mark Catesby. This two-volume large folio contained 200 important copper engravings which were considered to be exceptionally fine in their day. They represented the first comprehensive collection of coloured plates of North American natural history. Albin had included the cardinal and turkey in his *Birds*, and as far back as the last two decades of the 15th century John White had produced a series of watercolours of American birds, fish, insects, crabs and reptiles, but these had remained unpublished.

As stated earlier Catesby was trained principally as a botanist, and flowers figure prominently in his compositions. In most cases he

"The Fieldfare of Carolina" from Catesby's *Natural History*. . . . This is the dead robin on a stump of snake-root.

chose vegetation that was typical of the habitat of the species, an artistic approach that broke completely with the tradition of placing the bird on a stylized trunk or branch. It was an innovation that anticipated the full ecological settings of Audubon a century later.

Catesby had done his own field research and had taken copious notes, but the actual specimens after a twenty year interval preserved in kegs of dark navy rum had become faded and hopelessly deformed. Some birds, therefore, appear quite stiff; others are "stiff" for other reasons. His meadowlark will be dismissed by the casual observer as being a faithful reproduction of advanced *rigor mortis*. But anyone who has ever watched a meadowlark will recognize that they actually do stand and stretch just as Catesby portrayed them. In his plate of the blue jay, the bird is tilting forward on his branch, screaming at an unseen intruder, perhaps Catesby himself, or more likely at the end of Catesby's gun. Here, for the first time we see not only the birds, but the bird doing something that is typical or characteristic of a particular species.

There were several very colourful plates of fish, which from an ichthyological point of view were important. His snake plates, on the other hand, were among the most beautiful natural history paintings of all time, proving, as Léo-Paul Robert of Switzerland did with his caterpillars from the Jura Mountains in the 1920's, that great animal art is independent of the species being depicted. As examples of layout they are Catesby masterpieces of form and colour, and understandably served as the basis for scores of later plagiarisms. The wonderously blue wampum snake had only one problem: it had never been seen before, nor since, but it appeared in the major encyclopedic works for almost another century. In some of his animal

"The Large Lark" from *The Natural History of Carolina, Florida, and the Bahama Islands* by Mark Catesby. London, 1771.

plates he successfully tried foreshortening, a radical departure from the traditional strict scientific profile, but overall Catesby's larger animals are best forgotten.

Perhaps Catesby's greatest plate was his portrait of a robin. The bird was totally dead. The pose was appreciated by the 18th century English for whom "Who killed Cock-Robin?" was a favorite nursery rhyme. The ability to capture death as well as life, the overall accuracy, the expressions and characteristics of specific species, the sure sense of composition, and the use of environmental elements all represent a break with the past and anticipate the future. Quite rightly, Catesby is known today as "the Colonial Audubon."

A contemporary and friend of Catesby was George Edwards (1694-1773), known as "The Father of British Ornithology" partially because of the more than 900 zoological watercolours that he painted for Sir Hans Sloane (Founder of the British Museum), and partially for his two extensive publishing ventures, *A Natural History of Uncommon Birds* (1017) in 1743-1751, and *Gleanings of Natural History* (1017) from 1758 to 1764, two of the rarer of the great zoological works. In retrospect he deserves the title more because of his art which represents a significant dividing line between the basically ill-informed primitives and the scientifically more correct artists that would follow in the next generation.

Although trained as a businessman, Edwards travelled extensively throughout the Continent observing wildlife in its native haunts.It was obvious to him that all previous bird art failed in what he described as his goal: "I have endeavoured to finish my figures with such exactness both as to drawing and colouring that the prints themselves may give a tolerable idea of the subjects exhibited on them." Edwards paid such attention to accuracy that he claimed to have drawn three or four plates of each species before choosing the best. He did his own engraving (after learning the process from his friend, Mark Catesby) and even his own colouring, thereby justifying his claim that each print was "an original."[5]

Edward's approach was scientific rather than strictly artistic, so that his style has a very "engraved" look about it, with every line clearly showing through the rather transparent colours. The multitude of lines was perhaps better suited to fur than feathers, but nevertheless if identification of the species was the objective, Edwards succeeded admirably so that even laymen had no difficulty in recognizing the birds he depicted. His series of warblers from North America in *Gleanings* was extensive, and therefore ranks among the most important of the early American avifauna plates. Without doubt, his most famous plate was of the Dodo, copied from the original by Collins. Ironically, the last Dodo disappeared in 1693 the year before Edwards' birth.

"The Whooping Crane," hand-coloured copper engraving from George Edwards' *A Natural History of Uncommon Birds*. London, 1743-64, v. III.

The next major artistic jump took place in 1766 when Thomas Pennant produced his great folio, *British Zoology* (807, 1812 ed.) The title page called for 107 hand-coloured copper plates, but the complete set contained 132. And what plates they were! Almost all were drawn by Peter Paillou and engraved by P. Manzell, with the individual species including capercaillie, the largest member of the grouse family, being presented life-size. The birds of prey and the game birds were the most successful, along with the larger water birds. Paillou experimented with what in grandiose art terms is known as "contraposto," the twisting of the body on different planes. The resulting poses are more life-like than anything that had come before. The actual figures have a sense of three-dimensionality that is quite unprecedented and incidentally, the twisting allowed the larger birds to be incorporated within the page size without reduction. Paillou's animals were perhaps less successful than his major birds, but they too represent a quantum leap beyond the primitives in terms of size, pose and texture. The book itself is a monument to chauvinism, being an attempt to show the rest of Europe just how superior the English breeds were. Pennant gave credit for this to the marvelous English weather which, to this day, foreigners fail to appreciate.

British Zoology has been called the last great "old" book. The wonderful black and red lettering on the title page, the tactile quality of the broad sheets of paper, the lavishly extravagant wide borders, and the full calf bindings all speak of an age that was rapidly drawing to a close. The transition into the modern age was complete by the time the 4th edition appeared in 1812. The "f" and "s" confusion in the typeface had been resolved, but more importantly, the engraved title pages which combined text and illustration were of extraordinary beauty. For students of page design, the title pages of the 4th edition rank among the absolute classics of all time. But historically, from an artistic and scientific point of view, it was the great folio that remains supreme. The size and splendor of the plates and the broad range of animals that were covered had international impact.

In 1770, within four years of Pennant's great folio, Cornelis Nozeman began producing the five-volume foliǒ *Nederlandsche Vogelen* (1171) with 250 hand-coloured copper engravings. Science and art were subordinated to a Rococco flamboyance that make the result a delightful period piece. Its success underlines the broadening interest in such sumptuous undertakings.

Here we must digress for a moment to introduce a key personage who, along with two others, has had great impact on the popularization of natural history. He was not an artist, nor a true scientist, but rather a keenly observant rural cleric, the Rev. Gilbert White, who wrote a series of letters to Thomas Pennant describing the life around his parish. The letters were pulled together in the 1789 great pastoral

Hand-coloured copper engraving from *Nederlandsche Vogelen* by Cornelius Nozeman. Amsterdam, 1770-1829.

classic, *The Natural History and Antiquities of Selborne* (RBC) and (274). So far there have been more than eighty editions, and it has seldom been out of print.

The Rev. White recorded every fact he could—the arrival of the swallows, the habits of the field mice, the nesting of the various birds. He described everyday events with a timelessness and gentleness that made even simple observations seem scientifically important. He prepared the groundwork for the absolutely fabulous growth of interest in wildlife that produced the environment in which art could and did flourish. By his influence, White ushers in the golden age of popular books published during the last decade of the 18th and first forty years of the 19th centuries.

The second half of the 18th century was a period of great colonial expansion. Britannia ruled the waves, and therefore waived the rules with an immunity that was guaranteed by her ships of the line. The Seven Seas became the stage for the Five C's; in the name of conquest, conversion, colonization, commerce and cartography, Great Britain expanded an empire on which, it was rightly said, the sun never set. Centuries earlier the outlines of the continents had been fairly well established, but penetration into the land masses beyond the coasts tended to be minimal other than along major waterways. Much of North America and Australia remained *terrae incognitae*, and Africa remained truly the Dark Continent. By the 18th century, however, the pace of inland exploration—regardless of its motives—quickly accelerated. Among the leisure classes, one of the most popular pastimes was to haunt the bustling docks of the East India Company and other great trading monopolies to greet the ships returning with an incredible array of new species, some well preserved, others besodden with dark navy rum, but all fascinating. Never before in the history of man had there been such interest in wildlife. Collecting became an obsession with every aristocratic and new captain of industry vying to add "unique" items to their "cabinets." Unscrupulous dealers supplied the truly unique—"hummingbirds" from New Zealand and other species from equally inaccurate locations. From the publishing end, encyclopedic works sprang up in great profusion. The subjects covered the whole range of natural history, but the emphasis was on new discoveries, which, if new to science were accompanied by letter-press in both English and Latin. Several members of the parrot family from Australia were thus introduced to the European public.

So, too, were introduced several less significant species which had been discovered through the relatively new toy, the microscope. The world could share the enthusiasm evident in Leeuwenhoek's famous letter to the Royal Society from October 9, 1678 "...that no more pleasant sight has ever yet come before my eye than these many thousands of living creatures, seen all alive in a little drop of water..."

It is tempting to say that the introduction of the microscope brought natural history books back full circle to a medical tradition similar to that of the early botanical books, but the term "art" would have to be stretched beyond acceptable limits. Nevertheless, the impact of the microscope cannot be overstated. We do not know who first invented the instrument, but we can trace its name to John Faber who, called it a "micro" scope because it permits a view of minute things. It was Robert Hooke (1635-1703) who opened the eyes of Europe to the possibilities of the instrument in his *Micrographia* which was published in 1665, and contained fifty-seven illustrations by Hooke himself of such wonders as the compound eye of a fly. New insights in science were supplied by Jan Swammerdam (1637-1680) in whose *Historia Insectorum Generalis* (850) the intricate anatomy of insects was finally revealed, along with the proof that their reproductive systems were biologically consistent, not, as had been previously assumed, based on spontaneous generation from rotting meat and assorted slimes. Nehemiah Grew, a London physician, produced a prolific number of drawings from studies and observations with his microscope. Although most noted for his expertly illustrated *The Anatomy of Plants* (398) published in 1682, Grew also contributed much to the knowledge of comparative animal anatomy with his *Comparative Anatomy of Stomachs and Guts* (212) which was subjoined with *Musaeum Regalis Societatis* (212) and published in 1681.

While microscopic investigation led to a whole new branch of natural history, the main thrust of natural history illustration continued to be along more conventional lines. Volumes by the hundreds and plates by the tens of millions were published between 1790 and the 1840's. One of the most significant names of this period, and in fact of the entire history of animal art, was that of Edward Donovan, collector, naturalist, artist and author, whose encyclopedic interests resulted in close to fifty volumes containing more than one thousand plates on subjects ranging from the insects of India and China to a comprehensive study of British fauna.

Donovan is by far best known for his *Natural History of British Birds* which began publication in 1799. From a scientific draftsmanship point of view, the birds were somewhat stiff, but adequate. It was their hand-colouration that set them apart from anything that had appeared before or since. Donovan's plates are the most meticulously coloured engravings ever produced. Unlike others which relied heavily on engraved lines and cross-hatchings for shading and colour variation, those by Donovan achieved their jewel-like quality through the use of lavish amounts of different pigments. The actual engraving underneath was so faint that it often became obliterated after the paint had been applied. Under a magnifying glass the details of the individual feathers can be seen to have been applied with a single-hair brush.

In Donovan's more than twenty volumes of insects (204, 692-695) he carried realism to an even higher degree. Artistically his layouts had an Oriental quality that raised them above the artistic level of his contemporaries—and his floral settings were beautiful. Predictably his colouration was phenomenal. He was able to achieve a powdery effect on the wings of his moths and butterflies which, when seen in the originals, is palpable. When called for, he used real gold paint for the metallic effect on certain insect bodies and wings; and for the

Engraving showing microscopic detail from Nehemiah Grew's *Anatomy of Plants*. London, 1682.

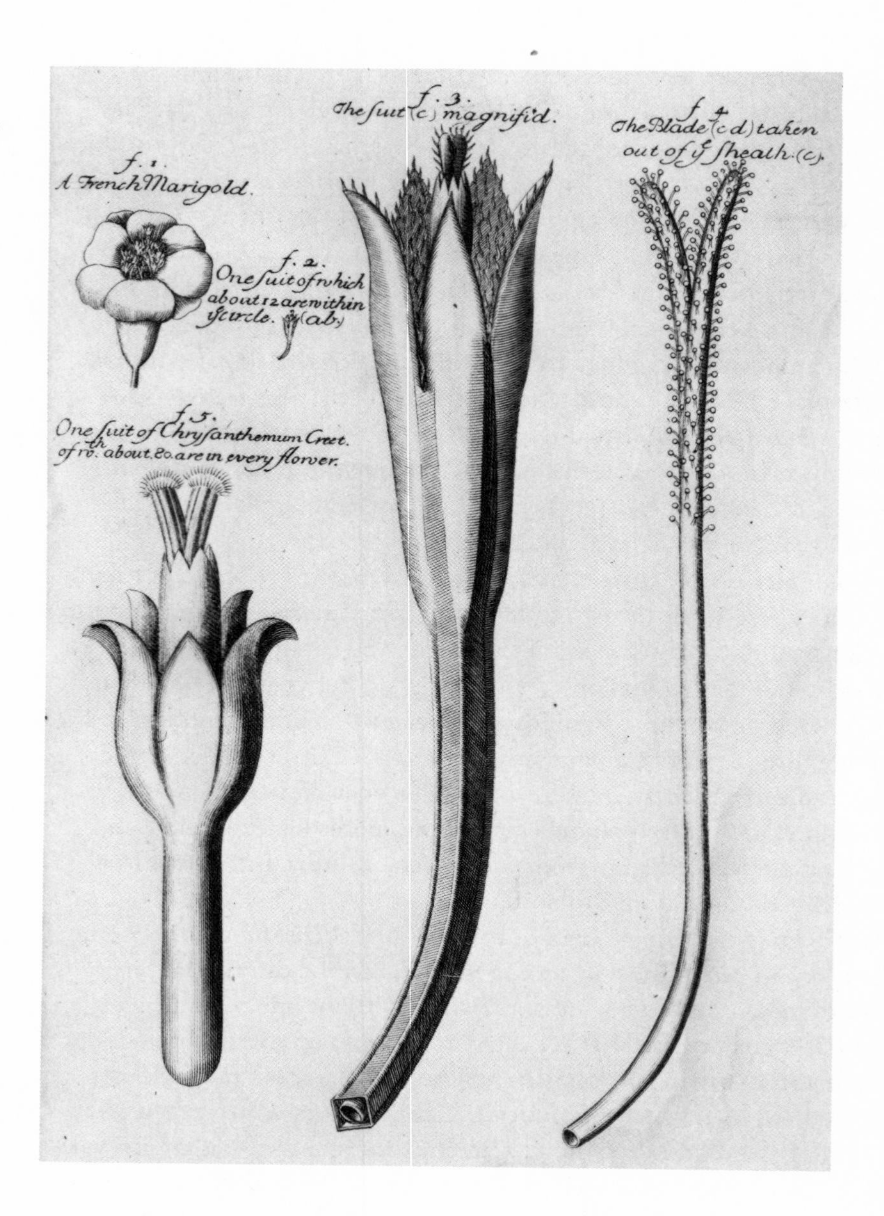

shimmering iridescence of dragonfly wings he used a substrate of colour with an overlay of clear varnish, the results of which cannot be reproduced by even the most modern printing techniques.

Donovan's shells surpassed those of his time as well. Subtle but complex rules of mathematical perspective came into play in rendering a sense of depth, and colouration again played a dominant role. His *Natural History of British Fishes* (692) contained some of the most detailed portraits of all time, down to the last scale. The colouration was brilliant, but artistically he usually preferred the safe route of strict scientific profile. But sometimes his imagination went berserk with flying fish actually flying hundreds of feet above the ocean among a flock of geese.

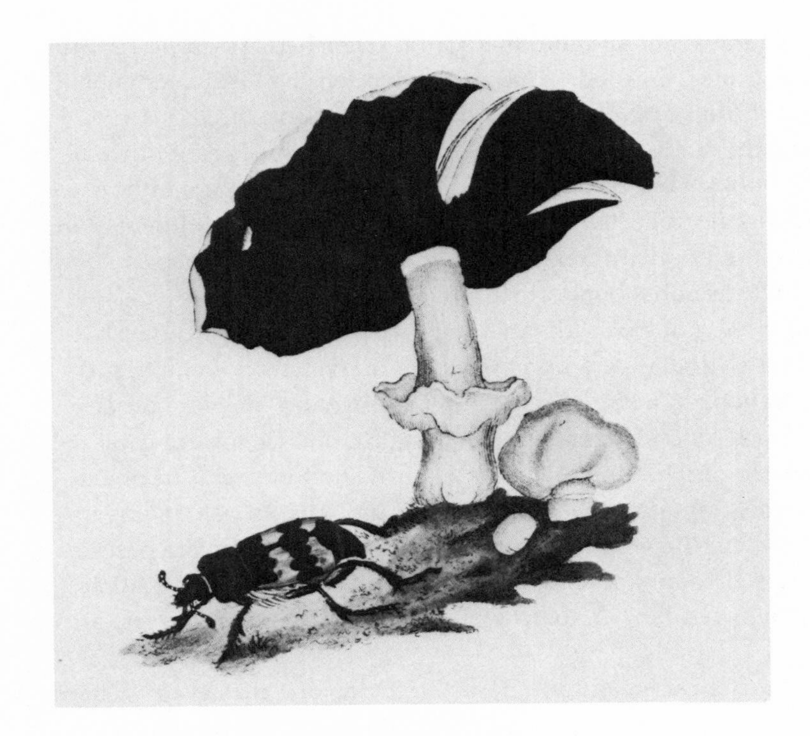

Engraved hand-coloured illustration from *The Natural History of British Insects* by Edward Donovan. London, 1793-1802, v. 1.

Unfortunately, Donovan felt obliged to try his hand at mammals. Facial expression, muscle tone, foreshortening, the breaking of fur, and the lack of familiarity with the living subjects all brought Donovan's limitations sharply into focus. Far better to remember him for his birds, insects, fish and shells.

Donovan's small octavos during their day were popular and went through a few editions, mostly for the profit of others. By 1833 by his own piteous account, he claimed that greed and dishonesty on the part of his publishers and booksellers had illegally deprived him of £70,000, and he was reduced to begging. His once vast collections of natural history objects from around the world had long since been sold off. Today, Donovan is forgotten, largely because of the scientific weakness of his texts, but also because his laborious colouration put the price of his books beyond the means of the masses. Each plate was, in effect, an original work of art and therefore not well-suited to efficient publishing. As the cost of such meticulous work proved prohibitive, and in view of the fact that the time consumption could not be sustained in a more hurried age, one of the most remarkable publishing ventures of all time died. The legacy of more than one thousand hand-coloured engravings had negligible lasting impact on the history of natural history art.

It was Donovan's great contemporary, Thomas Bewick, whose work was to mark the turning point in the history of animal art and in the popularization of natural science. Along with Gilbert White and Roger Tory Peterson, Bewick, it can be argued, is one of the three most influential naturalists of all time. In 1790 he published *A General History of Quadrupeds* (634, 1800 ed.), and in 1797 added the first volume of his *Figures of British land birds* (927) dealing with land birds, water birds being the subject of the second volume which appeared in 1800 (930). The books went through many major editions—both authorized and pirated—and collectively they rank among the most influential in broadening the popular interest in the natural world. Their greatness lay in the illustrations, both artistically and technically.

Bewick, while not truly "discovering" wood engraving, certainly was the first to understand and exploit its marvelous possibilities for book illustration. In a stunningly simple innovation, Bewick turned the woodblock on end, thereby presenting a drawing surface composed entirely of the miniscule ends of each shaft of grain. In practical terms, the grain disappeared, allowing the artist to use engravers' tools rather than the cruder gouges used in woodcuts for the preceding centuries. The new freedom permitted the most detailed feather work, and allowed Bewick to do wonders with trees and grasses, and even water.

This technological breakthrough in the hands of a master artist permitted the finest natural history art to be inexpensively produced for mass distribution because the illustrations could be integrated into the body of text and printed as a unit, something that had not been feasible with copper or lithography.

More than just a brilliant engraver, Bewick was also an artist of towering proportions. He was one of the very first to really go directly to nature for his subjects whenever possible, so his birds are usually scientifically correct. Accuracy and artistry are not necessarily synonymous but in Bewick's case they tended to be. As an example of layout, the tree-creeper demonstrates the artist at his best. The white breast is contrasted against the dark shadow of the tree, and the tawny back against the highlight of the upper bark. Two leaves add balance, with three-dimensionality achieved through a white highlit curl at the top, a medium tone for the main surface, and a muted intensity for the shadowed underleaf. All this was achieved with black and white lines of different widths and depths, and all in a masterpiece less than two inches across.

Typically, Bewick's birds and quadrupeds were superimposed on a highly detailed background—forests, rivers, fields with scarecrows, hunting scenes, or farms alive with other animals. For instance in his famous engraving of the rooster, the bird towers over his barnyard with six chickens, two other roosters, four chicks, two thatched buildings, a tiled wall, a ladder, assorted pebbles and clumps of grass—all in about an inch and a quarter. In his engraving of a mastif there are more than a dozen other dogs of recognizable species, and under the belly of the hartebeest there is a perfect miniature lion less than a quarter of an inch long.

Bewick's miniature masterpieces had a sense of acceptance of Nature's ways without betraying any anthropomorphism, editorializing or moralizing. His cur fox killed the chicken because that is what foxes do. There is no criminal, no innocent victim, just two players in Nature's pageant.

At the end of most chapters in his books, Bewick inserted a vignette—a tail piece—depicting some scene he himself had witnessed during his weekly walk of twenty miles or so, in all kinds of weather, to visit his parents. He recorded everything with compassion, accuracy, and sometimes with the ribald humor that was so much a fixture of rural life in the late 18th, early 19th centuries. Bewick captured his Age without resorting to the crude and gross characterizations of William Hogarth, Thomas Rowlandson, or John Leach. Artistically, he was the first to blur the spokes of a turning wheel, and to show all four feet of a galloping horse off the ground at the same time. His vignettes were monumental art conceived on a miniature scale—a beautiful record of a lifestyle that is gone.

Bewick, who was also the first to recognize that his thumb print was truly "His Mark" and who used an engraving of it as a receipt, left his mark not just on art, but also on people. In a touching description of their first meeting, Audubon wrote: "When I parted from Bewick that night, I parted from a friend. "When the master from Newcastle-upon-Tyne died in 1828, all of England went into mourning for they all felt that they too had parted from a friend. Seldom has anyone enjoyed such widespread affection during their lifetime, and seldom has posterity been indebted so much to one man.

Bewick's students and imitators carried on and illustrated a wide variety of books not only in his style but also in his format. Among the best known are the three volumes of William Yarrell's *A History of*

"The Mastiff," engraving on wood by Thomas Bewick from *A General History of Quadrupeds*. Newcastle-upon-Tyne, 1800.

British Birds (1284) which, with its subsequent editions and revisions, became the standard text between 1837 and 1843. Most of the 550 wood engravings were by J. Thompson, a skilled engraver who, nevertheless, lacked Bewick's spontaneity.

Bewick planned but never began a history of fishes, although some of the engravings for the intended project actually appeared in the later editions of the *Birds* and *Quadrupeds*. In Bewick's absence, Yarrell undertook the task and produced a masterpiece, *A History of British Fishes* (874). Most of the fish were expertly engraved by Henry White, one of Bewick's better pupils, and the format was copied from Bewick's books, using vignettes as tailpieces at the end of the chapters. These were for the most part drawn and engraved by Ebenezer Landells, perhaps Bewick's most successful pupil, who for many years was to be the leading engraver for the *Illustrated London News* and *Punch*. Some of the drawings were done by such notables as the great landscapist, John Constable, and by Edward Lear of *Nonsense Book* fame, about whom we shall hear more presently.

Bewick's influence spread rapidly across the sea. An American edition of his *Quadrupeds* appeared in 1831, with the drawings re-engraved by a Mr. A. Anderson. Much more successful was Thomas Nuttall's *A Manual of the Ornithology of the United States and of Canada* (1172) illustrated by Hale. This work first appeared in 1832, and the text was reprinted well into the 1900's. Many of the engravings by Hale were based on those of Bewick, or in some cases were copied directly. The raven, osprey, magpie, chickadee and shrike are straight plagiarisms. In numerous other plates he borrowed decorative elements: the twigs and berries in the American robin plate are taken from Bewick's engraving of the closely related European Fieldfare. Hale also stole from Audubon and Wilson.

Thanks to Bewick and his successors, popularly priced books of excellent quality were available to a greatly expanding reading public. Artistically the engravings were blessed with the introduction of Whatman-type papers with surfaces that were far smoother than the traditionally manufactured chain-laid paper. Clarity of printing was greatly enhanced. It was Whatman paper that was to be used by Audubon for his *Birds of America*.

The evolution of animal art took many turns. Here we must leave England and pick up the thread on the Continent. Along with England's John Gould, France's Jean Le Vaillant produced some of the most comprehensive sumptuous folios on rare and exotic birds (1123, 1124). One of his chief draftsmen was Jacques Barraband from 1767-1809, a flower and bird specialist who worked for the pottery firm at Sevres and for several tapestry weavers while at the same time teaching drawing at the Academy of Art at Lyons.

The 344 folio plates which Barraband drew for Le Vaillant were en-

"The Cur Fox" from Bewick's *General History of Quadrupeds.*

graved on steel, printed in colour, and then finished by hand. Even the elegance of the books, the remarkable technical achievement of the printer, Langlois, (who was also responsible for the matchless flower books of Pierre-Joseph Redouté) cannot obscure the fact that Barraband translated most faithfully the inert feeling of his (badly) stuffed models onto the plates. His technical sureness of line, however, placed him in such high esteem that he was selected to draw the largest bird plates ever produced in France to do justice to the glorious scientific discoveries made during Napoleon's Egyptian Expedition (RBC). These plates, published in 1805, assured Citizen Barraband a lasting fame in France. Worthy of mention, too, are the comparable plates in the 1802 folio of Audebert's *Oiseaux Dorés* (886).

As decorative pieces, Barraband's birds are striking, and the detailed feather work and bold colouring—especially of his parrots and birds of paradise—set the style for French bird art at the beginning of the 19th century. French critics still claim, with some merit, that Barraband's birds achieve a certain charm *because* of their dried-out stiffness. In his defense, they point out that he only had stuffed specimens from which to work, and that the exigencies of scientific accuracy prevented him from exercising his creative talents to a degree which non-scientific freedom would have allowed. This longed-for freedom was hinted at in his birds of paradise, perhaps his finest work. And perhaps this hint was why William Swainson called Barraband the greatest of perhaps three artists who had ever been able to draw a bird before Audubon.

Barraband, regardless of chauvinistic attempts to inflate his reputation, had only passing impact on nature art. In his case, the printing of Langlois was more important than Barraband's artistry. The main developments were to come from American influences.

In the early 1800's an unsuccessful Scottish poet from Paisley went over to America where he embarked on a project to paint all the spe-

Steel engraving by Jacques Barraband from *Oiseaux Dorés* by J. B. Audebert and L. P. Vieillot. Paris, 1802 v. 1.

cies of North American birds. The plates for Alexander Wilson's *American Ornithology* (1281) from 1808 to 1814 were the first coloured engravings produced in the United States, and for this reason alone they would be numbered among the most important bird plates of all time.

His birds were scientifically correct despite their lack of imaginative presentation. The compositions were cluttered with several species thrown together to fill a page. Most of the birds were shown in profile, but he did occasionally try some quite difficult foreshortening in his flying birds. Most of the combinations of species made little or no sense, but sometimes with a macabre twist of humor he combined a predator with several prey in one plate.

Wilson was clearly better than anyone who had come before. His champions pointed with pride to such plates as the Ruffed Grouse, and tended to overlook the gaggle of waterfowl that more closely typifies the Wilson style. Audubon himself had to struggle against the entrenched backers of Wilson, especially his publisher, Ord, for recognition by the learned societies in America. Ord—and his lithographer Lawson—saw a catastrophic threat in Audubon's work to their own vested interest in posthumous editions of Wilson. In retrospect, Wilson's greatness lay in his beautiful prose; he had been, after all, a poet. His descriptions of the American birds in the wilderness setting still are very readable, accurate for the most part, and certainly evocative. For this reason above all others Alexander Wilson rightly deserves his title as "The Father of American Ornithology."

After Wilson's death, Prince Charles Lucien Bonaparte published the "continuation" of Wilson, *American Ornithology; or, the Natural His-*

Page from Alexander Wilson's *American Ornithology* lithographed by Alexander Lawson. Philadelphia, 1808-14.

Alexander Wilson's Ruffed Grouse.

tory of Birds Inhabiting the United States not given by Wilson (941) in three folio volumes from 1825 to 1828. In style it was the artistic continuation of Wilson as well. The birds are workmanlike but stiff, and with little rationale behind the layout. One plate, however, is of fundamental importance. In 1826, Bonaparte published the "Great Crow Blackbird," our boat-tailed grackle. The bird in the foreground, the male, by A. Rider was stiff, thick in the body, arbitrarily attached to the stylized branch, and generally exhibiting an inaccurate reconstruction of skeleton and muscle. The female in the background, by contrast, was very much alive. The muscle tone is there, and the sureness of line and understanding of anatomy is apparent. The bird was painted by John James Audubon, his first published work.

Audubon arrived in England in 1826, and soon was introduced to the large folios produced by the then leading British artist, Prideaux John Selby, *Illustrations of British Ornithology* (1231) published over a period of thirteen years from 1821 to 1834, and reissued by Bohn in 1841. The numbering and dating of the various plates and editions is notoriously confused, with watermarks on the Whatman paper from 1846 being difficult to explain.

Selby had drawn most of his specimens life size, an innovation only in magnitude, not in concept, for Peter Paillou had portrayed birds up to the larger grouse and sea gulls life size for Pennant's *British Zoology* back in the 1760's. The birds of prey were undeniably powerful even though Audubon in his diary had sniffed that he thought his son, John Woodhouse, could have done a better job. The plates had been rather coarsely engraved by William Home Lizars of Edinburgh, the best-known engraver in the British Isles at the time. It was Selby, and the great naturalist Sir William Jardine, who introduced Audubon to Lizars, whose firm was to engrave the first ten plates for the *Birds of America*. Selby's folio remains the largest British bird book ever.

The other leading British artist of the period was William Swainson, who from his self-appointed Olympian heights had previously dismissed Donovan as being "garrish" and amateurish. Swainson was admittedly a far better artist than Donovan, but his major work, *Zoological Illustrations* was also admittedly very dull from an artistic point of view. Its main significance lay in its being the first book to make use of lithography in animal art. His scientific profile of a toucan on an anonymous branch can hardly be called "art" but it was perhaps the best plate in the first series of three volumes. The second series of three was produced *after* Swainson had met Audubon, and the draftsmanship greatly improved. By 1831, Swainson had also done the lithographs for the volume on birds in the classic *Fauna Boreali Americana* of Sir John Richardson. By then he had blossomed into a superior artist, as can be seen in his beautiful magnolia warblers and evening grosbeak. The birds were very much alive, and the poses were much more imaginative, showing to advantage the field marks of the species. What had happened to Swainson in the interim was quite simply what had happened to the entire world of natural history and of art—Swainson had seen the art of John James Audubon, and had actually taken a few lessons from the unknown American.

Audubon is perhaps the most important name in the history of animal art. From the wilderness of America this self-taught genius burst onto the European scene without warning and without precedent. The course of animal art changed for all time when he displayed his paintings in Edinburgh in 1827. As an artist and a naturalist his contribution cannot be overstated. His perception was so astonishing that he saw and was able to record what no one else had mastered before. Writing for *Le Monde*, Philarète-Chasles enthused: "Imagine! A landscape wholly American, trees, flowers, grasses, even the tints of the sky and waters, quickened with a life that is real, peculiar, trans-Atlantic...It is a real and palpable vision of the New World, with its atmospheres, its imposing vegetation, and its tribes which know not yet the yoke of man."[6] This sophisticated Frenchman could not come right out and say, "It is great bird art."

That, too, would have been incomplete, for through his birds Audubon had painted the reality of a land which was, to the European of the 1800's, totally unfamiliar and almost unbelievable. Thousands of miles and years of travel through the wilderness had given Audubon an unparalleled exposure to the birds and animals in their living environment, allowing him to place them in their proper ecological settings, an approach which has been considered the norm ever since. Others had anticipated this style; Catesby, for instance, placed his birds in appropriate floral settings, but the use of such sizes of the flowers or plants and the birds or animals were not always in scale, sometimes not even remotely so. James Bolton as early as 1794 had used proper floral settings in his wonderfully decorative and personal *Harmonia Ruralis; or, an Essay towards a Natural History of British Song Birds* (940), one of the rarest and most delightful colourplate books. Another precursor of Audubon's approach was Patrick Syme, who in 1823 anonymously published the exceedingly rare *Treatise on British Song Birds* whose hand-coloured plates engraved by R. Scott are enlarged versions of Bewick-inspired portraits. And of course, Bewick himself used environmental backgrounds, but many of his trees, flowers and bushes are too stylized to be identified. Audubon's use of environmental elements was on a scale and of a complexity and accuracy that represents a radical departure from the past, even though antecedents can be found. It is also arguable that such earlier examples most probably were not available to him when he was finishing his major portraits in Henderson, Kentucky; Mill Grove, Pennsylvania; or Bayou Sarah in Louisiana.

Some critics were not so generous. They pointed with glee at Audubon's tree-climbing snakes with re-curved fangs, and to the numerous "distortions." "There is no truth here," huffed Thomas Lawson (Wilson's printer and engraver in Philadelphia), dismissing Audubon's work as inferior rather than admitting that his own skills were far too modest to engrave the magnificent plates. Audubon's originals for him were "too soft, too much like oil paintings."

Audubon himself anticipated the criticism when he wrote, in the Introductory Address of his *Ornithological Biography* (903) of 1831: "The positions may, perhaps, in some instances appear *outré*; but such supposed exaggerations can afford subject of criticism only to persons unacquainted with the feathered tribes; for, believe me, nothing can be more transient or varied than the attitudes or position of birds."

High speed photography and expanded field observation have confirmed that many of the so-called distortions are indeed accurate; other poses, especially those of the skins brought back from the west by Lewis and Clarke, underline the fact that Audubon was not familiar with the living bird.

There was also the nagging matter of the plagiarisms from Wilson.

Ruffed Grouse from John James Audubon's *Birds of America*. London, 1827-38.

Without doubt, at least seven of Audubon's plates contain birds that were absolute copies of Wilson. The engraver, Havell, it appears added two himself—but we assume with Audubon's approval—and in the case of the bald eagle, I am convinced Audubon purposely aped Wilson's poses to show how infinitely superior were his artistic talents. It seems extraordinary that as recently as 1961, Robert Cantwell could claim in his biography of *Alexander Wilson: Naturalist and Pioneer* (191) that Wilson's *American Ornithology* "remains unrivalled for the fidelity of the paintings," defending this absurdity with the convoluted logic that Audubon's purpose was to create "great art, rather than great birds." This is the same author who could write that "Catesby's wonderful bird paintings, challenged those of Audubon in their delicacy and freshness." Perhaps intentionally, Cantwell chose a Wilson plate for the dustjacket which contained the Everglades Kite, one of the birds that Audubon plagiarized.

Audubon's monumental *Birds of America* (892) was the largest bird book ever published, containing 435 double elephant folio plates measuring 29 1/2 by 39 1/2 inches, depicting more than 400 species, and containing no less than 1,065 life size figures, thousands of bo-

Bewick's Wren from Audubon's *Birds of America.*

tanical portraits, and vast numbers of insects, shells, and other appropriate ecological elements. The backgrounds contain views of identifiable cities, rivers, mountains, as well as glimpses of American rural architecture, farming, and Indian life. The original watercolours were etched on copperplates through a process known as aquatinting, first by William Lizars for ten plates, and then for the remainder by

Robert Havell, Jr. whose ability to re-create Audubon's paintings in all their detail (and to add more in some cases upon Audubon's written instructions) is one of the great technical achievements of all time. The plates were then hand-coloured under Audubon's supervision and sold on a subscription basis. Fewer than 200 complete sets were produced. The cost, however, was so prohibitive that even such subscribers as the Rothschilds felt obliged to cancel. Waldemar H. Fries did an in-depth analysis of Audubon's masterpiece in his book *The Double Elephant Folio* (1042).

It was immediately obvious that a wider audience was waiting. Audubon published an octavo edition with 500 hand-coloured plates lithographed by Bowen in Philadelphia (LA COLL). The reduced format and the softer texture of the lithography does diminish the artistic impact, and the size forced major revisions of some of the folio plates, the most obvious example being that of the Carolina Parakeet. Nevertheless, if the octavo edition had been the only edition—that is, if the double elephant folio had never been produced—then the so-called miniature Audubon would still have been a major turning point in the history of bird art.

Far too few realize that Audubon was also one of the great animal painters. Following the success of his *Birds* he undertook an almost equally ambitious project, *The Viviparous Quadrupeds of North America* in conjunction with the Rev. John Bachman of South Carolina who was to be the future father-in-law for Audubon's two sons. With life

Lithograph from *The Quadrupeds of North America*, by John James Audubon, 1849, v. 2

size being out of the question, Audubon chose the imperial folio format of approximately 20 by 26 inches. The softness of lithography suited the texture of fur most admirably, and so it was used rather than copper engraving or aquatinting. Many of the 150 animal plates are stupendous, the animals being accurate for the most part, and like the birds, being placed in their proper settings. Audubon's life of vicissitudes had begun to take its toll, and halfway through the enterprise he turned his brushes over to his talented son, John Woodhouse Audubon who was responsible for at least seventy-one of the plates, and probably a few more. History has denied the son the fame accorded to the father. Despite a real talent and a technical mastery that could capture the inquisitiveness of a ferret or the short fur of an ocelot, John Woodhouse lacked his father's creative spark. His best work was conceived by and totally influenced by the senior Audubon. It was fitting, therefore, that in John Woodhouse's outstanding portrait of an obviously mortally wounded deer, the young Audubon painted a miniature hunter which, upon close inspection, is unquestionably a portrait of his father.

As had been the case in the smaller edition of the *Birds*, the plates in the octavo editions of the *Quadrupeds* (622-624) suffered from the reduction in size. All 155 plates were re-drawn by John Woodhouse Audubon or William Hitchcock using a camera lucida to capture a remarkably good likeness of the original folio lithographs. Just how vital had been the father's influence becomes clear in the five "bonus plates" created by John Woodhouse after his father's death; the animals are woefully inferior. It is perhaps justice that Audubon slipped from childlike senility into peaceful death before the work was completed. Despite its shortcomings, the octavo edition of the *Quadrupeds* certainly ranks as one of the major works on North American mammals, and a significant milestone in the history of animal art.

Audubon also never lived to see his son's attempt to reproduce in 1860 the folio *Birds* through the still unperfected process of chromolithography. In the so-called Bien edition, (the chromolithographer was J. Bien of New York) some of the plates, printed in colour and then touched up by hand, were quite good, but the poor register of the printing process gave many of the plates an out-of-focus appearance. For various reasons, but mostly because of the Civil War, the project was abandoned after the completion of 150 plates, and remains, consequently, scarcer than the original.

John James Audubon had raised the level of natural history illustration to that of true art by any criteria. His sense of vitality, the magnificent layouts, the accuracy and the drama all set new standards for the modern era of nature painting. Once he had broken new ground, others followed rapidly. The talent had been there, waiting, as it were, in the wings, and could now burst forth to profit from the artis-

Wounded deer by John Woodhouse Audubon from *The Quadrupeds of North America.*

tic breakthroughs to satisfy the public demand for outstanding art.

One of the greatest names in the history of ornithological art scarcely ever completed a painting. He did the original rough sketches, and left the detailed drawing, lithography and hand-colouring to others. And yet this in no way diminishes the importance of John Gould. The testimony to his life's work is a series of 43 imperial folios, containing 3,159 hand-coloured lithographs, many of which rank individually among the finest nature portraits of all time. For scope, consistency and elegance, nothing rivals the opus of John Gould.

To Gould's factory came birdskins from all over the world. After laborious study and classification, he turned his sketches over to a series of talented artists who completed the final drawings, transferring them onto stone, and finishing the lithographs in wonderful hand-colouring. After the death of his first assistant and wife, Elizabeth Gould, he employed W. Hart and H.C. Richter whose work deserves greater fame. So much were they guided by Gould, however, that their styles never became individualized, and thus their paintings have tended to be lumped under the general catch-all of "Gould plates." Two other Gould artists, however, established reputations independently, Edward Lear and Joseph Wolf.

Gould and his artists produced regional works on a vast scale—the *Birds of Europe* (1055), *Birds of Australia* (1054)—or on more local areas—*A Century of Birds from the Himalaya Mountains* (1057), and *The Birds of Great Britain* (1056). Monographs on *Hummingbirds* (1058) and *Trogons* (1063) and *Partridges of America* (1061) dealt with specific families. Typically, the male, female and frequently the young of the various species were depicted together, set in floral and geological surroundings that were excellently handled. His *Mammals of Australia* (724) carried the style into an expanded zoological area. The feeling was more relaxed than was the case with the almost Gothic gore of some of the Audubon plates. In fact, criticism for a lack of tension has been levelled against the Gould approach.

Few people realize that Edward Lear was an outstanding bird artist, although in his day he was lionized, not only as a Gould artist, but also for the extraordinary series of paintings in his 1830-1832, *Illustrations of the family of Psittacidae, or parrots* (1115) which he published when only 18 years old. The folio of hand-coloured lithographs appeared without text, an avian counterpart of the florilegia, and has remained unchallenged in the parrot field until Joseph Forshaw and William Cooper's *Parrots of the World* (1037) almost a century and a half later in 1973.

Lear is far better known throughout the English-speaking world for his *Book of Nonsense* which he compiled as a diversion for the children of Knowsley Hall where he had been invited to paint the birds and animals in the private menagerie of the Earl of Derby. Some of the natural history drawings appeared in *Gleanings from the Menagerie and Aviary at Knowsley Hall* (726) in 1846. Assurance of Lear's reputation as a scientific draftsman in the natural history community, had by then already been well-established, and he was asked to contribute many excellent plates for Gould, as well as for the volume on pigeons in Jardine's *The Naturalist's Library* (223, 1097), a publishing venture about which more will be said later. By 1837, deteriorating health forced Lear to move to Italy. He returned to England for a brief visit in 1845 and found that his reputation was still so high that he was invited to give private drawing lessons to Queen Victoria. By the 1850's he had broken completely from his bird art and was publishing a series of *Journals of a Landscape Painter* in *Southern Calabria*, *Albania*, and *Corsica* through to 1870. The soft lithographs in these and other works show just how underserved is the relative eclipse of Lear's artistic reputation. He was a great painter in a broad field of art.

The best of Gould's artists—in fact, one of the best animal painters of all time—was Joseph Wolf. Archibald Thorburn called his art "faultless" and Sir Edwin Landseer stated that he was "without exception the best all-round animal painter that ever lived." When scarcely 20 years old he supplied the matchless portraits—especially of the Greenland falcon—for Schlegel's epic folio *Traité de Fauconnerie* (1385). When John Gould first saw Bulwer's Pheasant he said that the only man in the world who could do justice to such a splendid creature was Joseph Wolf. During his lifetime from 1820 to 1899 Wolf became acknowledged as the leading wildlife painter of his day.

"We can see distinctly only what we know thoroughly,"[7] said Wolf,

"Picus Mahrattensis" from *A Century of Himalayan Birds*. "Drawn from Nature & on Stone by E. Gould. Printed by C. Hullmandel." John Gould, London, 1832.

Hand-coloured lithograph by Edward Lear from *Illustrations of the Family of Psittacidae, or Parrots*, 1832.

who spent his days absorbing as much natural history as any man that ever lived. He mastered landscape, realizing the importance of the settings to heighten the effect of the animals he so endlessly observed. Watercolour, oil, pencil and above all charcoal, were Wolf's preferred media. In his sketches the eye saw far more than the lines, but when called for his unmatched powers of observation and technical skill permitted the most detailed rendering of feathers, scales, muscles, fur or botanical elements. This correctness is only appreciated by a handful of people who can tell at a glance the right foot of a bird from the left, if it is not attached to the body. Whereas the other artists often slavishly copied "the distortions of the bird-stuffers," Wolf never suggested a stuffed specimen.

Wolf was resolutely determined not to become a mere scientific draftsman. For generations science and art suffered from mutual misunderstanding, even contempt. As Wolf said, "some ornithologists don't recognize nature—don't know a bird when they see it flying. A specimen must be well dried before they recognize it...It is impossible, for instance, for a mere museum man to know the true colour of the eyes."[8] Many well-known artists, on the other hand, boasted of having absolutely no knowledge of natural history. Wolf was both artist and naturalist, and brought as no man before, art and science together.

The artistic superiority of Wolf and other Gould artists over many of their contemporaries is obvious when his paintings are compared to those of, for example, Jean Théodore Descourtilz, whose monographs or national avifauna folios from Brazil are of more historic than aesthetic importance. An apparently unique copy of Descourtilz' *Oiseaux Remarquables du Brésil* (1003) from 1843 is of specific interest as an indication of how thick hand-colouring obliterates the finer work of the lithographer. It nevertheless demonstrates the ambitious work undertaken in such isolated centers as Rio de Janeiro. It is quite interesting to note that Gould's *Hummingbirds* (1058) was being produced at the same time.

To appreciate Wolf fully it is best to look at those paintings in which he had a free hand. Such are the two folios of plates he did for the London Zoo entitled *Zoological Sketches* (869) published in 1867. Here he proved he was at home with any kind of animal, whether it was a green tree boa slumbering ponderously under the draped coils of its sated body, or a fennec fox with ears palpably alert. He chose moonlit scenes for elephants and storks, and convincing African or desert locales for indigenous species.

In his day Wolf was appreciated on two levels. On the popular side his works were copied, engraved on wood, lithographed, and generally badly mauled by very inferior craftsmen. The public was delighted despite reproductions that Wolf himself called a fiasco. On the

Leopard drawing by Edward Lear published in *Gleanings from the Menagerie and Aviary at Knowsley Hall,* 1846.

The African Elephant by Joseph Wolf from *Zoological Sketches*, Made for the Zoological Society of London. 1861-67.

higher plane his true greatness was understood by a circle of experts to which it was no small honor to belong.

It is fitting that we leave Wolf with the appearance of Daniel Giraud Elliot's great folio, *A Monograph of the Phasianidae or family of the pheasants* (1022, 1872 ed.) from 1864 and 1865, considered by many to be the most beautiful bird book ever produced. Wolf did the free and easy charcoal sketches, and left the final detailed drawings to a young newcomer of whom Wolf admiringly said, "that young man really observes his birds." This protégé of Wolf's was John Keulemans. Inexplicably, Elliot, in his efusive preface fails to mention Keulemans while praising Wolf and the lithographer, Joseph Smit. Later, Keulemans was to supply the beautiful lithographs for the second edition of Elliot's *Pittidae* or Ant Thrushes (Elliot himself having supplied the plates for the first folio edition when the French artist Oudart died before commencing the task).

Every generation has one or two artists who dominate the world of ornithological art. From the late 1860's through to the mid 1890's virtually every major book contained the hand-coloured lithographs of John Keulemans. Like his master, Joseph Wolf, he laboured unceasingly to capture the exact configuration of the feather patterns, structure, and sense of life. To an extraordinary degree he succeeded, despite the fact that almost without exception he worked from museum skins.

The list of works illustrated by Keulemans is a compendium of the acknowledged classics of the twilight years of the great birdbooks: Elliot's *Hornbills* (1021), Blaauw's *Cranes* (937), Buller's *Birds of New Zealand* (963), Sclater's *Puff Birds and Jacamars*, Rowley's *Ornithological Miscellany* (1210), Dresser's *Rollers* and *Bee-Eaters* (1008), Seebohm's *Sandpipers* (1229), Shelley's *Sun-birds* (1238), and the monumental nine-volume *Birds of Europe* by H. F. Dresser. In all, Keulemans was responsible for well over two thousand fully finished bird and animal portraits during a working career that spanned over half a century. Keulemans was the greatest traditionalist in the 19th century mode of having the bird totally dominating the picture in a lifelike but still more or less static pose.

Hand-colouring obviously had limitations as far as mass-produced books were concerned. Combinations of mechanical colour plus touching up by hand were used effectively by the end of the 18th century. It was not until the late 1820's, however, that truly successful examples of this approach are to be found in non-botanical books. Starting in 1828, Edward Griffith added up-dated material to Baron Georges Cuvier's *The Animal Kingdom* (676) in an elegant 16-volume edition printed for Whittaker, which contained hundreds of engraved plates, most of which were printed in black and then coloured by hand. Many, however, were inked in two colours on different portions of the copper, and then touched up.

Chromolithography, in its earlier experiments tended to produce garish and out-of-focus plates, but it was a popular—and inexpensive—medium for books targeted at the lower end of the economic scale. Lloyd's *Natural History*, a 16-volume set printed in London in 1897 was contemporary with Lord Lilford's exquisite seven volume *Coloured Figures of the Birds of the British Islands* (1125), and the difference in quality is remarkable. The former can be described as washed-out, the latter reached a peak of perfection never before seen, truly challenging hand-colouration for subtlety of tone. John Millais' three folio volume from 1904 on *The Mammals of Great Britain and Ireland* (789) contained similarly wonderful chromolithographs of the paintings by Thorburn, and William Beebe's classic four folios *A Monograph of the Pheasants* (920) again reached sublime heights. Such quality was not universal. As late as 1895 Jacob Studer permitted truly inferior chromolithographs for the plates in his pretentious *Birds of North America* which he claimed were "drawn and coloured from nature." A greater crime was committed by the Swiss lithographer Thurwanger and French printer Lemercier who combined to distort the original

Gulls by J. G. Keulemans from *A History of New Zealand* by Walter Lawry Buller. London, 1873.

paintings of Léo-Paul Robert in the undated (ca. 1879) folios, *Les Oiseaux dans la Nature*. The wood engravings, however, of the 1933 posthumous edition of *Les Oiseaux de Chez Nous* clearly indicate that Switzerland's Léo-Paul Robert a national treasure at home, although little known abroad, was easily one of the top two or three bird artists of the 20th century.

Modern printing methods alone could do justice to the best artists;

Photogravure from *A Monograph of the Pheasants* by Charles William Beebe.

chromolithography could not, and died a natural death, especially when photography blurred its way onto the scene in the 19th century. A word in passing about photography; the photogravures of the natural habitats in Beebe's *Pheasants* (921) have a tactile quality and a tonality that far surpass anything that has been achieved since. Photogravure was an artistic medium that should never have been allowed to die.

The other main aspirant for quality colour work was an outgrowth of wood engraving. Just as each colour in chromolithography required a separate drawing on a separate stone prior to over-printing, each colour in chromoxylography required that a separate engraving be made before the various blocks were printed sequentially. One firm stands out above all others as the exponent of this method, that of Benjamin Fawcett, Printer and Engraver from Driffield, England. Fawcett, and his chief artist, Richard Alington, were reponsible for the more than 350 plates for the Rev. Francis Orpen Morris' *History of British Birds* (1155) mentioned earlier as the standard British bird book for over half a century. The plates were engraved on wood, printed in two colours, and touched up by hand. The same plates were used in

Chromoxylograph of fishes by A. F. Lydon from William Houghton's *British Fresh-water Fishes*. London, 1879.

literally dozens of other books. Fawcett also engraved the chromoxylographs for Beverley R. Morris' *British Game Birds and Wildfowl* (1154) which also enjoyed enormous popularity over the last decades of the 19th century.

Without question, the zenith of this printing technique was reached in F.G. Dutton's and William Thomas Greene's three-volume royal octavo *Parrots in Captivity* (1074). It was entirely produced by Fawcett, to whom it was dedicated. The colour printed plates, in most cases finished by hand, were drawn by A.F. Lydon who was associated with Fawcett for about 30 years. Of equal excellence was Rev. William Houghton's two folio volume *British Fresh-Water Fishes* (741), first published in an undated edition from 1879. The 41 full page plates were also by Lydon whose skill, coupled with Fawcett's, created plates with such subtle colouration that most biographers assume them to be fine chromolithographs rather than chromoxylographs. Lydon was also responsible for the wood engravings of river scenes interspersed among the pages of text. With the fine boards decorated with gold and black, the original publisher's binding is considered a classic of the genre, and the book itself is thought by some to be the most beautiful example of the bookmaker's art of the Victorian period.

Chromolithography and chromoxylography were mechanical responses to a market need, and as such were successful. Artistically, however, only Fawcett for coloured wood engraving, and the Berlin firm of Greve for chromolithography did justice to the artists' input.

A new dawn was breaking, and, happily for the art historian, the last of the old and the first of the new were combined in one book, Lord Lilford's *Coloured Figures of the Birds of the British Islands* (1125) published between 1885 and 1897. Of the 421 plates, sixty-odd were hand-coloured lithographs supplied by Keulemans; the others were chromolithographs by the brilliant Archibald Thorburn. The fact that modern catalogs generally list all 421 plates as being chromolithographs might suggest that chromolithography by such firms as Greve & Co. in Berlin had finally equalled the quality of hand-colouring.

Hand-coloured lithograph by J. G. Keulemans from Lord Lilford's *Coloured Figures of the Birds of the British Islands*. London, 1891-97, v. III.

Although Thorburn was only 23 years old when he received his first commissions from Lilford, it was immediately obvious that he had surpassed Wolf and Keulemans. Thorburn's approach was totally new. He thought in terms of a total composition rather than in terms of a bird with a composition around it. He approached his work as an artist first and lover of nature second. He introduced accurate dynamics of floating bodies, reflections and wave patterns, and a host of other facets of the natural world which he effortlessly incorporated into his deceptively simple masterpieces. Above all, Thorburn's paintings achieved a remarkable sense of depth. More is implied by this

Chromolithograph of Wood-Pigeon by Archibald Thorburn from Lord Lilford's *Coloured Birds of the British Islands*, v. IV.

"The Wild Cat" by Archibald Thorburn from *The Mammals of Great Britain and Ireland* by John G. Millais. London, 1904-06, v. I.

than mere perspective which can be reduced to a mathematical formula easily mastered by a draftsman. Using sophisticated techniques he would build up his pictures in a series of planes with elements of the birds and backgrounds leading the viewer into the picture, trailing away to a hazy horizon. Implicit is an understanding of the role of atmosphere and light—direct, diffused or reflected.

Such refinements showed Thorburn to be a consummate artist. In the work he did for Millais' *The Mammals of Great Britain and Ireland* (789) and his own *British Birds* (1263) and *British Mammals* we can see that animal art in the academic sense had finally come of age.

Some great artists from other fields now felt comfortable about turning to bird art. One such was Charles Whymper, one of that class of indefatigable Englishmen of the second half of the 19th century who resolved to become reincarnations of the Renaissance Man. Whymper was the first to climb the Matterhorn and dozens of other Alpine peaks, and undertook extended climbs among the active volcanoes in the Cotopaxi region of Ecuador. His two most famous books, *Scrambles Amongst the Alps* from 1860-1869 and *Travels Amongst the Great Andes of the Equator* from 1892, are the classics of mountaineering, but they are also filled with acute observations on the local

Barn-owl by Charles Whymper from his *Egyptian Birds.* London, 1909.

geology, weather, flora and fauna. The precision of his comments is largely due to the fact that Whymper was also a first-rate artist, supplying the illustrations for his own books. His contribution to bird art, although largely unknown was, nevertheless, significant, for in his *Egyptian Birds* (1279) from 1909 he represented the various species most likely to be seen by tourists in the most touristy locations—tombs, harbours, oases, ruins, and along the low-lying banks of the Nile. This is the obverse of the coin used by artists such as Audubon who placed their birds in their so-called natural settings.

With the encroachment of man, natural history artists began to incorporate the man-made elements of civilization into their work. Of particular note was Whymper's barn owl perched atop a fallen stone which is engraved with an ancient hieroglyph, ironically depicting an owl. Back in the 1700's Edwards had placed a kangaroo rat in the desert with the pyramids in the background, but this was an isolated affectation, not a coherent approach. Whymper's portrait of the gulls in the harbor of Alexandria is as much an interpretation of the Lesser Black-backed and Black-headed Gulls as a mood piece of sunset over a mosque along the river. The plate of the cattle egrets attending their host buffalo gives a tantalizing hint of what a brilliant wildlife painter

Whymper would have been had he chosen to scale the summits of this career instead of the highest peaks in the Alps and Andes.

In the 20th century animal art progressed at a fantastic pace, much of it destined for publication. In 1909 a new dilemma was posed by the brilliantly talented but slightly possessed Abbott Thayer, with his *Concealing Coloration in the Animal Kingdom* (853). Birds and animals generally, said Thayer, were coloured the way they were so that you would not see them. Therefore, for Thayer, animal paintings should stress the obliterative. Carried to its extreme, the object would blend so perfectly with its surroundings that it disappeared altogether! Copperheads in fall leaves and a famous ruffed grouse in an autumn forest were cases in point. Thayer gave up a promising career as one of America's leading impressionists to pursue his obsession with camouflage to *reductio ad absurdem*.

Nevertheless, the impact of Thayer must be stressed because his pupil, Louis Agassiz Fuertes, was to become the greatest painter of his time. At the height of his powers, Fuertes died in a tragic mishap when his car was hit by a train. He had just completed the brilliant portfolio of *Abyssinian Birds and Mammals* (715) which were published posthumously in 1930 by the Field Museum in Chicago.

During his career Fuertes was always tugged in two directions. His mentor, Abbott Thayer, was forever complaining that his birds stood out too much, while his editors, Gilbert Grosvenor and Frank M. Chapman, publishers respectively of the *National Geographic Magazine* and of *Bird Lore*, the fore-runner of *Audubon Magazine*, who paid Fuertes' salary did not hesitate to complain that his illustrations were too camouflaged. The inevitable compromise was somewhere in between—a bird that Thayer thought stood out too much, one that the publishers felt stood out too little. The only people truly satisfied were those who instinctively recognized the work of a genius. Fuertes faced, as had no other artist before, the problem of *how* to paint a bird. Either it was a feathered map with every plume in place, every colour seen in indirect light, or it was a living entity, bathed in light and shadow, surrounded by the softening influence of atmosphere. By showing the range of possibilities, Fuertes defined the problem that today's artists must confront. For this reason alone Fuertes would rank among the greats of animal art.

Fuertes was much in demand for the numerous new books that were published on many of the local avifauna. Less than perfect printing and colour fidelity did not do justice to the originals and Fuertes' reputation went into temporary eclipse. Only with the resurrection of his original works have contemporary artists and historians accorded him his rightful place in wildlife art.

Two other significant artists were at work in North America during the first third of this century—Canadians Ernest Thompson Seton and Colonel Allan Brooks, both artists of extraordinary talent.

The Nile Helmet Shrike by Louis Agassiz Fuertes from an *Album of Abyssinian Birds and Mammals*.

Seton, a naturalist and artist, was born in Scotland in 1860, but spent many years in the Canadian wilderness converting his experiences into writing and art. Many of his 8,000 drawings and paintings are permanently on display at the Seton Museum in Cimarron, New Mexico. As wonderful as his paintings were, Seton's greatest contribution was from two very modest black and white illustrations in his *Two Little Savages* (RBC) from 1903 of the profiles of floating river and sea ducks. As he wrote, "Now if I can put their uniforms down on paper I'll know the Ducks as soon as I see them on a pond a long way off."[9] This was the kind of woodlore appropriate from someone who would later be the co-founder of the Boy Scouts in America. The use of simple patterns as a means of identification struck a young Roger Tory Peterson as being a fine idea, and led directly to the field-guide system that has revolutionized natural history today. The "mere museum men" about whom Joseph Wolf had complained in the 1880's have become field naturalists as a result of that system.

Colonel Brooks also anticipated Peterson, but in addition deserves far greater recognition as a fine artist and ornithologist. He was responsible for the coloured plates of transcendent beauty in Miriam Bailey's 1928 *Birds of New Mexico* (905) with the full environmental settings associated with Fuertes (who supplied the text illustrations and some plates for this very book), and the atmospheric light we associate with Archibald Thorburn. It was also in this book that Brooks included two black and white plates of soaring hawks and vultures specifically designed to show the field marks. It was as if he had taken to heart the prophetic passage of Henry David Thoreau who, as a voice in the wilderness, wrote in his Journal for March 15, 1860: "I get a very fair sight of a (hen-hawk) sailing overhead. What a perfectly regular and neat outline it presents! an easily recognized figure anywhere. Yet I never see it represented in books. I do not believe that one can get as correct an idea of the form and color of the undersides of a hen-hawk's wings by spreading those of a dried specimen in his study as by looking up at a free and living hawk soaring above him in the fields."[10] The birth of the true field guide was inevitable.

Along with Gilbert White and Thomas Bewick it can be said that Roger Tory Peterson has had the most profound impact on the history of natural history. Through his successful adaptation and exploitation of the field guide concept—rapid identification through pattern—19th century precursors such as W.J. Gordon's *Our Country's Birds and How to Know Them* (1051) were merely pocket-sized books confusingly illustrated. Peterson brought the natural world out of the museum drawers and the exclusive purview of museum men into the realm of the layman. Since they were first published in 1934, Peterson's *Field Guides* have been the most successful wildlife books ever produced, with sales now numbering in the millions. In the intervening fifty years, field guides of everything from grasses to stars, and of the birds and mammals from virtually every known region on earth have appeared.

In our own time when we are witnessing another rapid expansion of scientific knowledge, it can be said that the 1970's ushered in a new "golden age" of fine natural history illustration with monographs such as Elizabeth Butterworth and Rosemary Low's *Parrots and Cockatoos* (1183) and *Amazon Parrots* (883) and John Harrison's *Birds of Prey* (1079). In addition, the recent interest in wildlife has prompted the publication of excellent facsimiles of rare classics of the past such as Schlegel and Wulverhorst's *Traité de Fauconnerie* (1386), published in

Gambel Quail by Allan Brooks reproduced in Florence Augusta Bailey's *Birds of New Mexico.* Santa Fe, 1928.

1979 by Pion Ltd. of London; Gesner's *Historia plantarum* (391) 1972 reprint; Redouté's *Les Roses* (529) reprinted in 1974-78 by De Schutter; and Jardine's *British Salmonidae* (752) reprinted by Decimus of London in 1979.

Throughout this essay many artists have been discussed who, through their stylistic and scientific contributions, represent the major turning points in the evolution of natural history art. It is relatively simple to identify these outstanding individual artists; less obvious are the contributions that have been made over the centuries by individuals who invented, improved or perfected processes for reproduction, printing, paper-making, for colour-fast paints and inks. Perhaps over-looked or too easily taken for granted has been the role of colourists who, before the introduction of mechanically applied pigments, played a vital role in transmitting the artists' works faithfully

Painting by Allan Brooks showing field marks from *Birds of New Mexico.*

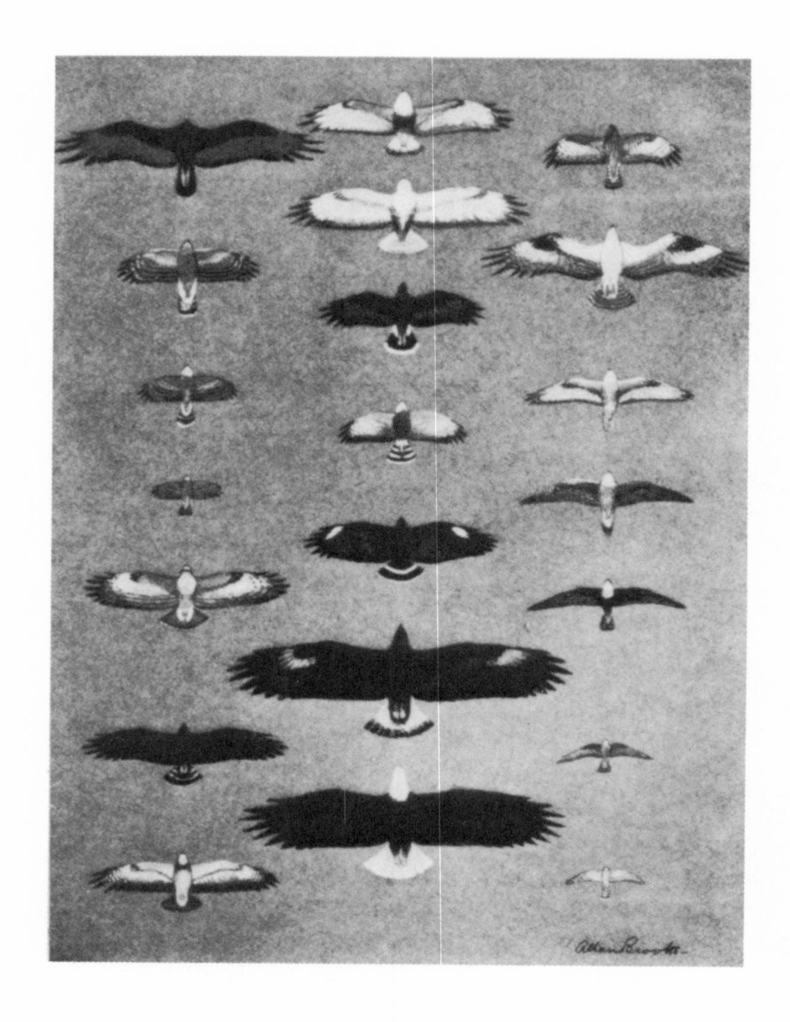

to countless copies of books containing their art. Some are named; we know from the Preface of R. Bowdler Sharpe's *A Monograph of the Hirundinidae or Family of Swallows* that the colouring of the elegant lithographs by Claude W. Wyatt was "executed by Miss Bertha Sharpe and Miss Dora Sharpe, with the occasional help of their sisters Emily and Eva," the daughters of the author. And Daniel Giraud Elliot in his *Monograph of the Phasianidae* (1022) from 1872 specifically credits a Mr. J.D. White with the colouring of what must rank among the most complex family of featheration and colouration.

Most colourists, however, will forever remain unknown. The scale of their anonymous contribution to animal art can be appreciated in the context of one of the most ambitious and successful publishing ventures of the mid-third of the 19th century. In 1833 there appeared the first of what would ultimately comprise 40 volumes of *The Naturalists' Library* (223) in which there were more than 1400 hand-coloured copper engravings after the work of such famous artists as Edward Lear, Audubon, William Swainson, and Prideaux John Selby to name but a few. With an average of 4000 copies per volume, and with at least six subsequent editions, it can be estimated that over 20 million hand-coloured plates were produced. In a Publisher's Advertisement, Sir William Jardine wrote an astonishing yet reassuring paragraph: "Altogether independent of the gratification which these plates have given the public, the publication has opened up a source of agreeable, permanent, and profitable employment, to a very numerous class of most deserving and industrious persons in Edinburgh, whose rank in society and whose education precluded them from applying themselves readily to any other occupation than that of colouring."

Thus the legacy of natural history illustration was created by talented persons from diverse occupations and stations in life. So too, special collections such as the E.A. McIlhenny Natural History Collection invite study and enjoyment by an equally diverse population of users. There is a challenge to satisfy this range of interests in such a collection. Meeting this challenge is a formidable task, but continual additions to the already rich variety of works in the McIlhenny Collection, which are significant for both their scientific and artistic value, will all the more enrich a fine natural history collection.

REFERENCES

1. Allen, David Elliston. *The Naturalist in Britain.* London: Pelican Books, 1976, p.16.
2. *Ibid.*
3. *Audubon and His Journals.* By Maria R. Audubon. With zoological and other notes by Elliott Coues. 2 vols. New York: Dover, 1960. Vol. I, pp.326-330.
4. Thomas, Keith. *Man and the Natural World: a history of the Modern Sensibility.* New York: Pantheon Books, 1983, pp.144-150.
5. Allen, Elsa Guerdrum. "The History of American ornithology before Audubon." *Transactions of the American Philosophical Society.* New Series, volume 41, part 3, 1951, p.483.
6. Herrick, Francis Hobart. *Audubon the Naturalist: a History of His Life and Time.* 2 volumes, illustrated. New York: D. Appleton and Company, 1917. Vol. I, pp.359-60.
7. Palmer, A.H. *The Life of Joseph Wolf, Animal Painter.* London and New York: Longmans, Green & Co., 1895, p.99.
8. *Ibid*, pp.57-58.
9. Seton, Ernest Thompson. *Two Little Savages.* New York: Grosset & Dunlop, 1903, p.386.
10. Thoreau, Henry D. *The Journal of Henry D. Thoreau.* Edited by Bradford Torrey and Francis H. Allen. Boston: Houghton Mifflin Co., 1949. Vol. XIII, p.194.

The Catalogue

This *Catalogue* is divided into nine subject divisions or parts. Entries are alphabetical within each part. All entries are consecutively numbered throughout the *Catalogue*. Elements in the *Catalogue* are referred to by entry number in the Index to the *Catalogue*.

An asterisk by the entry number indicates those titles which were part of the original McIlhenny gift or which have subsequently been purchased through the generosity of the benefactor of the collection.

In the compilation of the *Catalogue* each book was examined to verify and amplify the existing library cataloging (which is in a variety of styles and rules). A format was designed for a clear, non-repetitive bibliographical entry containing a standardized collation and descriptive notes. Much information not present in the existing cataloging was added. For multi-volume sets or serials LSU's exact holdings are given. Binding descriptions and provenance are included. Citations to major natural history catalogues and bibliographies in which each work is listed appear as a separate line after the collation and notes. These bibliographies are briefly listed below with their *Catalogue* entry number and the abbreviations used in the citations.

Author/Title	Abbreviation	Entry Number
Anker, Jean. *Bird Books and bird art*	Anker	2
British Museum. *Catalogue of the books, manuscripts, maps and drawings in the British Museum (Natural History)*. Vols. 1-5. Supplement. Vols. 6-8.	BM	7
Catalogue of the Edward E. Ayer Ornithological Library of the Field Museum of Natural History Chicago	Ayer	16
Catalogue of botanical books in the collection of Rachel McMasters Hunt	Hunt	24
Nissen, Claus. *Die botanische Buchillustration...*	Nissen BBI	37
———. *Die illustrierten Vogelbucher*	Nissen IVB	38
———. *Die zoologische Buchillustration*	Nissen ZBI	39
Plesch, Arpad. *Mille et un livres botaniques*	Plesch	40
Pritzel, Georg August. *Thesaurus literaturae botanicae...*	Pritzel	41, 42
Yale University. Library. *Ornithological books...*	Yale	55

PART I

Reference and Bibliography

Page from *A New & Complete Dictionary of the Arts & Sciences*. London, 1815.

1

AGASSIZ, LOUIS (1807-1873). *Bibliographia zoologiæ et geologiæ. A general catalogue of all books, tracts, and memoirs on zoology and geology.* Corrected, enlarged, and edited by H.E. Strickland. London: The Ray Society, 1848-54.

4 v. 22 cm. (The Ray Society Publications, v.13, 18, 22, 26). Vol.4 edited by H.E. Strickland and Sir William Jardine. Bindings vary.

BM I p.8.

2

ANKER, JEAN (1892-1957). *Bird books and bird art; an outline of the literary history and iconography of descriptive ornithology based principally on the collection of books containing plates with figures of birds and their eggs now in the University library at Copenhagen and including a catalogue of these works.* Written and compiled by Jean Anker, issued by the University Library, Copenhagen, to commemorate the inauguration of the new building of the library. Copenhagen: Levin & Munksgaard, 1938.

xviii, 251 p. col. front., 1 illus., xii plates (part col.) 31 cm. Each plate accompanied by guard sheet with descriptive letterpress. Blue library binding.

——— *Reprint edition.* Levin and Munksgaard, Copenhagen. The Hague: W. Junk, 1973, c1938.

xviii, 251 p., [24] leaves of plates. illus. 31 cm. Green cloth binding.

*3

BLAKE, SIDNEY FAY (1892-1959). *Geographical guide to floras of the world, an annotated list with special reference to useful plants and common plant names.* By S.F. Blake and Alice C. Atwood. Washington: For sale by the Superintendent of Documents, U.S. Government Printing Office, 1942-61.

2 v. 24 cm. (U.S. Dept. of Agriculture. Miscellaneous publication no.401, 797.) Green library binding.

4

BONAPARTE, CHARLES LUCIEN JULES LAURENT, PRINCE DE CANINO (1803-1857). *Conspectus generum avium.* Lugduni Batavorum: apud E.J. Brill, 1850-57.

2 v. 25 cm. Vol.1, green cloth binding, spine taped; v.2, black library binding.

BM I p.194. Yale p.34. Zimmer p.68.

5

BOURLIÈRE, FRANÇOIS (1913-). *Éléments d'un guide bibliographique du naturaliste.* Macon: Protat frères, imprimeurs, 1940. Paris: P. Lechevalier, 1941.

ix, 302 p., 1 l. 28 cm. Grey library binding.

——— *Suppléments I et II.* Paris: P. Lechevalier, 1941. ([303]-368 p., 1 l. 28 cm.) Bound with main piece.

*6

BRITISH MUSEUM. DEPT. OF PRINTED BOOKS. *A catalogue of the works of Linnaeus (and publications more immediately relating thereto) preserved in the libraries of the British Museum (Bloomsbury) and the British Museum (Natural History) (South Kensington).* 2d edition. London: Printed by order of the Trustees of the British Museum, 1933.

xi, 246, 65, [2]p. VII plates (incl. front. (port.) facsims.) 29 cm. Green cloth binding.

——— *An index to the authors (other than Linnaeus) mentioned in the catalogue of the works of Linnaeus preserved in the libraries of the British Museum.* 2d edition, 1933. London: Printed by order of the Trustees of the British Museum, 1936. (59 p. 28.5 x 22 cm.) "The index is the work of Dr. C. Davies Sherborne." Green paper wrappers.

7

BRITISH MUSEUM (NATURAL HISTORY). LIBRARY. *Catalogue of the books, manuscripts, maps and drawings in the British Museum (Natural History).* "Compiled and edited by Mr. B.B. Woodward, assisted by representatives from the several departments." London: Printed by order of the Trustees, 1903-1915.

5 v. 29 x 22 cm. Blue cloth binding.

——— *Supplement.* London: Printed by order of the Trustees, 1922- . (v. 28 x 22 cm.) Compiled and edited by Mr. B.B. Woodward, assisted by representatives from the several departments. (v.6-8). Black cloth binding.

8

BRITISH MUSEUM (NATURAL HISTORY). LIBRARY. *Serial publications in the British Museum (Natural History) Library.* 1968- . London: British Museum (Natural History), 1968-

v. 30 cm. (Publication no.673, 778). Title varies slightly: *List of serial publications in the British Museum (Natural History). Library.* Blue cloth binding.

*9

BURNS, FRANKLIN LORENZO (1869-). *A bibliography of scarce or out of print North American amateur* *and trade periodicals devoted more or less to ornithology*. [n.p.], 1915.

32 p. 23 cm. Caption title. Originally appeared as a supplement to *The Oölogist*, vol.32, no.7, whole no.336, July 15, 1915. Paper wrappers.

10

CANBERRA, AUSTRALIA. NATIONAL LIBRARY. *Checklist to the Mathews Ornithological Collection*. Canberra: National Library of Australia, 1966.

v, 309 p. port. 27 cm. Grey library binding.

* 11

CHALMERS-HUNT, J.M. *Natural history auctions, 1700-1972; a register of sales in the British Isles*. With articles by S. Peter Dance, Peter G. Embrey, W.D. Ian Rolfe, Clive Simson, William T. Stearn, J.M. Chalmers-Hunt, Alwyne Wheeler. London: Sotheby Parke Bernet, 1976.

xii, 189 p. 31 cm. Green cloth binding, dustjacket.

* 12

CHOATE, ERNEST ALFRED. *The dictionary of American bird names*. Boston: Gambit, [c1973].

xviii, 261 p. illus. 19 cm. Black cloth binding, dustjacket.

* 13

CHRISTIE, MANSON & WOODS INTERNATIONAL INC. *Highly important natural history books and autographs including Audubon's "Birds of America" and an extensive collection of his manuscripts, the properties of the heirs of Grace Phillips Johnson, the trustees of Deerfield Academy, the Grolier Club of New York and other owners which will be sold on Thursday May 26, 1977 at 8 p.m. precisely*. New York: Christie, Manson & Woods International Inc., 1977.

115 p., [3] leaves of plates, illus. (some col.) 26 cm. "Estimates": [3]p. laid in. Printed paper over boards, ms. facsimiles on endpapers.

* 14

COBRES, JOSEPH PAUL VON. *Deliciae Cobresianae*. J.P. Cobres Büchersammlung zur Naturgeschichte. Amsterdam: A. Asher, 1966.

2 v. (xxviii, 956 p.) 23 cm. Each vol. has added title page, engraved with date 1782. Blue cloth binding.

15

COUES, ELLIOTT (1842-1899). *American ornithological bibliography*. Introduction by Keir B. Sterling. New York: Arno Press, 1974.

239-1066 p. 24 cm. (Natural sciences in America). Consists of the 2d and 3d installments of American ornithological bibliography reprinted from the 1870-80 edition issued by the U.S. Govt. Printing Office. Originally published in the *U.S. Geological and Geographical Survey of the Territories*. Bulletin, v.5, no.2, pp.239-330 and no.4, pp.521-1066. Red cloth binding.

* 16

FIELD MUSEUM OF NATURAL HISTORY, CHICAGO. EDWARD E. AYER ORNITHOLOGICAL LIBRARY . *Catalogue of the Edward E. Ayer Ornithological Library*. By John Todd Zimmer. Chicago, 1926.

2 v. fronts. (v.1: col. port.), plates, facsims. 25 cm. (Field Museum of Natural History. Publication 239, 240. Zoological series. Vol. XVI). Printed paper covers.

* 17

FREEMAN, RICHARD BROKE. *British natural history books, 1495-1900; a handlist*. Folkestone, Kent: Dawson; Hamden, Conn.: Archon Books, 1980.

437 p. 23 cm. Green cloth binding, gilt lettering.

* 18

GREGORY, GEORGE (1754-1808). *A new & complete dictionary of arts & sciences: including the latest improvement & discovery and the present state of every branch of human knowledge*. London: S.A. Oddy, 1815.

2 v. illus. 140 leaves of plates (some col.) 29 cm. LSU copy imperfect: plates 59, 67, 70, 77, 86-87, 108, 112, 123, 125, 130-131, 133-134, 136-137, 139-140 wanting. Plate 129 duplicated. Quarter red morocco, marbled boards.

* 19

GRUSON, EDWARD S. *Words for birds: a lexicon of North American birds with biographical notes*. New York: Quadrangle Books, 1972.

xiv, 305 p. illus. 25 cm. Blue cloth binding, dustjacket.

20

HAGEN, HERMANN AUGUST (1817-1893). *Bibliotheca entomologica. Die Litteratur über das ganze Gebiet der Entomologie bis zum Jahre 1862...* Leipzig: W. Engel-

mann, 1862-1863.
2 v. in 1. 23 cm. Half calf/dark red cloth binding, marbled endpapers.
BM II p.762.

* 21

HALLER, ALBRECHT VON (1708-1777). *Bibliotheca botanica: qua scripta ad rem herbariam facientia a rerum initiis recensentur.* Bologna: Forni, 1967.
2 v. 25 cm. Reprint of the 1771-1772 edition published by Crell, Gessner, Fuessli, Tiguri. Brown cloth binding.
BM II p.775 (1771 ed.) Pritzel 3727 (1771 ed.)

* 22

HENREY, BLANCHE (? -1983) *British botanical and horticultural literature before 1800: comprising a history and bibliography of botanical and horticultural books printed in England, Scotland, and Ireland from the earliest times until 1800.* London, New York: Oxford University Press, 1975.
3 v. illus. (some col.) 28 cm. Grey cloth binding, slipcase.

23

HORBLIT, HARRISON D. *One hundred books famous in science; based on an exhibition held at The Grolier Club.* New York: The Grolier Club, 1964.
449 p. illus., facsims. 29 cm. Quarter blue/grey cloth binding.

24

HUNT, RACHEL MCMASTERS (MILLER). *Catalogue of botanical books in the collection of Rachel McMasters Miller Hunt.* Pittsburgh: Hunt Botanical Library, 1958-61.
2 v. in 3. illus., plates, port. 26 cm. Vol.1, compiled by Jane Quinby. Vol.2, compiled by Allan Stevenson. Green cloth binding.

25

JACKSON, BENJAMIN DAYDON (1846-1927). *Guide to the literature of botany; being a classified selection of botanical works, including nearly 6000 titles not given in Pritzel's Thesaurus.* London: Pub. for the Index Society, by Longmans, Green, 1881.
xl, 626 p. 22 x 17 cm. (Index Society. Publications, 1880, VIII). Full red morocco, in slip case.
BM II p.915. Plesch p.277.

* 26

JUNK (W.) FIRM, PUBLISHERS. *Rara historico-naturalia et mathematica/Wilhelm Junk (1866-1942).* Editio stereotypa, praefatione tabulisque aucta, curante F.H. Schwarz. Amsterdam: A. Asher, 1979.
3 v. in 1 (vi, 314 p.) 24 cm. Reprint of the 1900-1939 edition published by W. Junk, Berlin and Den Haag. Vol.1, no.1-19 reprinted from *Bibliographische Zeitschrift für Naturwissenschaften und Mathematik (formerly Laboratorium & Museum).* Vols.2-3 have title: *Rara historico-naturalia.* Vol.3 is the *Supplementum.* Blue cloth binding.

27

KANSAS UNIVERSITY LIBRARIES. *A catalogue of the Ellis collection of ornithological books in the University of Kansas Libraries.* Compiled by Robert M. Mengel. Lawrence, 1972.
v. 26 cm. (University of Kansas publications. Library series, 33). Printed paper covers.

* 28

KEYNES, GEOFFREY, SIR (1887-). *Dr. Martin Lister; a bibliography.* Godalming, Surrey, Eng.: St. Paul's Bibliographies, 1981.
xii, 52 p. illus., facsims. 23 cm. (St. Paul's bibliographies, no.3). "Appeared originally in *The Book Collector,* Winter 1979 and Spring 1980 issues — p.[iv]." "Edition limited to 350 copies." Green cloth binding with gilt decoration.

29

KNIGHT, DAVID M. *Natural science books in English, 1600-1900.* New York: Praeger, [c1972].
x, 262 p. illus. 26 cm. Green cloth binding, dustjacket.

30

KURODA, NAGAMICHI (1889-). *A bibliography of the duck tribe, Anatidae, mostly from 1926 to 1940, exclusive of that of Dr. Phillip's work.* Tokyo: Herald Press, 1942.
852 p. 22 cm. Title and editor's name also in Japanese on title page. Quarter grey cloth/printed boards.

* 31

LISNEY, ARTHUR A. *A bibliography of British Lepidoptera, 1608-1799.* London: Chiswick Press, 1960.
xviii, 315 p. ports., facsims. 26 cm. Green cloth binding.

* 32

MALLALIEU, HUON. *The dictionary of British watercolour artists up to 1920.* Woodbridge, England: Antique Collectors' Club, [c1976-1979].

2 v. 29 cm. Blue cloth binding, dustjacket.

33

MEISEL, MAX (1892-). *A bibliography of American natural history; the pioneer century, 1769-1865; the rôle played by the scientific societies; scientific journals; natural history museums and botanic gardens; state geological and natural history surveys; federal exploring expeditions in the rise and progress of American botany, geology, mineralogy, paleontology and zoology.* Brooklyn, N.Y.: The Premier publishing co., 1924-29.

3 v. 23.5 cm. Original red cloth binding.
BM VII p.821.

* 34

MULLENS, WILLIAM HERBERT (1866-). *A bibliography of British ornithology from the earliest times to the end of 1912, including biographical accounts of the principal writers and bibliographies of their published works.* By W.H.Mullens and H. Kirke Swann. London: Macmillan and co., limited, 1917.

xx, 673 p., 675-691 numbered leaves. 23 cm. The numbered leaves are printed on one side only. Issued in 6 parts, 1916-17.

——— *Supplement.* By H. Kirke Swann. London: Wheldon & Wesley, ltd., 1923. LSU lacks supplement. Green library binding.
BM VII p.880. Zimmer p.446.

35

MULLENS, WILLIAM HERBERT (1866-). *A geographical bibliography of British ornithology from the earliest times to the end of 1918, arranged under counties, being a record of printed books, published articles, notes and records relating to local avifauna.* By W.H. Mullens, H. Kirke Swann, and Rev. F.C.R. Jourdain. London: Witherby & co., 1920.

viii, 558 p., 1 l. 23 cm. Issued in 6 parts, 1919-20. Red library binding.
BM VII p.880. Zimmer p.447.

* 36

The Naturalists' directory. Containing names, addresses and special subjects of study of professional and amateur naturalists of North and South America, etc., and a list of periodicals dealing with the subjects of natural history; also a list of natural history museums and dealers announcements. [1st]-ed.; 1877-[19]. Salem, Mass.: The Cassino Press, 1877-[19].

v. 18.5 x 23 cm. LSU owns 1898. Grey linen binding.
BM III p.1399. Yale p.203.

37

NISSEN, CLAUS (1901-). *Die botanische Buchillustration, ihre Geschichte und Bibliographie.* Stuttgart: Hiersemann, 1951-66.

3 v. in 2. 31 cm. Vol.1 and 2 bound together. Vol.3: 2d edition. Green cloth binding.
Plesch p.348.

38

NISSEN, CLAUS (1901-). *Die illustrierten Vogelbücher: ihre Geschichte und Bibliographie.* Stuttgart: Hiersemann verlag, 1953.

222 p. illus. 31 cm. Yellow cloth binding.

39

NISSEN, CLAUS (1901-). *Die zoologische Buchillustration. Ihre Bibliographie und Geschichte.* Stuttgart: A. Hiersemann. 1969-

2 v. 31 cm. Issued in parts. Blue cloth binding.

* 40

PLESCH, ARPAD. *Mille et un livres botaniques; répertoire bibliographique de la bibliothéque Arpad Plesch.* Arcade, Bruxelles: 1973.

517 p. illus., 33 mounted col. plates, col. port. 31 cm. Preface and introduction in English and French. Green cloth binding, pictorial endpapers.

41

PRITZEL, GEORG AUGUST (1815-1874). *Thesaurus literaturae botanicae omnium gentium inde a rerum botanicarum initiis ad nostra usque tempora, quindecim millia operum recensen: Curavit G.A. Pritzel.* Lipsiae: F.A. Brockhaus, 1851.

3 p.l., viii, 547 p. 26.5 x 21.5 cm. Quarter brown morocco/marbled boards, decorated endpapers.
BM IV p.1617. Jackson p.3. Plesch p.369.

*42

PRITZEL, GEORG AUGUST (1815-1874). *Thesaurus literaturae botanicae omnium gentium, inde a rerum botanicarum initiis ad nostra usque tempora, quindecim millia operum recensens.* Editionem novam reformatam curavit G. A. Pritzel. Lipsiae: F.A. Brockhaus, 1871-1877. Koenigstein: Otto Koeltz Antiquariat, 1972.

576 p. 32 cm. Blue cloth binding.

43

QUARITCH, FIRM, BOOKSELLERS, LONDON. *Catalogue of works on natural history, offered at the net prices affixed by Bernard Quaritch.* London: Quaritch, 1907.

234 p. 25 cm. Red cloth binding.

*44

READING, ENGLAND. UNIVERSITY LIBRARY. *The Cole Library of early medicine and zoology: catalogue of books and pamphlets, by Nellie B. Eales.* Oxford: Alden Press [for] the Library, 1969-

v. plate, facsim. 26 cm. (Reading University Library Publications, 1). LSU owns v.1. Red cloth binding.

45

RONSIL, RENÉ. *Bibliographie ornithologique française. Travaux publiés en langue française et en latin en France et dans les Colonies françaises de 1473 à 1944.* Préf. de Marcel Legendre. Paris: P. Lechevalier, 1948-49.

2 v. in 1. facsim. 25 cm. (Encyclopédie ornithologique, 8-9). Red library binding.

46

ROYAL SOCIETY OF LONDON. *Catalogue of scientific papers* (1800-1900). Compiled and published by the Royal Society of London... London: George Edward Eyre and William Spottiswoode, 1867-77; J. Murray, 1879; C.J. Clay and sons, 1891-

19 v. 29 x 22.5—28 x 21.5 cm. Compiled under the supervision of Henry White, Herbert McLeod, H. Forster Morley, and others. V.2-5, half brown morocco/cloth; other vols., red cloth binding.

——— *Subject Index.* Cambridge: University Press, 1908-1943. (3 v. in 4). Brown cloth binding.
BM IV p.1755.

*47

Scientific books, libraries and collectors: a study of bibliography and the book trade in relation to science. By John Thornton and R.I.J. Tully. 3d revised edition. London: Library Association, 1971.

ix, 508 p., 17 plates, facsims. 23 cm. Brown cloth binding, dustjacket.

48

SITWELL, SACHEVERELL (1897-). *Fine bird books, 1700-1900.* By Sacheverell Sitwell, Handasyde Buchanan and James Fisher. London, New York: Collins & Van Nostrand, 1953.

viii, 120 p. illus., col. plates. 50 cm. Half burnt orange cloth/marbled boards, illustrated endpapers.

49

SOCIETY FOR THE BIBLIOGRAPHY OF NATURAL HISTORY, LONDON. *The Journal of the Society for the Bibliography of Natural History.* v.1- ; Oct. 1936- London, [1936-].

v. illus., plates, facsims. 26 cm. Irregular. Numbering very irregular. Now renamed: Society for the History of Natural History — publication also renamed: *Archives of Natural History.* Green library binding.

*50

SOTHEBY, FIRM, AUCTIONEERS, LONDON. *The magnificent botanical library of the Stiftung für Botanik, Vaduz, Liechtenstein, collected by the late Arpad Plesch, which will be sold by auction by Sotheby & Co....Monday 16th June, 1975...* [n.p., 1975].

3 v. illus. 29 cm. "Prices and buyers names," part 1-[3]: [12] leaves laid in v.1. Printed paper over boards.

*51

TRINITY COLLEGE (HARTFORD, CONN.) LIBRARY. *Ornithology books in the library of Trinity College, Hartford: including the library of Ostrom Enders.* Prepared by Viola Breit and Karen B. Clarke. Hartford: The Library, 1983.

270 p. front. 27 cm. [Limited edition of 1500 numbered copies.] LSU owns copy no.161. Dark blue cloth binding, dustjacket.

52

WHITTELL, HUBERT MASSEY (1883-). *The literature of Australian birds; a history and a bibliography of Australian ornithology*. Perth, Western Australia: Paterson Brokensha, 1954.

xi, 788 p. plates (1 col.) ports., facsims. 25 cm. Green library binding.

*53

WILLEY, BASIL (1897-). *The eighteenth century background; studies on the idea of nature in the thought of the period*. Boston: Beacon Press, [1961].

viii, 301 p. 21 cm. "First published in 1940 by Chatto and Windus, Ltd." Paperbound.

54

WOOD, CASEY ALBERT (1856-1942), ed. *An introduction to the literature of vertebrate zoology; based chiefly on the titles in the Blacker library of zoology, the Emma Shearer Wood library of ornithology, the Bibliotheca Osleriana and other libraries of McGill University, Montreal*. Compiled and edited by Casey A. Wood. London: Oxford University Press, H. Milford, 1931.

xix, 643 p. col. front. 29 cm. (McGill University publications. Series XI (Zoology), no.24). Blue cloth binding.
BM VIII p.1451.

*55

YALE UNIVERSITY LIBRARY. *Ornithological books in the Yale University Library including the library of William Robertson Coe*. Compiled by S. Dillon Ripley and Lynette L. Scribner. New Haven: Yale University Press, 1961.

338 p. port. 26 cm. Beige cloth binding.

PART II

Natural History and the Arts

Plate from *Flora's Gems* with "poetical illustrations by Louisa Anne Twamley" and "twelve bouquets, drawn and coloured from nature, by James Andrews." London, 1837.

*56

BAIN, IAIN. *The watercolours and drawings of Thomas Bewick and his workshop apprentices.* Introduced and with editorial notes by Iain Bain. London: Gordon Fraser, 1981.

2 v. illus. (some col.), ports. 23 x 26 cm. Grey cloth, in slipcase.

57

BEWICK, THOMAS (1753-1828). *My life.* Edited and with an introduction by Iain Bain; with numerous wood-engravings and watercolors by the author. London: Folio Society, 1981.

192 p. front. illus. (some col.) 24 cm. Quarter brown cloth/gold paper over boards, illustration on front cover.

*58

BEWICK, THOMAS (1753-1828). *Thomas Bewick, Vignettes: being tail-pieces engraved principally for his General history of quadrupeds & History of British birds.* Edited with an introduction by Iain Bain. London: Scolar Press, 1979.

25 p. 167 leaves of plates, chiefly illus. 18 cm. Linen spine, paper over boards, dustjacket.

59

Bird paintings of the Ch'ien Lung period, 1736-1796. With an introduction and notes by J. Longridge. [London]: Holland Press, [1967].

[35]p. 8 col. plates. 34.5 x 42.5 cm. Pictorial cloth binding.

*60

BOOS, JOHN E. *The philosopher of Woodchuck Lodge, John Burroughs.* Albany, N.Y.: Boos, [1941?].

102 leaves. illus. (some col.) 24 cm. Signed in ms.: John E. Boos, 1941. Includes notes in ms. with typescript of notes preceding those in ms. and holograph letters by John Burroughs, Julia Burroughs, and A. Agassiz. "The summit of the years," by John Burroughs: [9]p. at end. Quarter blue cloth, paper boards.

*61

BUCHANAN, HANDASYDE. *Nature into art: a treasury of great natural history books.* 1st American edition. New York: Mayflower Books, c1979.

220 p. illus. (some col.) 31 cm. Green cloth binding, dustjacket.

*62

CALMANN, GERTA. *Ehret, flower painter extraordinary.* Boston: New York Graphic Society, c1977.

160 p. 95 illus. (some col.) 35 cm. Grey cloth binding, dustjacket.

*63

[*Collection of ten hand colored plates; mainly birds, four to eight to a sheet.*] London: Published as the act directs, by Harrison & Co., 1784-85.

10 col. plates in portfolio. 37 cm.

*64

COSENTINO, FRANK J. *Boehm's birds; the porcelain art of Edward Marshall Boehm.* With an appreciation by John D. Morse. New York: Frederick Fell, [1966, c1960].

202 p. plates (part col.) ports. 28 cm. Blue cloth binding, dustjacket.

*65

COSENTINO, FRANK J. *Edward Marshall Boehm, 1913-1969.* Chicago: Printed by The Lakeside Press, c1970.

264 p. illus. (part col.), ports. (part col.) 32 cm. Red/beige cloth binding, decorated endpapers.

66

Curious woodcuts of fanciful and real beasts; a selection of 190 sixteenth-century woodcuts from Gesner's and Topsell's natural histories. [By] Konrad Gesner. New York: Dover Publications, [1971].

111 p. illus. 31 cm. (Dover pictorial archive series). Paperbound covers with plastic coating.

*67

DANCE, S. PETER. *The art of natural history; animal illustrators and their work.* [Edited, designed, and photographed by Ian Cameron]. Woodstock, N.Y.: Overlook Press, c1978.

224 p. illus. (some col.) 35 cm. Brown cloth binding, dustjacket.

*68

DAVIDS, ARLETTE. *Plantes grasses, dessinées.* Par Arlette Davids. Préface by Henry de Montherlant.

Edited by André Gloeckner. Paris: Éditions Hypérion, [c1939].
[6]p. 40 col. plates. 37 cm. Each plate accompanied by leaf with descriptive letterpress. Paper boards, linen spine. Nissen BBI 2: 452n.

*69

DAVIDS, ARLETTE. *Rock plants.* Drawn by Arlette Davids. Preface by Henry de Montherlant. Translated from the French by S.P. Skipwith. London, New York: Hyperion Press, [c1939].
3 p.l. 40 col. plates, 1 l. 47 cm. (Her Flowers) "This album, edited by André Gloeckner, was first published in December MCMXXXIX by the Hyperion Press, Paris." Blocks engraved by Les Arts Photographiques appliques. Plates printed by L. Delaporte, text by G. Desgrandchamps. Binding by Joseph Taupin, Paris." Grey linen binding.

*70

DOUGHTY, DOROTHY (d. 1962). *The American birds of Dorothy Doughty.* A critical appreciation by George Savage. The plates described by Dorothy Doughty. Preface by Joseph F. Gimson. Worcester, England: Worcester Royal Porcelain Co., [c1962].
x, 203 p. 10 col. mounted illus., 70 (i.e.71) col. mounted plates, 3 mounted ports. 35 cm. "One thousand five hundred copies of this book have been printed, each of which has been numbered, and signed by the artist." LSU mounted copy no.748. Full tan goatskin binding in paperboard case.

*71

DOUGHTY, DOROTHY (d. 1962). *The British birds of Dorothy Doughty.* Preface and description by George Savage. Worcester, England: Worcester Royal Porcelain Co., [c1965].
51 p. illus. 20 cm. Blue cloth binding.

72

EDE, BASIL. *Birds of town and village.* Paintings by Basil Ede. Text by W.D. Campbell. Foreword by H.R.H. The Prince Philip, Duke of Edinburgh. London: Country Life Ltd., [c1965].
154 p. 36 col. plates, illus. 30 cm. Green cloth binding, dustjacket.

*73

EVANS, HENRY HERMAN. *Botanical prints with excerpts from the artist's notebooks.* Foreword by Wilfrid Blunt. San Francisco: W.H. Freeman, c1977.
64 p. illus. (some col.) 32 cm. Beige linen cloth binding, decorated endpapers, dustjacket.

*74

EVERITT, CHARLES (1906-). *Birds of the Edward Marshall Boehm aviaries; major portion devoted to softbills.* [Trenton, N.J.: E.M. Boehm, c1973].
297 p. illus. 32 cm. Brown cloth binding, with photograph on cover, marbled endpapers. In beige cloth slipcase with Boehm photograph.

75

FEASEY, PEGGY. *Rubies & roses; gems portrayed in flowers.* Rutland, Vt.: C.E. Tuttle Co., [1970, c1969].
128 p. illus. (part col.) 27 cm. Red cloth binding.

*76

Feathers to brush: the Victorian bird artist, John Gerrard Keulemans, 1842-1912. By Tony Keulemans and Jan Coldewey. Epse, The Netherlands: Privately published by the authors, c1982.
xvii, 94 p. illus. (some col.), facsims., ports. 34 cm. "Limited edition of 500 copies..." LSU copy no.137. Brown half leather, green linen boards.

*77

Flowers in art from east and west. By Paul Hulton and Lawrence Smith. London: British Museum Publications, 1979.
x, 150 p. [16] leaves of plates. illus. (some col.) 29 cm. Blue paper over boards, dustjacket.

*78

GRAPE-ALBERS, HEIDE. *Spätantike Bilder aus der Welt des Arztes: medizinische Bilderhandschriften der Spätantike und ihre mittelalterliche Überlieferung.* Wiesbaden: G. Pressler, c1977.
203 p. illus. 35 cm. Based on thesis, Vienna. Brown paper over boards, vellum spine.

*79

HUYSUM, JACOBUS VAN (ca.1687-ca.1740). *The twelve months of flowers; in the collection of the Hon. Henry Rogers Broughton.* With a commentary by Maurice Harold Grant. Leigh-on-Sea, England: F.

Lewis, [1950].

65 p. 12 col. plates. 45 cm. "This edition de luxe is limited to five hundred copies." LSU owns copy no.392. Blue cloth binding.

*80

JACKSON, C.E. *Bird illustrators: some artists in early lithography.* London: H.F. & G. Witherby, 1975.

133 p. [12] leaves of plates, illus. (some col.) 26 cm. Black cloth binding, dustjacket.

*81

JAQUES, FLORENCE (PAGE) (1890-1972). *Francis Lee Jaques: artist of the wilderness world.* Foreword by Roger Tory Peterson. A treasury of prose writings from the six Jaques books. Three new appraisals of the artist's work. Memorials to Francis Lee Jaques by Sigurd F. Olson and to Florence Page Jaques by Harriet Buchheister. With 64 paintings and dioramas in full color and 100 drawings in scratchboard, pen and ink, and pencil. Garden City, N.Y.: Doubleday, 1973.

xxi, 370 p. illus. (part col.) 30 cm. Green cloth binding with gilt decoration, illustrated endpapers; illustrated paperboard box.

*82

KIRBY, WILLIAM (1759-1850). *On the power, wisdom and goodness of God as manifested in the creation of animals and in their history, habits and instincts.* London: W. Pickering, 1835.

2 v. 20 plates (incl. front.) 23 cm. In ms. on fly leaf of each vol.: Enoch Ellis Jones, Rhostyllen, 11 Oct. 1899. Grey paper over boards, spine labels.
BM II p.983.

*83

KITAGAWA, UTAMARO (1753?-1806). *Utamaro: a chorus of birds.* With an introduction by Julia Meech-Pekarik, and a note on kyoka and translations by James T. Kenney. New York: Metropolitan Museum of Art; Viking Press, 1981.

[48] leaves. col. illus. 26 cm. Folded accordion style. Green linen binding, in paperboard case.

*84

LANK, DAVID M. *Once-upon-a-Tyne: the angling art and philosophy of Thomas Bewick.* Montreal: Published by the Antiquarian Press for the Atlantic Salmon Association, 1977.

xiv, 105 p. illus. 18 x 28 cm. Edition limited to 1,000 signed and numbered copies. Quarter beige linen/brown paper over boards.

85

LANK, DAVID M. *Paintings from the wild; the art and life of George McLean.* Special introduction by Bob Kuhn. Toronto: Brownstone Press, distributed by J. Wiley and Sons Canada, 1981.

141, [5]p. illus. (some col.) 31 cm. Quarter green cloth/ green paper over boards, pictorial dustjacket.

*86

LEWIS, FRANK. *A dictionary of British bird painters.* The Tithe House, Leigh-on-Sea, England: F. Lewis Publishers, [c1974].

47 p. 48 plates. 30 cm. Blue cloth binding.

*87

LOATES, MARTIN GLEN. *The art of Glen Loates.* Text by Paul Duval. Scarborough, Ontario: Cerebrus Publishing Co., [c1977].

189 p. (6 fold.) illus. (chiefly col.) 36 cm. Imitation leather, dustjacket.

*88

LOCKHART, JAMES L. *Portraits of nature; paintings, drawings, & text by James Lockhart.* New York: Crown Publishers, [c1967].

[25]p., 34 col. plates. illus. 46 cm. Illustrated paper over boards, spiral bound.

*89

MATHEWS, FERDINAND SCHUYLER (1854-1938), comp. *The golden flower chrysanthemum.* Verses by Edith M. Thomas, Richard Henry Stoddard [and others]... Collected, arranged and embellished with original designs by F. Schuyler Mathews. Illustrated with reproductions of studies from nature in water color by James & Sidney Callowhill, Alois Lunzer and F. S. M. Boston: L. Prang & co., [c1890].

2 p.l., 8 p. 23 leaves, col. front., illus., 15 col. plates. 30 cm. Quarter imitation vellum, green cloth binding. Gilt decoration and edges.

*90

MERIDITH, LOUISA ANNE (TWAMLEY) (1812-1895). *Flora's gems; or, The treasures of the parterre.* Twelve bouquets, drawn and coloured from nature, by James Andrews. With poetical illustrations by Louisa Anne Twamley. London: Charles Tilt, [1837].

[45]p. 15 col. plates. 37 cm. Green cloth binding. BM V p.2156.

*91

MORRIS, FRANK THOMSON. *Pencil drawings, 1969-78.* Foreword by David Dridan. Melbourne: Lansdowne Editions, 1978.

[219]p. illus. 52 cm. Limited edition of 500 copies. LSU copy no.278, signed by the artist. Full brown calf, gilt ornamentation.

*92

NORELLI, MARTINA R. *American wildlife painting.* New York: Watson-Guptill Publications, 1975.

224 p. chiefly illus. (some col.) 32 cm. Black cloth binding, dustjacket.

93

NORTON, CHARLES ELIOT (1827-1908). *The poet Gray as a naturalist. With selections from his notes on the "Systema naturae" of Linnaeus and facsimiles of some of his drawings.* Boston: Charles E. Goodspead, 1903.

[1]-66 p. facsims. 21.9 cm. "A limited edition of five hundred copies of this book was printed on hand-made paper, by D.B. Updike, The Merrymount Press, Boston, November, 1903. This is copy no.172." Green library binding. BM VII p.939. Yale p.211.

94

PENNANT, THOMAS (1726-1798). *The literary life of the late Thomas Pennant, esq. By himself...* London: Benjamin and John White, and Robert Faulder, 1793.

3 p.l., 144 p. front. (port.), plates. 26 x 19 cm. Half brown morocco/marbled boards.
Yale p.224.

95

The Pictorial Museum of Animated Nature. London: Charles Cox, [1844].

2v. col. fronts., illus. 35.5 cm. Original red cloth binding, gilt decoration.

*96

Pictures of animals; with eighteen coloured plates and twenty-five illustrations. London, New York: G. Routledge, [1897].

64 p. illus. (some col.) 27 cm. Red pictorial cloth binding.

97

ROBINSON, ALAN JAMES. *Gamebirds & waterfowl.* Ten etchings, hand water-colored. [Easthampton, Mass.: Cheloniidae Press, c1980].

[10] leaves of plates. col. illus. 36 cm. "Edition of one hundred. Twenty-six copies of each print have been reserved for sale as a special suite, numbered with the letters of the alphabet ...signed by the artist." LSU owns copy C. Beige linen portfolio.

*98

ROGERS-PRICE, VIVIAN. *John Abbot in Georgia: the vision of a naturalist artist (1751-ca.1840). Catalogue of an exhibition at the Madison-Morgan Cultural Center, September 25-December 31, 1983.* Madison, [Ga.]: Madison-Morgan Cultural Center, 1983.

149 p. illus. 26 cm. Paperbound.

*99

RONSIL, RENÉ (1908-). *L'Art Français dans le livre d'oiseaux; éléments d'une iconographie ornithologique française.* Paris: [n.p.], 1957.

136 p., [20] leaves of plates. illus. (some col.) 28 cm. (Memoires de la Societe Ornithologique de France et de l'Union Francaise, no.6). *L'Oiseau et la Revue Francaise d'Ornithologie.* Supplement, v.27, pt.3. Paper wrappers. Unbound, in original paper covers.

*100

SETON, ERNEST THOMPSON (1860-1946). *Studies in the art anatomy of animals; being a brief analysis of the visible forms of the more familiar mammals and birds.* Designed for the use of sculptors, painters, illustrators, naturalists, and taxidermists. Illustrated with one hundred drawings by the author. Philadelphia: Running Press, 1977.

ix, 96 p. 47 leaves of plates. 33 cm. Reprint of the 1896 edition published by Macmillan, London. Paperbound. Nissen ZBI 3822 (orig. ed.)

* 101

SKIPWITH, PEYTON. *The great bird illustrators and their art,* 1730-1930. London, New York: Hamlyn, 1979.

176 p. col. illus., ports. 31 cm. Brown linen cloth binding, dustjacket.

* 102

SOUTHERN, JOHN. *Thorburn's Landscape: the major natural history paintings.* London: Elm Tree Books, 1981.

120 p. illus. (chiefly col.) 23 x 31 cm. Brown paper over boards, illustrated dustjacket.

* 103

STERN, HAROLD P. *Birds, beasts, blossoms, and bugs: the nature of Japan.* New York: H.N. Abrams, [1976].

496 p. illus. 26 cm. Red brocade with silver gilt, dustjacket.

104

STIX, HUGH. *The shell; five hundred million years of inspired design.* By Hugh and Marguerite Stix and R. Tucker Abbott. Photographs by H. Landshoff. New York: Abrams, [1968?].

[106]p. 188 plates (part col.) illus. 28 x 30 cm. In ms. on half title: "Pal, Here we are again another year to add to the 500,000,000. Love Johnny & Marge 1969." Blue cloth binding with gilt decoration.

* 105

THOMAS, BILL. *The swamp.* Special consultants, Gary Hendrix, Don Whitehead. 1st edition. New York: W.W. Norton, 1976.

222 p. col. illus. 29 cm. Green cloth binding, dustjacket.

* 106

TIRSCH, IGNACIO (b. 1733). *The drawings of Ignacio Tirsch, a Jesuit missionary in Baja California.* Narrative by Doyce B. Nunis, Jr. Translation by Elsbeth Schulz-Bischof. Los Angeles: Dawson's Book Shop, 1972.

125 p. (chiefly plates, part col.) 23 cm. (Baja California travels series, 27). Captions in English and German. "900 copies printed by Grant Dahlstrom at the Castle Press, Pasadena. Color plates by Prago Press, Prague. Bound by Bela Blau, Los Angeles." Red cloth binding.

* 107

TWIGDEN, BLAKE L. *Pisces tropicani: an artist's collection of portraits of twenty-six species of coral reef fishes.* Commentary by Roger Lubbock and foreword by Sir Peter Scott. Melbourne: Lansdowne Editions, 1978.

91 p. 26 plates. illus. 42 cm."Limited edition of 350 signed & numbered copies." LSU owns copy no.278. Bluering angel-fish plate, numbered & signed, laid in. Half blue morocco/cloth binding.

* 108

VAN GELDER, PAT. *Wildlife artists at work.* New York: Watson-Guptill, 1982.

175 p. illus. (some col.) 32 cm. Brown cloth binding, pictorial dustjacket.

109

VÉLINS DU MUSEUM; peintures sur vélin de la collection du Muséum national d'histoire naturelle de Paris. Exposition, Bibliothèque Royale Albert Ier. Catalogue. Bruxelles (bd. de l'Empereur 4): Bibliothèque Royale Albert Ier, 1974.

xxiv, 31 p., [21] leaves of plates. illus. (some col.) 26 cm. Errata slip inserted. Paperbound with plasticized coating.

110

WO, HENRY YUE-KEE. *Chinese paintings of Henry Wo Yue-kee.* Alexandria, Va.: Artland Studio, 1980-

v. illus. (part col.) 25 cm. Colophon title. English and Chinese. "Originally published in 1972 under the title *'Chinese Watercolours'.*" Illustrated paper wrappers.

* 111

The World of Owen Gromme. Introduction by Roger Tory Peterson. Biography by Michael Mentzer. Commentaries by Judith Redline Coopey. Madison, Wis.: Stanton & Lee Publishers, Inc., 1983.

240 p. illus. (chiefly col.), ports. 28 x 34 cm. "The limited edition of *The World of Owen Gromme* accompanied by a signed and numbered print *'Bobwhites'* is published by Stanton & Lee Publishers in association with Wild Wings in a limited edition of nine hundred and fifty copies." LSU owns copy no.506. Full brown leather with brown leather slipcase.

PART III

Travel and Scientific Expeditions

From *Expédition dan les parties centrales de l'Amérique du Sud, de Rio de Janeiro à Lima, et de Lima au Para;* exécutée par ordre du gouvernement Français pendent les années 1843 à 1847 sous la direction de Francis de Castelnau. Paris, 1850-59. Part 2, Vues et scènes.

* 112

AUDUBON, JOHN JAMES (1785-1851). *Scènes de la nature dans les États-Unis et le Nord de l'Amérique.* Ouvrage traduit d'Audubon par Eugène Bazin. Avec préface et notes du traducteur. Paris: P. Bertrand, 1857.

2 v. 22 cm. In original printed paper covers.

* 113

BARTRAM, WILLIAM (1739-1823). *Travels through North and South Carolina, Georgia, east and west Florida.* A facsimile of the 1792 London edition embellished with its nine original plates, also seventeen additional illustrations and an introduction by Gordon DeWolf. Savannah: Beehive Press, c[1973].

xx, 534 p. illus. 24 cm. Quarter green cloth binding, yellow paper over boards.
BM I p.105 (other eds.)

114

BATES, HENRY WALTER (1825-1892). *The naturalist on the River Amazons; a record of adventures, habits of animals, sketches of Brazilian and Indian life, and aspects of nature under the Equator, during eleven years of travel.* 5th edition. London: John Murray, 1884.

x, 394 p. illus. 20 cm. Bookplate of E.C. Freeland. Original gold cloth binding with gilt palm tree on front cover, spine mended.
BM I p.109.
Gift of E.C. Freeland.

* 115

BECKER, LUDWIG (1808-1861). *Ludwig Becker: artist & naturalist with the Burke & Wills expedition.* Edited and with an introduction by Marjorie Tipping. Carlton, Australia: Melbourne University Press on behalf of the Library Council of Victoria; Forest Grove, Oregon, distributed by International Scholarly Book Services, 1979.

xiii, 224 p. illus. (some col.) 31 cm. Grey linen cloth binding, dustjacket.

116

BEEBE, CHARLES WILLIAM (1877-1962). *The Arcturus adventure; an account of the New York zoological society's first oceanographic expedition.* With 77 illustrations from colored plates, photographs and maps, published under the auspices of the Zoological society. New York, London: G.P. Putnam's sons, c1926.

3 p.l., xix, 439 p. col. fronts., plates (part col.) maps. 24.5 cm. Green library binding.
Yale p.25.

117

BEEBE, CHARLES WILLIAM (1877-1962). *Galapagos, world's end.* With 24 coloured illustrations by Isabel Cooper, and 83 photographs by John Tee-Van. Published under the auspices of the New York Zoological Society. New York & London: G.P. Putnam's sons, 1924.

xxi, 443 p. col. front., plates (part col., 1 double), ports., maps. 27 cm. Original dark blue cloth binding.
Yale p.24.

* 118

BEEBE, CHARLES WILLIAM (1877-1962). *Pheasant jungles.* With 60 illustrations from photographs by the author. New York: G.P. Putnam, 1927.

xiii, 248 p. illus. 23 cm. ports. Original blue cloth binding, pictorial endpapers.
Yale p.25.

119

BEEBE, MARY BLAIR (RICE). *Our search for a wilderness; an account of two ornithological expeditions to Venezuela and to British Guiana.* By Mary Blair Beebe and C. William Beebe. Illustrated with photographs from life taken by the authors. New York: Henry Holt and company, 1910.

xix, 408 p. front., illus., map. 23 cm. Green library binding.
Yale p.25. Zimmer p.50.

120

BOYDELL, JOHN (1719-1804). *An history of the River Thames.* London: Printed by W. Bulmer, for John and Josiah Boydell, 1794-96.

2 v. illus., 76 col. plates, 2 fold. maps. 48 cm. In ms. on p.[2] of cover of each vol.: James Wood. Each vol. in original boards in brown cloth box.
Gift of Dr. and Mrs. A. Brooks Cronan.

* 121

CASTELNAU, FRANCIS, COMTE DE (1812-1880). *Expédition dans les parties centrales de l'Amérique du Sud, de Rio de Janeiro à Lima, et de Lima au Para; exécutée par ordre du gouvernement français pendant les années 1843 à 1847, sous la direction de Francis de Castelnau*. Paris: P. Bertrand, 1850-59.

7 pts. in 15 v. plates (part col.) maps, charts. 22.5- 31 cm. Each part, except pt.1, has also special title page with date in some cases differing from that of general title page.

Part 1. *Histoire du voyage*. 1850-51. 6 v. fold. map, tables. 22.5 cm. LSU copy Leipzig: R.W. Hiersemann, 1922.

Part 2. *Vues et scènes*. 1853. 4 p.l., [5]-16 p. 60 plates (part col.) 31 cm.

Part 3. *Antiquités des Incas et autres peuples anciens...* 1854. 4 p.l., 7 p., 60 plates (part col.) 31 cm.

Part 4. *Itinéraires et coupe géologique à travers le continent de l'Amérique du Sud...* 1852. 4 p.l., [v]- vii, iiip., 76 double charts. 46.5 cm.

Part 5. *Géographie, des parties centrales de l'Amérique du Sud...* 1854. 4 p.l., [v]-x., 1 l., 30 double maps. 46.5 cm.

Part 6. *[Botanique] Chloris andina. Essai d'une flore de la région alpine des Cordillères de l'Amérique du Sud...* 1855-57. 2 v. 90 plates. 31 cm.

Part 7. *[Zoologie] Animaux nouveaux ou rares recuellis pendant l'expédition...* 1855-59. 3 v. in 5. 174 plates (part col.) 32 cm.

Half red morocco, marbled boards.

BM I p.325. Nissen ZBI 837. Yale p.54. Zimmer p.125.

122

CHATEAUBRIAND, FRANÇOIS AUGUSTE RENÉ, VICOMTE DE (1768-1848). *Travels in America and Italy*. London: H. Colburn, 1828.

2 v. 24 cm. Translated from the French. Stamped on verso of title page of each vol.: Sion College Library. Stamped on fly leaf of each vol.: Withdrawn from Sion College Library. Grey boards with red buckram spine.

Gift of Dr. and Mrs. A. Brooks Cronan.

* 123

COLVOCORESSES, GEORGE MUSALAS (1816-1872). *Four years in the government exploring expedition; commanded by Captain Charles Wilkes*. By Lieut. George M. Colvocoresses. Fifth edition. New York: J.M. Fairchild & co., 1855[c1852].

371 p. front., plates. 20 cm. Red library binding.

* 124

CROUCH, NATHANIEL (1632?-1725?). *Het Britannische ryk in Amerika: zynde eene beschryving van de ontdeking, bevolking, inwoonders, het klimaat, den koophandel, en tegenwoordigen staat van alle de Britannische Coloniën, in dat gedeelte der wereldt. Met eenige nieuwe kaarten van de voornaamste kusten en eilanden. Uit het Engelsch; als mede een omstandig berecht aangaande de koffy e koffy-plantery, uit het fransch vertaald*. Amsterdam: R. en G. Wetstein, 1721.

2 v. in 1. illus., maps. 23 cm. Modern hand-tooled leather binding.

* 125

DE BRAHM, JOHN GERAR WILLIAM (1717-ca.1799). *Report of the general survey in the southern district of North America*. Edited and with an introduction by Louis De Vorsey, Jr. Columbia: University of South Carolina Press, c1971.

xvi, 325 p. illus., maps. 26 cm. (Tricentennial edition, no.3). Transcribed from Kings mss. 210 and 211 in the British Museum. Imitation blue leather binding, dustjacket.

* 126

FERNÁNDEZ DE OVIEDO Y VALDÉS, GONZALO (1478-1557). *De la natural hystoria de las Indias*. Chapel Hill: University of North Carolina Press, [1969].

xvi, 116 p. illus., port. 23 cm. (University of North Carolina. Studies in the Romance languages and literatures, no.85). "A facsimile edition issued in honor of Sterling A. Stoudemire." Brown cloth binding.

* 127

FRANKLIN, JOHN, SIR (1786-1847). *Narrative of a journey to the shore of the Polar Sea in the years 1819-20-21-22*. 2d edition. London: J. Murray, 1824.

2 v. 3 maps (some col.) 23 cm. On title page: "Published by authority of the Right Honourable the Earl Bathurst." In ms. on half title page of v.2: Caroline Narg (?) 1824. Red imitation leather binding.

BM II p.613. Nissen ZBI 1419. Yale p.101.

128

Galapagos: the flow of wildness. Photographs by Eliot Porter. Introduction by Loren Eiseley. Edited by Kenneth Brower. Foreword by David Brower. San Francisco: Sierra Club, [1968].

2 v. 138 col. illus., map. 36 cm. (Sierra Club exhibit format series, 19-20). Red cloth with gilt lettering, pictorial endpapers, green cloth slipcase.

129

GREAT BRITAIN. CHALLENGER OFFICE. *Report on the scientific results of the voyage of H.M.S. Challenger during the years 1873-76 under the command of Captain George S. Nares and the late Captain Frank Tourle Thomson, R.N. Prepared under the super-intendence of the late Sir C. Wyvill Thomson and now of John Murray...* Published by order of Her Majesty's government. [Edinburgh]: Printed for H.M. Stationery Off. [by Neill and company], 1880-95.

40 v. in 44. illus., plates, maps, plans, tables, diagrs. and atlas. 6 v. 33 x 25 cm. Contents: v.I. Narrative (2 v. in 3. 1882, 1885); v.II. Physics and chemistry (2 v. 1884-89); v.III. Deep-sea deposits (1 v. 1891); v.IV. Botany (2 v. 1885-86); v.V. Zoology (32 v. in 40. 1880-89); v. VI. Summary (1 v. in 2. 1895). Green cloth bindings.
BM II p.716-17. Nissen ZBI 4754.

* 130

GUILLEMARD, FRANCIS HENRY HILL (1852-1933). *The cruise of the Marchesa to Kamschatka & New Guinea. With notices of Formosa, Liu-Kiu, and various islands of the Malay Archipelago.* With maps and numerous woodcuts drawn by J. Keulemans, C. Whymper, and others, and engraved by Edward Whymper. London: J. Murray, 1886.

2 v. illus., 14 maps (5 fold. col.), 2 col. plates. 23 cm. Armorial bookplate of Frances Evelyn, Countess of Warwick, in each vol. Stamped on fly leaf of each vol.: Hatchards, 187 Piccadilly. Half brown morocco, marbled endpapers.
Anker 192. BM II p.750. Nissen ZBI 1761. Zimmer p.277.

131

HARRIS, EDWARD (1799-1863). *Up the Missouri with Audubon; the journal of Edward Harris.* Edited and annotated by John Francis McDermott. Norman: University of Oklahoma Press, [c1951].

xv, 222 p. illus., ports. 24 cm. (American exploration and travel 15). Grey cloth binding.
Yale p.123.

* 132

HENDERSON, GEORGE, M.D. *Lahore to Yārkand. Incidents of the route and natural history of the countries traversed by the expedition of 1870, under T.D. Forsyth.* By George Henderson and Allan O. Hume. [Edited by H.W. Bates]. London: L. Reeve, 1873.

xiv, 370 p. illus. (part col.), fold. map. 26 cm. "The plates of birds [32 plates] are all by Keulemans." Half brown morocco, marbled boards. Binding by William A. Payne.
Anker 64. BM II p.822. Nissen ZBI 1885. Zimmer p.297.

133

HOOKER, JOSEPH DALTON, SIR (1817-1911). *Himalayan journals; or, Notes of a naturalist in Bengal, the Sikkim and Nepal Himalayas, the Khasia mountains, &c.* London, New York: Ward, Lock, and Bowden, 1891.

xxxi, 574 p. illus., front., maps (part fold.) (The Minerva library of famous books). Green cloth binding.
BM I p.869 (1854 ed.)

134

INTERNATIONAL POLAR EXPEDITION, 1882-1883. *Report of the International polar expedition to Point Barrow, Alaska, in response to the resolution of the U.S. House of Representatives of December 11, 1884.* Washington: Government Printing Office, 1885.

8 pts. in 1 v. 695 p. col. front., plates, fold., map, plan. 80 cm. U.S. Signal office. Arctic series of publications, no.1. Issued also as House ex. doc. 44, 48th Cong., 2nd sess. (v.23, serial no.2298). LSU lacks fold. map, plan. In ms. on fly leaf: Thomas Mason-1st. Lt. U.S.R.M. recd. from G.P.O. Feby 16th, 1886. Label on back fly leaf: Free Public Library, East Orange, N.J. Original brown cloth binding.
BM V p.272. Yale p.200.

* 135

KOTZEBUE, OTTO VON (1787-1846). *Entdeckungs-Reise in die Süd-See und nach der Berings-Strasse zur Erforschung einer nordöstlichen Durchfahrt. Unternommen in den Jahren 1815, 1816, 1817, und 1818, auf kosten sr. erlaucht des...Grafen Rumanzoff auf dem Schiffe Rurick unter dem Befehle des Lieutenants der russisch-kaiserlichen Marine Otto von Kotzebue...* Weimar: Gebrüder Hoffmann, 1821.

3 v. in 1. col. front., col. plates, fold. maps. 29 cm. Manuscript notes on p.22 of v.2. Dark blue quarter leather, marbled boards, endpapers, and edges.

* 136

LAET, JOANNES DE (1593-1649). *Novus orbis, seu Descriptionis Indiae Occidentalis, libri XVIII. Novis tabulis geographicis et variis animantium, plantarum fructuumque iconibus illustrati.* Cum privilegio Lugd. Batav., apud Elzevirios, A., 1633.

16 p.l., 690, [18]p. illus., 14 maps. 37 cm. LSU copy imperfect: pp.105-204 omitted in numbering only, 480 incorrectly numbered 470, 627 incorrectly numbered 613, 654 incorrectly numbered 648. Clipping containing bibliographic information and label (Domus Aquensis Societatis Jesu) in envelope laid in. Contemporary calf rebacked.
Nissen ZBI 2359.

* 137

LINNÉ, CARL VON (1707-1778). *Lachesis lapponica; or, A tour in Lapland, now first published from the original manuscript journal of the celebrated Linnaeus; by James Edward Smith.* London: Printed for White and Cochrane, by R. Taylor, 1811.

2 v. illus. 24 cm. Signed in ms. on lining paper of each vol.: Dorothy Roddick. Original boards.
BM III p.1129.

*138

LOPES, DUARTE (fl. 1578). *Relatione del reame di Congo et delle circonvicine contrade, tratta dalli scritti & ragionamenti di Odoardo Lopez, Portoghese, per Filippo Pigafetta, con dissegni vari di geografia, di piante, d'habiti, d'animali, & altro.* Roma: Appresso B. Grassi, [1591].

4 p.l., 82 p., 1 l. 8 fold. plates, 2 fold. maps. 23 cm. Vellum binding.
BM III p.1177 (other eds.)

*139

LYELL, CHARLES, SIR, BART. (1797-1875). *Travels in North America; with geological observations on the United States, Canada, and Nova Scotia.* London: J. Murray, 1845.

2 v. illus., plates (part fold.) maps (part fold.) facsims. 20 cm. Label with stamped no.239 D mounted on fly leaf of each vol. Clipping with bibliographical information mounted on p.[2] of cover of each vol. Original green cloth binding.
BM III p.1199.

*140

PALLAS, PETER SIMON (1741-1811). *Voyages du Professeur Pallas, dans plusieurs provinces de l'empire de Russie et dans l'Asie septentrionale.* Traduits de l'allemand par le C. Gauthier de la Peyronie. Nouvelle édition, revue et enrichie de notes par les CC. Lamarck...et Langlès. Paris: Maradan, l'an II de la République, [1794].

8 v. 21 cm. and atlas (9 plates). 34 cm. Vols.6-7 and atlas also edited by J.B.L.J. Billecocq. Two bookplates (with names erased) inserted in each vol. Contemporary calf, marbled endpapers.
BM IV p.1505. Nissen ZBI 3076. Yale p.219 (German ed.) Zimmer p.480 (German ed.)

*141

PERNETY, ANTOINE JOSEPH (1716-1801). *Histoire d'un voyage aux isles Malouines, fait en 1763 & 1764; avec des observations sur le detroit de Magellan, et sur les Patagons, par Dom Pernetty...* Nouvelle édition. Refondue & augmentée d'un discours préliminaire, de remarques sur l'histoire naturelle, &c. Paris: Chez Saillant & Nyon; Delalain, 1770.

2 v. and 16 plates (incl. maps) in 1 v. 21 cm. Armorial bookplate in each volume. Stamped throughout: Cambridge University Library, No.3, 1925. Stamped on title page of vols. 1 & 2 and fly leaf of plates: Massiliae, Soc. Jes., JHS. Full contemporary calf restored. Plates vol. modern full calf.
BM IV p.1549 (English ed.) Yale p.224.

*142

PRINCETON UNIVERSITY EXPEDITIONS TO PATAGONIA, 1896-1899. *Reports of the Princeton University Expeditions to Patagonia, J.B. Hatcher in charge.* Edited by William B. Scott. Princeton, N.J.: The University Press (J. Pierpont Morgan Publication Fund); Stuttgart: Schweizerbart, 1901-03—1932.

8 v. in 13. illus. (part col.) maps (part fold., part col.) tables. 33 x 26 cm. Vol.1 published 1903. Vols. 2 (5 pts.), 3 and 8 (each in 2 pts.), paged continuously. Vols. 1, 3-8 library binding; v.2 in original blue paper covers.
BM IV p.1615. Nissen ZBI 4758. Yale p.233.

*143

SLOANE, HANS, SIR, BART. (1660-1753). *A voyage to the islands Madera, Barbados, Nieves, S. Christophers and Jamaica, with the natural history of the herbs and trees, four-footed beasts, fishes, birds, insects, reptiles, &c. of the last of those islands; to which is prefix'd an introduction, wherein is an account of the inhabitants, air, waters, diseases, trade, &c. of that place, with some relations concerning the neighbouring continent, and islands of America. Illustrated with the figures of the things described, which have not been heretofore engraved; in large copperplates as big as the life.* London: Printed by B. M. for the author, 1707-25.

2 v. XI, 274 fold. plates (incl. map). 35.5 cm. In ms. on title page of v.1-2, 939 F. Log., 2 vols. Bookplate of the Loganian Library mounted on lining paper of v.1-2. LSU copy imperfect: plates I, 1, 6, 143-144, 146, 207, and 259 wanting. Vol.2, p.195 incorrectly numbered 185; p.232 incorrectly numbered 322; pp.467-468 wanting?; p.472 incorrectly numbered 572. Title pages in red and black. Contemporary calf binding, rebacked with modern calf.
BM IV p.1939. Hunt 417. Jackson p.370. Nissen BBI 1854. Pritzel 8723.

*144

SMITH, HERBERT HUNTINGTON (1851-1919). *Brazil, the Amazons and the coast.* Illustrated from sketches by J. Wells Champney and others. New York: C. Scribner's Sons, 1879.

xv, 644 p. illus., fold. map. 24 cm. Stamped on title page and p.101: Ames Free Library, Easton, Mass. Label mounted on p.[2] of cover: The Ames Free Library, Easton, Mass., April 14, 1880, Shelf no. 337.27, accession no.1203.

In ms. on verso of title page: 1203. Original brown cloth decorated binding with gilt title.

* 145

The Story of the life and travels of Alexander von Humboldt. London, New York, etc.: T. Nelson and sons, 1879.

207 p. incl. front., illus. (map), plates. 17 cm. In ms. on fly leaf Eddie Sayles, Clay Station, Nov. 25, 1880. Original cloth decorated binding.

146

UNITED STATES NAVAL ASTRONOMICAL EXPEDITION, 1849-1852. *The U.S. Naval astronomical expedition to the southern hemisphere, during the years* 1849-'50-'51-'52. Lieut. J.M. Gilliss, superintendent. Washington: A.O.P. Nicholson, Printer, 1855-56.

4 v. front. illus. plates (part col.), maps (part fold.), plans. 30 x 23 cm. (33d Cong., 1st Sess. House Ex. doc. no.121). Vols. published as I, II, III, and VI. Vols. IV and V were not published. The material was later published by the U.S. Naval Observatory in its Astronomical and meteorological observations (Washington observations) for 1868, App. I. 1871 and Astronomical, magnetic, and meteorological observations (Washington observations) for 1890, app. I. 1895. LSU owns only v.II. In ms. on fly leaf: H.B. Bailey. Half calf/marbled boards.
BM V p.2166. Nissen ZBI: 1574. Yale p.54 (v.2 only).

* 147

WATERTON, CHARLES (1782-1865). *Excursions dans l'Amérique Méridionale, le Nord-Ouest des États-Unis et les Antilles, dans les années* 1812, 1816, 1820, *et* 1824: *avec des instructions totalement neuves sur la conservation des oiseaux/par Charles Waterton; suivies d'une notice sur les sauvages de l'Amérique Septentrionale; traduit de l'Anglais.* Paris: Lance, 1833.

xvi, 470 p. [1] leaf of plates, port. 22 cm. In ms. on half title page: C. Dupuis. Label mounted on p.[2] of cover: Librairie Ancienne & Moderne, a Rouen, chez Edet Jeune, Rue Beavoisine, no.9, près la Crosse, Libres Anglais, Italiens, &c. Brown calf binding, gilt decorated spine.
BM V p.2271 (English ed.)

148

WATERTON, CHARLES (1782-1865). *Wanderings in South America, the North-west of the United States, and the Antilles, in the years* 1812, 1816, 1820, *and* 1824. *With original instructions for the perfect preservation of birds, &c. for cabinets of natural history.* London: J. Mawman, 1825.

vii, 326 p. front. 28 cm. Clipping containing bibliographic information mounted on p.[3] of cover. Quarter brown leather, marbled boards, marbled end papers.
BM V p.2271.

149

WATERTON, CHARLES (1782-1865). *Wanderings in South America, the North-west of the United States, and the Antilles, in the years* 1812, 1816, 1820, *and* 1824. *With original instructions for the perfect preservation of birds, &c. for cabinets of natural history.* 4th edition. London: B. Fellowes, 1839.

iv, 307 p. front. 28 cm. Bookplate: Jules M. Burguieres Sugar Collection. On fly leaf in ms.: Jules Burguieres, June 1957, Sugar cane cultivation, 2; Sugar cane juice, 66; Sugar Plantation near Pernambuco, 90; Sugar in Demerara, 101-102. Half calf, gilt decorated spine, marbled endpapers and edges.
BM V p.2271. Yale p.306.

* 150

WIED-NEUWIED, MAXIMILIAN ALEXANDER PHILIPP, PRINZ VON (1782-1867). *Reise nach Brasilien in den Jahren* 1815 *bis* 1817; *von Maximilian, Prinz zu Wied-Neuwied.* Mit zwei und zwanzig Kupfern, neunzehn Vignetten und drei Karten. Frankfurt: H.L. Brönner, 1820-21.

2 v. 19 plates. 38 cm. and portfolio of 22 plates (5 col.), 3 maps. 52 cm. Full green morocco, gilt decorated spine and boards, gilt edges.
BM V p.2315.

151

WILKES, CHARLES (1798-1877). *Narrative of the United States exploring expedition during the years* 1838, 1839, 1840, 1841, 1842. Philadelphia: Lea & Blanchard, 1845.

5 v. illus. plates, maps (part double) and atlas of 5 fold. maps. 28 cm. Vols. 1, 3, 5, original brown cloth binding; vols. 2, 3 and atlas, brown library binding.
BM V p.2167. Nissen ZBI 434.

152

WOLF, THEODOR (1841-1924). *Geografía y geología del Ecuador; publicada por órden del supremo gobierno de la república por Teodoro Wolf.* Leipzig: F.A. Brockhaus, 1892.

xii, 671 p. illus., 12 plates (incl. front.) 2 fold. maps. 27.5 cm. Label on inside front cover: E.C. Freeland, 14 Highwood Ave., Larchmont, New York. Original brown cloth binding, gilt decoration.
BM V p.2349.

PART IV

General Natural History

Impish monkeys from Edward Donovan's *The Naturalist's Repository of Exotic Natural History*, 1822-24.

153

ACADEMY OF NATURAL SCIENCES OF PHILADELPHIA. *Journal*. v.1-8; 2d ser., v.1-16. Philadelphia, 1817-1918.

24 v. illus. (part col.) 22-36 cm. Bindings vary.
BM I p.7. Nissen ZBI 4650. Yale p.2.

154

ACHARD, LOUIS AMÉDÉE EUGÈNE (1814-1875). *The history of my friends; or, Home life with animals.* Translated from the French of Émile [!] Achard. New York: G.P.Putnam's Sons, 1891.

2 p.l., 193 p. plates. 20.5 cm. Green library binding.

155

ADAMS, ALEXANDER B. *John James Audubon; a biography*. New York: Putnam, [c1966].

510 p. illus., facsims., ports. 22 cm. Red cloth binding.

156

ALLEN, DAVID ELLISTON. *The naturalist in Britain: a social history*. London: Allen Lane, 1976.

291 p. [4] leaves of plates. illus. 23 cm. Original tan cloth binding.

157

Annales des sciences naturelles. t.1-30. Paris: Crochard [etc.], 1824-33.

30 v. plates (part col., part fold.) 21-29 cm. and atlases in 5 v. Continued in two sections: *Annales des sciences naturelles. Botanique et biologie végétale* and *Annales des sciences naturelles. Zoologie et biologie animale*. LSU lacks vols. 19-21, 23-27. Bindings vary.
BM I p.50. Nissen ZBI 4547. Pritzel 10683. Yale p.10.

158

ARTHUR, STANLEY CLISBY. *Audubon, an intimate life of the American woodsman*. New Orleans: Harmanson, 1937.

517 p. front., illus., plates, ports., maps (1 double). 26.5 cm. "This edition, printed on Hazelbourn satin antique deckle edge paper and signed by the author, is limited to 375 copies, of which only 350 are for sale." LSU owns copies nos.25, 39, and two unnumbered copies. Tan cloth bindings.
Yale p.14.

*159

AUDUBON, JOHN JAMES (1785-1851). *The art of Audubon; the complete birds and mammals*. With an introduction by Roger Tory Peterson. London: Macdonald, 1981.

xiii, 674 p. col. illus. 29 cm. Red cloth binding, dustjacket.

160

AUDUBON, JOHN JAMES (1785-1851). *Audubon and his journals*. By Maria R. Audubon. With zoölogical and other notes by Elliott Coues. New York: Dover Publications, [1960].

2 v. illus., ports. 21 cm. "An unabridged and unaltered republication of the first edition originally published...in 1897." Paperbound, illustrated covers.
Yale p.14 (1st ed.) Zimmer p.26 (1st ed.)

161

AUDUBON, JOHN JAMES (1785-1851). *Audubon in the west*. Compiled, edited, and with an introduction by John Francis McDermott. [1st ed.] Norman: University of Oklahoma Press, [c1965].

xi, 131 p. illus., facsims., port. 23 cm. "Letters from Audubon to his family." Grey cloth binding.

162

AUDUBON, JOHN JAMES (1785-1851). *Audubon's America; the narratives and experiences of John James Audubon*. Edited by Donald Culross Peattie; illustrated with facsimiles of Audubon's prints and paintings. Boston: Houghton Mifflin company, 1940.

vii, [1]p., 1 l., 328, [1]p. col. front. (port.), col. plates (part double) 29.5 cm. Map on lining-papers. Green library binding.

——— "Extra set of illustrations to accompany your copy of *Audubon's America*." Boston: Houghton Mifflin company, [1940]. 17 col. plates. 38.5 x 29.5 cm. In envelope in pamphlet binder.
Yale p.14.

163

AUDUBON, JOHN JAMES (1785-1851). *Journal of John James Audubon made during his trip to New Orleans in 1820-1821*. Edited by Howard Corning, foreword by Ruthven Deane. Boston: The Club of Odd Volumes, 1929.

ix, 234 p., 1 l. front. (port.), facsim. 25 cm. "This edition is limited to three hundred and one copies, of which twenty-five are reserved by the Club of Odd Volumes."

"An exact copy of the original manuscript which is in the Museum of Comparative Zoology of Harvard University; now published for the first time." Quarter linen/blue boards, slipcase.
Yale p.14.

164

AUDUBON, JOHN JAMES (1785-1851). *Journal of John James Audubon made while obtaining subscriptions to his "Birds of America" 1840-1843.* Edited by Howard Corning, foreword by Francis H. Herrick. Boston: The Club of Odd Volumes, 1929.

1 p.l., [ix], 179, [3]p. front. 25 cm. "This edition is limited to two hundred and twenty-five copies, of which twenty-five copies are reserved by the Club of Odd Volumes." "An exact copy, except for a few corrections, of the original manuscript which is in the Museum of Comparative Zoology of Harvard University; now published for the first time." Quarter linen/blue boards, slipcase.
Yale p.14.

165

AUDUBON, JOHN JAMES (1785-1851). *Letters of John James Audubon, 1826-1840.* Edited by Howard Corning. Boston: The Club of Odd Volumes, 1930.

2 v. 25 cm. "This edition is limited to two hundred and twenty-five copies, of which twenty-five copies are reserved by the Club of Odd Volumes." Quarter linen/blue boards.
Yale p.14

166

AUDUBON, JOHN JAMES (1785-1851). *The life of John James Audubon, the naturalist.* Edited by his widow. With an introduction by Jas. Grant Wilson. New York: Putnam, 1869.

443 p. port. 19 cm. Consists "in good part of extracts from...[Audubon's] *journals and episodes*... The following pages are substantially the...[*Life and adventures of John James Audubon*, ed. by Robert Buchanan] reproduced with some additions, and the omission of several objectionable passages." In ms. on fly leaf: Thomas A. Murray. Original green cloth binding.
Zimmer p.26.

* 167

AUSTRALIAN MUSEUM, SYDNEY. *Lord Howe Island. Its zoology, geology, and physical characters.* Printed by order of the trustees, E.P. Ramsay, curator. Sydney: C. Potter, government printer, 1889.

5 p.l., 132 p., 7 l. front. (fold. map), x plates (part fold., incl. 2 fold. maps, 1 col.) 24 cm. (The Australian Museum, Sydney. Memoirs, no.2). In ms. on cover of c.2: B.B. Woodward. "Note on the bibliography of Lord Howe Island," by R. Etheridge: [5]p. tipped in c.2. C.1, blue printed boards; c.2, green printed boards.
BM V p.2063 (as Memoirs). Yale p.15 (as Memoirs).

* 168

BALDNER, LEONHARD (1612-1694). *Vogel-, Fisch- und Thierbuch. Handschrift Ms. 2° phys. et hist. nat. 3 der Murhardschen Bibliothek der Stadt Kassel und Landesbibliothek.* Einfuhrung von Robert Lauterborn. Stuttgart: Müller und Schindler, [c1973-1974].

2 v. facsimile and 2 v. commentary. 20 x 32 cm. "Fotomechanischer Nachdruck aus: *Das Vogel- Fisch- und Thierbuch des Strassburger Fischers Leonhard Baldner aus dem Jahre 1666.*" Half leather/cork edition; commentary, green paper wrappers.

* 169

BANCROFT, EDWARD (1744-1821). *An essay on the natural history of Guiana, in South America. Containing a description of many curious productions in the animal and vegetable systems of that country. Together with an account of the religion, manners, and customs of several tribes of its Indian inhabitants. Interspersed with a variety of literary and medical observations. In several letters from a gentleman of the medical faculty during his residence in that country.* London: T. Becket and P.A. De Hondt, 1769.

2 p.l., iv, 402, [2]p. front. 20 cm. Full calf, raised bands, gilt decoration.
BM I p.95. Yale p.19. Zimmer pp.38-39.

* 170

BARBER, LYNN. *The heyday of natural history, 1820-1870.* Garden City, N.Y.: Doubleday, 1980.

320 p., [8] leaves of plates. illus. (some col.) 27 cm. Quarter brown cloth, paper over boards, dustjacket.

* 171

BAZIN, GILLES AUGUSTIN (d. 1754). *Observations sur les plantes et leur analogie avec les insectes: precedées de deux discours; l'un sur l'accroissement du corps humain, l'autre sur la cause pour laquelle les bestes nâgent naturellement, & que l'homme est obligé d'en étudier les moyens.* Strasbourg: J.R. Doulssecker, 1741.

1 p.l., [4], xvi, 134 p. 20 cm. Manuscript notes on fly leaves. Paper over boards.
BM I p.116. Plesch p.137. Pritzel 534.

172

Biologia Centrali-Americana; or, contributions to the knowledge of the fauna and flora of Mexico and Central America. Edited by Frederick Du Cane Godman and Osbert Salvin. London: Porter & Dulau, 1879-1915.

Nissen ZBI 4589.

1. *Zoology, botany, and archaeology.* Edited by Frederick Du Cane Godman and Osbert Salvin. Introductory volume by Frederick Du Cane Godman. 1915. (viii, 149 p., 2 port., 8 double maps. 32 cm.) Black library binding.
2. *Mammalia.* By Edward R. Alston. With an introduction by P.L. Sclater. 1879-82. (xx, 220 p. illus. 3 cm. Issued in parts.) Half red morocco, marbled boards, endpapers.

3-6. *Aves.* By Osbert Salvin and Frederick Du Cane Godman. 1879-1904. (4 v. 79 (i.e.84). col. plates. 31 cm. Issued in parts.) Quarter red morocco, marbled boards, marbled endpapers. Nissen IVB 811.

7. *Reptilia and Batrachia.* By Albert C.L. Gunther. 1885-1902. (xx, 326 p. 76 plates (part col.) 32 cm. Issued in parts.) Green cloth binding.

8. *Pisces.* By C. Tate Regan. 1906-08. (xxxii p., 1 l., 203 p. 26 plates, maps. 32 cm. Issued in parts.) Half blue cloth, marbled boards.

9. *Land and freshwater Mollusca.* By Eduard von Martens. 1890-1901. (xxviii, 706 p. 44 plates (part col.) 32 cm. Issued in parts.) Black library binding.

10-11. *Arachnida-Araneidea.* By Octavius Pickard-Cambridge and Frederick Octavius Pickard-Cambridge. 1899-1905. (2 v. plates (part col.) 32 cm. Issued in parts.) Black library binding.

12. *Arachnida. Scorpiones, Pedipalpi and Solifugae.* By Reginald Innes Pocock. 1902. (71 p., [12]l. 12 plates. 32 cm.) Red library binding.

13. *Arachnida-Acaridea.* By Otto Stoll. 1886-93. (xxi, 55 p. 21 plates (part col.) 32 cm. Issued in parts.) Half red morocco, marbled boards, marbled endpapers.

14. *Chilopoda and Diplopoda.* By Reginald Innes Pocock. 1895-1910. (217 p., 15 plates.32 cm. Issued in parts.) Black library binding.

15-32. *Insecta. Coleoptera.* 1880-1911. (7 v. in 18. plates (part col.) 32.5 cm. Issued in parts. Various bindings. BM VI p.380.

33-35. *Insecta. Hymenoptera.* By Peter Cameron and Auguste Forel. 1883-1900. (3 v. in 2. plates (part col.) 32 cm.) Issued in parts. Black library binding.

36-38. *Insecta. Lepidoptera-Rhopalocera.* By Frederick Du Cane Godman and Osbert Salvin. 1879-1901. (3 v. in 2. 112 (i.e.113) col. plates. 32 cm. Issued in parts.) Quarter red cloth, marbled boards.

39-42. *Insecta. Lepidoptera-Heterocera.* By Herbert Druce. 1881-1915. (4 v. in 3. illus. 32 cm. Issued in parts.) Quarter red cloth, marbled boards.

43-45. *Insecta. Diptera.* By C.R. Osten-Sacken, F.M. van der Wulp and S.W. Williston. 1886-1903. (3 v. in 2. col. plates. 32 cm.) Green cloth binding.

46-47. *Insecta. Rhynchota. Hemiptera-Heteroptera.* By William L. Distant and George C. Champion. 1880-1901. (2 v. plates (part col.) 32 cm. Issued in parts.) V.1, black library binding, v.2, quarter black cloth, marbled boards.

48-49. *Insecta. Rhynchota. Hemiptera-Homoptera.* By William L. Distant, W.W. Fowler, and T.D.A Cockerell. 1881-1909. (2 v. illus. col. plates. 31.5 cm. Issued in parts.) V.48, black library binding; v.49, half black cloth/marbled boards.

50. *Insecta. Neuroptera. Ephemeridae.* By A.E. Eaton. Odonata, by Philip P. Calvert. 1892-1908. (xxx p., 1 l., 420 p., 10 l. 10 plates (3 col.) map. 32 cm.) Quarter red morocco, marbled boards, marbled endpapers.

51-52. *Insecta. Orthoptera.* 1893-1909. (2 v. plates (part col.) 32.5 cm. Issued in parts.) Black library binding.

53-57. *Botany.* By William Botting Hemsley. 1879-88. (5 v. 110 (i.e. 111) plates (part col., incl. map) 33 x 27 cm.) Quarter brown morocco, cloth binding. BM VI p.380. Nissen BBI 846.

58. *Archaeology.* By Alfred Percival Maudslay. Appendix, [The archaic Maya inscriptions] by J.T. Goodman. 1889-1902. (5 v. in 1. 33 cm. Black library binding.)

59-62. *Plates.* 1889-1902. (4 v. in 16. illus. maps. 34 x 52 cm.) Paper boards.

* 173

BLUNT, WILFRID. *The compleat naturalist: a life of Linnaeus.* [By] Wilfrid Blunt; with the assistance of William T. Stearn. New York: Viking Press, [c1971].

256 p. illus. (some col.), facsims., maps, ports. (some col.) 26 cm. Green cloth binding.

174

BOITARD, PIERRE (1789?-1859). *Manuel d'histoire naturelle, comprenant les trois règnes de la nature; ou généra complet des animaux, des végétaux et des minéraux.* Paris: Roret, 1827.

2 v. 14 cm. Quarter black morocco, marbled boards.
Nissen ZBI 456a.
Gift of Blanche Duncan.

175

BOITARD, PIERRE (1789?-1859). *Manuel du naturaliste préparateur, ou l'art d'empailler les animaux et de conserver les végétaux et les minéraux.* Troisieme édition, revue, corrigee et entièrement refondue. Paris: A la Librairie encyclopédique de Roret, 1835.

306 p. illus. 15 cm. Full brown morocco, marbled endpapers and edges.
Yale p.34 (later ed.)

176

BONNET, CHARLES (1720-1793). *Oeuvres d'histoire naturelle et de philosophe de Charles Bonnet.* Neuchatel: Samuel Fauche, 1779-1783.

8 v. in 10. front. (v.1) illus., plates (part fold.) 26 cm. In ms. on title page of each vol.: Levi? Peirce. Some vols. inscribed to his sister Mary Peirce. Full brown modern morocco binding by William A. Payne.
BM I p.197. Nissen ZBI 461.

177

BOSTON SOCIETY OF NATURAL HISTORY. *Memoirs read before the Boston Society of Natural History.* v.1-9, no.3, 1863/69-1946. Boston: The Society, 1866-1946.

9 v. illus., plates (part col.) maps, tables, diagrs. 30 cm. Preceded by *Boston journal of natural history*, published 1834-63. V.1, marbled boards, vols.2-9, black library binding.
BM I p.207. Nissen ZBI 4684. Yale p.36.

178

BRADLEY, RICHARD (1688-1732). *A philosophical account of the works of nature. Endeavouring to set forth the several gradations remarkable in the mineral, vegetable, and animal parts of the creation. Tending to the composition of a scale of life. To which is added an account of the state of gardening, as it is now in Great Britain, and other parts of Europe: together with several new experiments relating to the improvement of barren ground, and the propagating of timber-trees, fruit-trees, &c. with many curious cutts.* London: W. Mears, 1721.

10 p.l., 194 p. XXVIII plates (1 fold.) 27.5 cm. Ivory imitation leather, hinge design, raised bands.
BM I p.220. Hunt 452. Nissen BBI 2: 221na. Pritzel 1077. Yale p.38.

179

BRITISH MUSEUM (NATURAL HISTORY). *The history of the collections contained in the Natural History departments of the British Museum.* London: Printed by order of the Trustees, 1904-12.

3 v. 22 cm. Green cloth binding, gilt decoration.
BM VI p.131. Yale p.42.

180

BRUSSELS. INSTITUT ROYAL DES SCIENCES NATURELLES DE BELGIQUE. *Annales.* t.1-14. Bruxelles, 1877-87.

14 v. illus., ports., maps. 40 cm. and atlases (plates). 40 x 58 cm. Red library binding.
BM I p.269. Nissen ZBI 4550.

181

BRYAN, WILLIAM ALANSON (1875-1942). *Natural history of Hawaii, being an account of the Hawaiian people, the geology and geography of the islands and the native and introduced plants and animals of the group. Illustrated with one hundred and seventeen full page plates from four hundred and forty-one photographs elucidating the ethnology of the native people, the geology and topography of the islands and figuring more than one thousand of the common or interesting species of plants and animals to be found in the native and introduced fauna and flora of Hawaii.* Honolulu: [Printed for the author by] the Hawaiian gazette co., ltd., 1915.

596 p. incl. illus., plates, port. 26 cm. Autographed presentation copy. Green cloth binding, gilt lettering.
BM VI p.152. Yale p.45.

* 182

BUENOS AIRES. MUSEO ARGENTINO DE CIENCIAS NATURALES "BERNARDINO RIVADAVIA." *Anales.* t.1-42; 1864—1946-47. Buenos Aires.

42 v. illus. ports., maps, plans, diagrs., tables. 27-37 cm. LSU holdings sparse and scattered. Various bindings.
BM I p.280. Yale p.47.

183

BUFFON, GEORGES LOUIS LECLERC, COMTE DE (1707-1788). *Histoire naturelle, extraite de Buffon et de Lacépède; quadrupèdes, oiseaux, serpents, poissons et cétacés, avec de nombreuses illustrations dans le texte.* Tours: Alfred Mame et fils, 1883.

368 p. front., illus. 27 cm. Red library binding.

184

BUFFON, GEORGES LOUIS LECLERC, COMTE DE (1707-1788). *Histoire naturelle, générale et particuliere...par le comte du Buffon...* Aux Deux-Ponts: Sanson & c, 1785-87.

43 v. front. (port.,tome 1) illus. (part col.) 17 cm. LSU lacks vols. 31 & 41. Full calf with gilt; many vols. with spine mended.

185

BUFFON, GEORGES LOUIS LECLERC, COMTE DE (1707-1788). *Histoire naturelle, générale et particuliere. Nouvelle édition, accompagnée de Notes, et dans laquelle les Supplémens sont inséres dans le premier texte, à la place qui leur convient. L'on y a ajouté l'histoire naturelle des Quadrupédes et des Oiseaux de converts depuis la mort de Buffon, celle des Reptiles, des poissons, des insectes et des vers; enfin, l'histoire des plantes dont ce grand naturaliste n'a pas eu le tems de s'occuper.* Redige par C.S. Sonnini. Paris: F. Dufart, [1801-1808].

127 v. illus. 20 cm. Quarter red leather, red paper over boards, gilt on spine, marbled endpapers.
BM I pp.281-82 (dif. eds.) Nissen ZBI 682. Yale p.47 (dif. eds.)

* 186

BUFFON, GEORGES LOUIS LECLERC, COMTE DE (1707-1788). *Natural history of birds, fish, insects, and reptiles. Embellished with upwards of two hundred engravings.* London: Printed for H.D. Symonds; and sold by Sherwood, Neely, and Jones, 1808-16. [v.1, 1815].

6 v. col. plates. 22 cm. On title page of v.6: "Vol. VI; or, Supplementary volume, containing a description of rare and curious birds, discovered since the death of Buffon, selected and arranged by Sonnini and J.J. Virey, and translated from the last edition of Buffon." Vol.6 has imprint: Printed for the proprietor; and sold by H.D. Symonds. Full red morocco, gilt decorated spine, marbled endpapers.
BM I pp.281-82 (dif. eds.) Yale p.47.

* 187

BUFFON, GEORGES LOUIS LECLERC, COMTE DE (1707-1788). *Oeuvres complètes de Buffon, avec des extraits de Daubenton et la classification de Cuvier.* Paris: Furne, 1844-48. [v.1, 1848].

6 v. 5 col. maps, 122 col. plates, port. 27 cm. Half green morocco, green marbled boards and edges, gilt decorated spine.

* 188

BUFFON, GEORGES LOUIS LECLERC, COMTE DE (1707-1788). *Per l'adolescenza ossia compendio dell'intera storia dei tre regni della natura, il Buffon; arrichita con la serie di tutte le necessarie figure incominciando con quella dell'uomo e proseguendo con le altre de' quadrupedi de' rettili, anfibj, pesci, insetti, e volatili, indi con quella delle piante dell' erbe, e de' fiori, e terminando con quella de' fossili, e de' minerali.* Napoli: N. Gervasi al Gigante, 1808.

2 v. col. illus. 22 cm. Paper covers, paper labels on spine.

189

BUFFON, GEORGES LOUIS LECLERC, COMTE DE (1707-1788). *The system of natural history, written by M. de Buffon, carefully abridged; and the natural history of insects; compiled, chiefly from Swammerdam, Brookes, Goldsmith, &c. embellished with sixty-four elegant copper plates; containing upwards of 284 figures, accurately described, and the page referred to, marked at each figure.* Edinburgh: Printed by J. Ruthven & sons, 1800.

2 v. illus. plates. 22 cm. Dark red library binding.

190

CALL, RICHARD ELLSWORTH (1856-). *The life and writings of Rafinesque.* Prepared for the Filson club and read at its meeting, Monday, April 2, 1894. Author's edition. Louisville, Kentucky: John P. Morton and company, 1895.

xii, 227 p. 2 port. (incl. front.) 3 facsims. 32.5 x 25.5 cm. (Filson club publications, no.10). Red library binding.
BM I p.300.

191

CANTWELL, ROBERT (1908-). *Alexander Wilson: naturalist and pioneer, a biography.* With decorations by Robert Ball. [1st edition]. Philadelphia: Lippincott, [c1961].

318 p. illus. 32 cm. Original green and black cloth binding.

192

CATESBY, MARK (1679?-1749). *The natural history of Carolina, Florida, and the Bahama islands: containing the figures of birds, beasts, fishes, serpents, insects, and plants; particularly, those not hitherto described, or incorrectly figured by former authors, with their descriptions in English and French. To which is prefixed, a new and correct map of the countries; with observations on their natural state, inhabitants, and productions.* Revised by Mr. George Edwards. To the whole is now added a Linnaean index of the animals and plants... [Title follows in French]. London: Printed for B. White, 1771.

2 v. fold. map. 54.5 cm. In ms. on preliminary leaf of both vols.: Joseph Jones, M.D., London, September 6, 1870. Bookplate of Jas. R. Hoffmann in both vols. Bookplate of Joseph Jones, M.D. in both vols. Full calf, spine restored, gilt edges.
Anker 94-95. BM I p.327. Hunt 486. Nissen ZBI 842. Pritzel 1602.

193

CHAMBERS, ROBERT (1802-1871). *Vestiges of the natural history of creation.* New York: Wiley and Put-

nam, 1845.
vi, [7]-291 p. 18 cm. Black library binding.
BM I p.334.

194

CLARK, AUSTIN HOBART (1880-). *The new evolution; zoogenesis.* Baltimore: The William & Wilkins company, 1930.
xiv, 297 p. illus. 22 cm. Blue library binding.

* 195

CLERCK, CARL ALEXANDER (1710-1765). *Nomenclator extemporaneus rerum naturalium: plantarum, insectorum, conchyliorum, secundum systema naturae Linnaeanum.* Stockholmiae, typis: L.L. Grefingii, 1759.
2 p.l., 67, [12]p. 21 cm. Half red morocco, red marbled boards.
Jackson p.14.

* 196

COMPANYO, LOUIS JEAN BAUDILE. *Histoire naturelle du département des Pyrénées-Orientales.* Perpignan: J.B. Alzine, 1861-64, [v.3, 1863].
3 v. 2 fold. plates. 23 cm. Vol.1 is author's autograph presentation copy. Original dark red cloth binding, spine labels.
BM I p.372.

197

COUSTEAU, JACQUES YVES. *The silent world.* By J.Y. Cousteau, with Frédéric Dumas. New York: Harper, [c1953].
xiv, 266 p. illus. 24 cm. Quarter red/blue cloth binding.
Gift of Jondon Shilg.

198

DARWIN, CHARLES ROBERT (1809-1882). *The descent of man and selection in relation to sex.* London: J. Murray, 1871.
2 v. illus. 19 cm. Original green cloth binding.
BM I p.423.

199

DARWIN, CHARLES ROBERT (1809-1882). *Journal of researches into the geology and natural history of the various countries visited by H.M.S. Beagle, under the command of Captain Fitzroy, R.N., from 1832 to 1836.* London: H. Colburn, 1839.
xiv, 629, [609]-615 p. maps (part fold.) 23 cm. "Addenda" (p.[609]-629) inserted between p.608 and [609]. In ms. on fly leaf: F.R.L. Henson, Epsom, May 6, 1920. Brown half-calf, marbled boards.
BM I p.422 (dif. eds.) Yale p.71.

200

DARWIN, CHARLES ROBERT (1809-1882). *On the origin of species by means of natural selection; or The preservation of favored races in the struggle for life.* 5th edition, with additions and corrections. New York: Appleton, 1871.
447 p. fold. diagr. 20 cm. Original red cloth binding.
BM I p.422 (various eds.) Pritzel 2057 (various eds.) Yale p.71 (various eds.)
Gift of Rietta Garland Albritton and Florence Garland-Brookes. Garland-Bullard Memorial Collection.

201

DARWIN, CHARLES ROBERT (1809-1882). *The variation of animals and plants under domestication.* London: John Murray, 1868.
2 v. illus. 23 cm. V.1, bookplate of Charles Atwood Kofoid. V.2, bookplate of E.G. Pretyman, Orwell Park. Original green cloth binding.
BM I p.422. Pritzel 2060.

202

DARWIN, CHARLES ROBERT (1809-1882). *The variation of animals and plants under domestication.* Authorized edition, with a preface by Professor Asa Gray. New York: Orange Judd & company, [1868].
2 v. illus. 19.5 cm. Green library binding.
BM I p.422 (Brit. ed.) Pritzel 2060 (Brit. ed.)

* 203

DELATTE, CAROLYN E. *Lucy Audubon; a biography.* Baton Rouge: Louisiana State University Press, 1982.
xiii, 248 p. illus. 24 cm. Southern biography series. Brown cloth binding, pictorial dustjacket.

204

DONOVAN, EDWARD (1768-1837). *The naturalist's repository of exotic natural history; consisting of seventy-two elegantly coloured plates, with appropriate scientific and general descriptions of the most curious, scarce and beautiful quadrupeds, birds, fishes, insects, shells, marine productions, and other interesting objects of natural history, the produce of foreign climates.* London: Printed for the author, and W. Simpkin and R. Marshall, [1822-24].

2 v. 72 hand col. plates. 26 cm. Each plate accompanied by descriptive letterpress. Half green morocco/marbled boards, marbled endpapers, gilt decorated spine.
BM I p.473. Nissen VBI 259.

205

EDWARDS, GEORGE (1694-1773). *Essays upon natural history and other miscellaneous subjects... To which is added, a catalogue, in generical order of the birds, beasts, fishes, insects, plants, &c. contained in Mr. Edwards' Natural history.* London: J. Robson, 1770.

231 p. front. 21 cm. Full contemporary calf, rebacked.
BM II p.510.

*206

EIFERT, VIRGINIA LOUISE (SNIDER) (1911-1966). *River world; wildlife of the Mississippi.* Illustrated by the author. New York: Dodd, Mead, 1959.

xiv, 1 l., 271 p. illus. 22 cm. Blue cloth binding, dustjacket.

207

ELLIOTT SOCIETY OF NATURAL HISTORY, CHARLESTON, S.C. *Proceedings of the Elliott Society of Natural History of Charleston, South-Carolina.* Vols. 1-2; Nov. 1853-Nov. 1890. Charleston: Russell & Jones, 1859-[91].

2 v. illus. plates. 24 cm. In November 1867, the name of the society was changed to Elliott Society of Science and Art. Each volume has a typewritten table of contents bound in. Half tan morocco/cloth binding.
BM II p.523.
Gift of Jim Bishop.

208

FAUJAS DE SAINT-FOND, BARTHÉLEMY (1741-1819). *Histoire naturelle de la Montagne de Saint-Pierre de Maestricht.* Paris: H.J. Jansen, An 7ème de la République François, [1799].

263 p. illus., map, plates (part fold.) 35 cm. Stamped on fly leaf: Mario Cermenat. Full calf, gilt decoration.
BM II p.557.

209

GEOFFROY SAINT-HILAIRE, ÉTIENNE (1772-1844). *Études progressives d'un naturaliste pendant les années 1834 et 1835...* Paris: Roret, 1835.

xiv, 1 l., 189 p., 1 l. 9 plates. 26 cm. With this is bound: His *Recherches sur de grands sauriens trouves a l'etat fossile.* Paris: Freres, 1831 (138 p., 1 plate). Marbled boards, marbled endpapers; spine mended.
BM II p.656 (1st title only).

*210

GOSSE, PHILIP HENRY (1810-1888). *Letters from Alabama, (U.S.) chiefly relating to natural history.* Mountain Brook, Alabama: Overbrook House, 1983. An annotated reprint edition of the original published London: Morgan and Chase, 1859.

2 p.l., [8], xiii, 306, [307-324]. Paperbound.
BM II p.697 (orig. ed.)

211

GRANDIDIER, ALFRED (1836-1921). *Histoire physique, naturelle et politique de Madagascar.* Paris: Imprimerie nationale, [1875-1954?].

39 v. in 36. plates (part col.) maps, tables. 30 cm. LSU lacks vols. 2, 7-8, 11, 24, 26, 28-29, 31-32, 37-38. Vols. 1, pt.2; 16, pt.2; 25, 33-36 lack title pages. Title page of v.3 A reads: v.4. Title page of v.13 corrected in pencil. Vols. numbered irregularly. 187 plates only partly available in v.39. No more published. V.1, pt.2, half red morocco, marbled boards, marbled endpapers. Other vols., tan library binding.
BM II p.705. Nissen BBI 746, ZBI 1676. Plesch p.241. Yale p.195. Zimmer pp.264-65.

212

GREW, NEHEMIAH (1641-1712). *Musæum regalis societatis; or, A catalogue & description of the natural and artificial rarities belonging to the Royal society and preserved at Gresham colledge. Whereunto is subjoyned the comparative anatomy of stomachs and guts.* By the same author. London: Printed by W. Rawlins, for the author, 1681.

6 p.l., 386, [4], 43 p. front. (port.) 31 plates (1 fold.) 31.5 cm. Full contemporary calf restored with new spine, gilt decorated.
BM IV p.1755. Nissen ZBI 1714. Pritzel 3558. Yale p.118.

*213

HARRIS, THADDEUS MASON (1768-1842). *The natural history of the Bible; or, A description of all the quadrupeds, birds, fishes, reptiles, and insects, trees, plants, flowers, gums, and precious stones, mentioned in the Sacred Scriptures.* Collected from the best authorities, and alphabetically arranged. Boston: Wells and Lilly, 1820.

xl, 476 p. 22 cm. Half calf, marbled boards.

*214

Henderson; a guide to Audubon's home town in Kentucky. Compiled by workers of the Writers' program of the Work Projects Administration in the state of Kentucky. Sponsored by Susan Starling Towles, librarian, Public Library, Henderson, Kentucky. Northport, L.I., N.Y.: Bacon, Percy & Daggett, [c1941].

120 p. front., plates, ports., maps. 22 cm. (American guide series). Original blue cloth binding.

*215

HERNÁNDEZ, FRANCISCO (1514-1587). *Nova plantarvm, animalivm et mineralivm Mexicanorvm historia a Francisco Hernandez...primum compilata, dein a Nardo Antonio Reccho in volvmen digesta, a Io. Terentio, Io. Fabre, et Fabio, Colvmna Lynceis notis, & additionibus longe doctissimis illustrata. Cui demum accessere, aliqvot ex principis Federici Caesii frontispiciis Theatri naturalis phytosophicae tabulae vna cum quamplurimis iccnibus ad octingentas, quibus singula contemplanda graphice exhibentur.* Romae: sumptibus B. Deuersini & Z. Masotti, typis V. Mascardi, 1651.

15 p.l., 950, [22], 90, [6]p. illus. 33 cm. Manuscript notes throughout text and index. Title page in red and black. Full calf binding, tooled.

BM II p.832. Hunt 32, 1949. Jackson p.376. Nissen ZBI 861. Pritzel 4000. Yale p.131.

216

HERRICK, FRANCIS HOBART (1858-). *Audubon the naturalist; a history of his life and time.* New York, London: D. Appleton and company, 1917.

2 v. fronts., plates (part col.) ports., facsims. 22.5 cm. Blue library binding.

Yale p.14. Zimmer p.300.

217

HOGG, JABEZ (1817-1899). *The microscope: its history, construction, and application, being a familiar introduction to the use of the instrument and the study of microscopical science.* Illustrated with upwards of five hundred engravings. 3d edition. London, New York: George Routledge, 1858.

xiv, 607 p. front., illus., tables, diagrs. 18 cm. Stamped on fly leaf and p.[vi]: Albert Godchaux. No.182 Esplanade, New Orleans, La. Half calf, marbled boards, endpapers, and edges.

BM II p.860 (dif. eds.)

Gift of Rietta Garland Albritton and Florence Garland-Brookes. Garland-Bullard Memorial Collection.

*218

HORREBOW, NIELS (1712-1760). *The natural history of Iceland: containing a particular and accurate account of the different soils, burning mountains, minerals, vegetables, metals, stones, beasts, birds, and fishes; together with the disposition, customs, and manner of living of the inhabitants. Interspersed with an account of the island, by Mr. Anderson. To which is added, a meteorological table, with remarks. Translated from the Danish original of Mr. N. Horrebow, and illustrated with a New General map of the island.* London: Printed for A. Linde, [etc], 1758.

xx, 207 p. front. (fold. map). 37 cm. Armorial bookplate of Jeremiah Milles mounted on verso of title page. Quarter black imitation morocco, blue cloth binding.

BM II p.876. Yale p.136.

219

HOUGHTON, WILLIAM. *Country walks of a naturalist with his children. Illustrated with eight coloured plates and numerous wood engravings.* London: Groombridge and sons, 1869.

vi, 154 p. illus., plates. 18.5 cm. Original green cloth binding with gilt decoration.

Yale p.137 (2nd ed.)

220

HUGHES, GRIFFITH (fl. 1750). *The natural history of Barbados.* In ten books. London: Printed for the author, 1750.

[2 p.l.], vii, [13], 314 (i.e. 318), [20]p. 43 cm. Four pages (*251-*254) inserted. Addenda 4 p. at end. Large paper edition. Bookplate: Jules M. Burguieres Sugar Collection. Half tan calf, marbled boards; raised bands, gilt decorated spine; binding by William A. Payne.

——— *Another copy.* [2 p.l.], vii, 314 (i.e. 318), [16]p. illus., 29 plates, fold. map. 43 cm. Four pages (*251-*254) inserted. Large paper edition. 2 plates both bearing no.24. Armorial bookplate with motto, "le bon temps viendra." Half calf, marbled boards.

BM II p.887. Hunt 536. Nissen BBI 950. Pritzel 4349. Yale p.140.

221

HUXLEY, THOMAS HENRY (1825-1895). *On the origin of species: or, The causes of the phenomena of organic nature. A course of six lectures to working men.* New York: D. Appleton and company, 1871.

150 p. illus. 21 cm. C.2, bookplate: John Dalton Shaffer. Ardoyne Plantation, Parish of Terrebonne, La. C.1, original red cloth, spine taped. C.2, original blue cloth binding.
BM II p.898 (dif. ed.)

* 222

JACQUIN, NIKOLAUS JOSEPH, FREIHERR VON (1727-1817). *Miscellanea Austriaca ad botanicam, chemiam, et historiam naturalem spectantia.* Vindobonae: Ex Officina Krausiana, 1778-81.

2 v. in 1. 44 hand col. plates (3 fold.) 24 cm. Half brown morocco, brown cloth, gilt decorated spine; marbled endpapers, mottled edges.
BM II p.918. Hunt 655. Nissen BBI 975. Pritzel 4367.

223

JARDINE, WILLIAM, SIR, BART. (1800-1874), ed. *The naturalist's library.* Edinburgh: W.H. Lizars, [1853-64].

40 v. fronts. (ports.), illus., col. plates. 18 cm. The volume numbering differs in various editions of this work. Vols.13, 25-26, 28 and 38 have imprint, London: H.G. Bohn, 1860-62. Original red cloth binding, spines mended.
BM III p.1399. Nissen VBI 471, ZBI 4708. Zimmer pp.326-27.

224

JEFFERIES, RICHARD (1848-1887). *The pageant of summer, with a preface by Thomas Coke Watkins.* Portland, Maine: Thomas B. Mosher, 1905.

xi, [1], 51, [1]p. 14 cm. Original paper covers printed in red and black.

225

KENTUCKY. DEPARTMENT OF PARKS. DIVISION OF MUSEUMS AND SHRINES. *John James Audubon Memorial Museum.* [n.p., c1965].

30 p. illus. 22 x 28 cm. *"John James Audubon's Quadrupeds"*: [3]p. laid in. Paper wrappers.

* 226

KINGSLEY, JOHN STERLING (1854-1929), ed. *The standard natural history.* Boston: S.E. Cassino, 1885-86 [v.1, 1886, c1884; v.6, 1885].

6 v. illus., map. 27 cm. Originally issued in 60 parts, 1884-85. Quarter green morocco/cloth, marbled endpapers.
BM II p.982.

227

KNAPP, JOHN LEONARD (1767-1845). *The journal of a naturalist.* Philadelphia: Gihon & Smith, 1846.

viii, 286 p., 1 l. illus., 2 plates (incl. front.) 19.5 cm. LSU copy imperfect: plate 1 wanting. In ms. on fly leaf: Jos. Jones, Philadelphia, Jan.-19th, 1851. Quarter brown cloth, paper boards.
BM II p.996 (earlier eds.) Pritzel 4737 (earlier ed.) Yale p.159

228

LANKESTER, EDWIN, SIR (1814-1874). *Half-hours with the microscope; being a popular guide to the use of the microscope as a means of amusement and instruction.* Illustrated from nature, by Tuffen West. A new edition. London: Robert Hardwicke, [18—?].

xx, 106 p. front. (col.), illus., plates (col.) 16.5 cm. Red cloth binding.

* 229

L'ÉCLUSE, CHARLES DE (1526-1609). *Rariorum ali-Atrebatis... Exoticorvm libri decem: quibus animalium, plantarum, aromatum, aliorumq̃ue peregrinorum fructuum historiae describuntur: item Petri Bellonii observationes, eodem Carolo Clusio interprete. Series totius operis post praefationem indicabitur.* [Antverpiae]: Ex officinâ Plantinianâ Raphelengii, 1605.

10 p.l., [823]p. illus. 38 cm. Various pagings. LSU copy imperfect: pp.341-344 incorrectly numbered 339-342; p.369 incorrectly numbered 367. Vellum over boards, raised bands, ties.
BM II p.1075. Hunt 182. Nissen BBI 369. Pritzel 1760.

230

LEDERMÜLLER, MARTIN FROBENIUS (1719-1769). *Versuch, bey angehender frühlings zeit die vergrösserungs werckzeüge zum nüzlich u. angenehmen zeitvertreib anzuwenden, von dem verfasser der Mikroskopischen vemüths u. augen ergözung... Essai d'employer les instruments microscopiques avec utilité et plaisir dans la saison du printems, par l'auteur des Amusements microscopiques tant des yeux que de l'esprit; traduit de l'Allemand par I.C. Harrepeter...* Nuremberg: A.L. Wirsing, 1764.

1 p.l., 48 p. 12 hand col. leaves. 46 x 31 cm. Double columns of text in German and French. Some signature numbers not printed. Title page hand-colored. Green library binding.
BM III p.1077. Nissen ZBI 2412.

231

LEEUWENHOEK, ANTHONY VAN (1632-1723). *Arcana naturae detecta, ed. novissima, auction and correction.* Lugduni Batavorum, apud J. Arnold: Langerak, 1722.

2 p.l., [14], 515, [11]p. front. illus., plates (part fold.) 20 x 16.5 cm. Stamped on verso of title page: Dr. Carl Rohrbach, Galberg II. Full calf, raised bands, gilt decoration, marbled endpapers.
BM III p.1080 (earlier ed.) Nissen ZBI 2417.

232

LINNÉ, CARL VON (1707-1778). Caroli Linnaei... *Systema naturæ per regna tria naturæ secundum classes, ordines, genera, species, cum characteribus, differentiis, synonymis, locis...* Ed.10, reformata... Holmi: impensis L. Salvii, 1758-59.

2 v. 20 cm. (v.2: 19 cm.) In ms. on fly leaf of both vols.: J.E. Getterstedt. In ms. on fly leaf of v.1: George Hallenberg. In ms. on fly leaf of v.1 description of the *Systema Naturae* signed G.U. Fernald. Bookplates in both vols.: Ex libris, C.U. Fernald. In v.1 three pages of manuscript notes inserted between pp.822-823. Full calf binding rebacked.
BM III p.1128. Yale p.172.

233

LINNÉ, CARL VON (1707-1778). Caroli Linnaei... *Systema naturae; sive, Regna tria naturae systematice proposita per classes, ordines, genera, & species.* Lugduni Batavorum, T. Haak; ex typ. J.W. de Groot, 1735. Facsimile. [Uppsala: Bokgillet, 1960].

[16]p. 31 cm. "The original size of 41 x 52.5 cm. has been reduced." Marbled paper over boards.
BM III p.1127 (orig. ed.) Pritzel 5404 (orig. ed.) Yale p.172 (dif. eds.)

234

Magazine of zoology and botany. Conducted by Sir W. Jardine, Bart., P.J. Selby, Esq. and Dr. Johnston... v.1-2. Edinburgh: W.H. Lizars; London: S. Highley, [etc.], 1837-38.

2 v. plates (part col.) 22.5 cm. United, in 1838, with W.J. Hooker's *Companion to the Botanical magazine* and continued as the *Annals of natural history*, later the *Annals and magazine of natural history*. Green library binding.
BM III p.1219. Nissen ZBI 4656. Yale p.182.

*235

MARCET, JANE HALDIMAND (1769-1858). *Conversations on natural philosophy in which the elements of that science are familiarly explained and adapted to the comprehension of young pupils.* Illustrated with plates. With additional illustrations and appropriate questions for the examination of scholars by Rev.J.L. Blake. 5th American edition. Hartford: G. Goodwin, 1823.

viii, [13]-256 p. 23 leaves of plates, illus. 19 cm. In ms. on fly leaf: F.S. Woods. Full contemporary calf.

236

MARGGRAF, GEORG (1610-1644). *Historia natural do Brasil; tradução de mons, dr. José Procopio de Magalhães; edição do Museu Paulista comemorativa do cincoentenário da fundação da imprensa oficial do estado de Sao Paulo.* Sao Paulo: Imprensa oficial do estado, 1942.

2 p.l., iv, 4 l., 293, [6], civ, [1]p. illus., facsims. 38.5 cm. Marggraf's *Historiae rervm natvralivm Brasiliae* was first published as pt.2 of Willem Piso's *Historia natvralis Brasiliae.* This edition reproduces in facsimile the added title page for pt.2 and the original engraved title page *Historia natvralis Brasiliae...* Lvgdvn. Batavorvm. apud F. Hackium, et Amstelodami, apud L. Elzevirium, 1648. Half calf, marbled boards, marbled endpapers; spine taped.
BM III p.1237, 1580 (orig. ed.) Hunt 280 (orig. ed.) Nissen BBI 1533. Pritzel 7157 (orig. ed.)
Gift of Mr. Wyeth A. Read in memory of his father, Dr. W.A. Read.

237

MARINE BIOLOGICAL ASSOCIATION OF THE UNITED KINGDOM. *Journal.* v.1-2, 1887-88; new ser. v.1- 1889- . Plymouth: The Association, 1887- .

v. illus. 24 cm. LSU Library holdings complete. V.1, 1887/88 and ns. v.1, 1889/90 in McIlhenny Collection. Bookplate of Rousoon, Devon. Half tan calf binding, marbled boards.
BM III p.1240.

*238

MARTIN, BENJAMIN (1704-1782). *The natural history of England; or, A description of each particular county, in regard to the curious productions of nature and art.* London: Printed and sold by W. Owen, 1759-63.

2 v. illus., fold. maps, plan. 21 cm. LSU owns only v.1 (iv, 410, [8]p.) Brown imitation leather, hinge decoration, gilt lettering.
BM III p.1249.

*239

MARTYN, WILLIAM FREDERICK. *A new dictionary of natural history; or, Compleat universal display of ani-*

mated nature, with accurate representations of the most curious and beautiful animals, elegantly coloured. London: Printed for Harrison, 1785.

2 v. in 1. illus. (part col.) 38 cm. LSU owns v.1-2 in which numerous illustrations from another source are mounted. The date, 1776, appears on two of the illustrations, one of which is a duplicate. In ms. on title page: E.A. Title page in red and black. Half brown morocco/cloth, marbled endpapers.
BM VII p.806. Nissen ZBI 2729.

* 240

MERRET, CHRISTOPHER (1614-1695). *Pinax rerum naturalium britannicarum: continens vegetabilia, animalia et fossilia in hac insula repperta inchoatus.* Londini: Impensis C. Pulleyn, typis F. & T. Warren, 1666.

221[i.e.223]p. 18 cm. Stamped on title page and p.[223]: Entomological Society of London. In ms. on verso of title page: William Spence. Manuscript notes on pp.198-199. LSU copy imperfect: pp.97-98 incorrectly numbered 87-88, 178-179 incorrectly numbered 162-163. Full calf, raised bands, tooled spine.
BM III p.1291.

* 241

MESMER, FRANZ ANTON (1734-1815). *Mémoire sur la découverte du magnétisme animal.* Geneve, Paris: P.F. Didot le jeune, 1779.

vi, 85 p. 17 cm. In ms. on fly leaf: Herbert G. Norman. Quarter calf, marbled boards.

* 242

MOUNTFORT, GUY. *The wild Danube; portrait of a river.* Illustrated by Eric Hosking. Foreword by Peter Scott. Boston: Houghton Mifflin, 1963.

206 p. illus. 25 cm. First published in London in 1962 under title: *Portrait of a river.* Blue cloth with gilt decoration, maps on endpapers, dustjacket.

243

La Naturaleza; periódico científico de la Sociedad Mexicana de Historia Natural. t.1-7, 1869/70-1885/86; 2. ser., t.1-3, 1887/90-1897/1903; 3. ser., t.1, 1910/12. Mexico: I. Escalante, 1870-1912.

11 v. plates (part col.) maps, diagrs. 30-34.5 cm. LSU holdings complete except for a few scattered numbers. Bindings vary.
BM III p.1296.

244

NEW YORK (STATE). NATURAL HISTORY SURVEY. *Agriculture of New-York; comprising an account of the classification, composition and distribution of the soils and rocks...together with a condensed view of the climate and the agricultural productions of the state.* By Ebenezer Emmons. Albany: Printed by C. Van Benthuysen & co., 1846-54.

5 v. front., illus., plates (part col.) 2 maps (1 fold.) 30 cm. (Natural history of New York...[div.5]). Original brown cloth binding, gilt decoration.

245

NEW YORK (STATE). NATURAL HISTORY SURVEY. *A flora of the state of New-York, comprising full descriptions of all the indigenous and naturalized plants hitherto discovered in the state; with remarks on their economical and medicinal properties.* By John Torrey. Albany: Carroll and Cook, printers to the Assembly, 1843.

2 v. plates (part fold., part col.) 28.5 x 23 cm. (Natural history of New York...[div.2]). Black library binding.
BM III p.1423, V p.2125. Jackson p.364. Nissen BBI 1973, ZBI 1064. Pritzel 9407.

246

NEW YORK (STATE). NATURAL HISTORY SURVEY. *Geology of New-York.* [by W.W. Mather, Ebenezer Emmons, Lardner Vanuxem, and James Hall.] Albany: Printed by Carroll & Cook, printers to the Assembly, 1842-43.

4 v. illus., plates (part fold., part col.) maps (part fold.) diagrs. 28.5 x 23 cm. (Natural history of New York... [div.4]). Original brown cloth binding, gilt decoration.
BM III p.1423. Nissen ZBI 1064.

247

NEW YORK (STATE). NATURAL HISTORY SURVEY. *Mineralogy of New-York; comprising detailed descriptions of the minerals hitherto found in the state of New-York, and notices of their uses in the arts and agriculture.* By Lewis C. Beck. Albany: Printed by W. & A. White & J. Visscher, 1842.

1 p.l., xxiv p., 1 l., 536 p., 2 l. VIII plates (incl. map, plans, diagrs.) 28.5 x 23 cm. (Natural history of New York...[div. 3]). Original brown cloth binding, gilt decoration.
BM III p.1423. Nissen ZBI 1064.

248

NEW YORK (STATE). NATURAL HISTORY SURVEY. *Palaeontology of New York...* By James Hall. Albany:

Printed by C. Van Benthuysen, 1847-94.

8 v. in 13. illus., plates (part fold.) 29 cm. (Natural history of New York...[div.6]). Some vols. original brown cloth binding; some vols. black library binding.
BM III p.1423. Nissen ZBI 1064, 1804.

249

NEW YORK (STATE). NATURAL HISTORY SURVEY. *Zoology of New York, or the New York fauna; comprising detailed descriptions of all the animals hitherto observed within the state of New York, with brief notices of those occasionally found near its borders, and accompanied by appropriate illustrations.* By James De Kay. Albany: Printed by Carroll and Cook, printers to the Assembly [etc.], 1842-44.

6 pts. in 5 v. plates (part col.) 28.5 x 23 cm. (Natural history of New York...[div.1]). Bookplate in v.1.: "Presented to the State of Louisiana from the State of New York..." Some vols. in original brown cloth binding with gilt decoration on front. Others in black library binding.
BM III p.1423. Nissen ZBI 1064. Yale pp.74, 208. Zimmer p.164.

250

NICOLSON, DOMINICAN. *Essai sur l'histoire naturelle de l'isle de Saint Domingue.* Avec figures en taille-douce. Paris: Chez Gobreau, 1776.

2 p.l., iii-xxxi, 374, [2]p. 10 fold. plates. 20.5 cm. Bookplate, c.2: Jules M. Burguieres Sugar Collection. In ms. on fly leaf of c.2: Sugar-3, Sugar, Commerce-4, Sugar, pp.22-31. In ms. on title page of c.2: By P. Nicholson. C.1., full calf, spine mended, marbled endpapers. C.2, quarter calf, marbled boards.
BM III p.1433. Nissen ZBI 2973. Pritzel 6695.

251

Nouveau dictionnaire d'histoire naturelle appliquée aux arts, principalement à l'agriculture et à l'économie rurale et domestique; par une société de naturalistes et d'agriculteurs: avec des figures tirées des trois règnes de la nature. Paris: Deterville, 1803-04.

24 v. plates. 20.5 cm. Vol.2: plate A 10, facing p.56, xerox copy. Plates out of order: v.6, plate B 31 facing p.251, not p.184; v.16 plate M 5 facing p.368, not p.312, v.17, plate M 3 facing p.134, not p.132; v.19, plate P 13 facing p.70, not p.90. Half leather, marbled boards, mottled edges.
BM I p.458. Nissen ZBI 4615. Yale p.211.

252

PEATTIE, DONALD CULROSS (1898-   ). *A Prairie Grove: a naturalist's story of primeval America.* New York: Simon and Schuster, 1938.

3 p.l., 289, [1]p. front., illus. 23.5 cm. [Life in America series]. Wood engravings by Paul Landacre. In ms. on fly leaf: Florinell L. Morton. Original green cloth binding, dustjacket.
Gift of Florinell L. Morton.

253

PHILP, ROBERT KEMP (1819-1882). *The reason why. Natural history; illustrating the natural history of man and the lower animals.* By the author of *Enquire within upon everything, etc., etc.* One hundred and thirty illustrations, Twelfth thousand. London: Houlston & Wright, [1869].

372 p. 18 cm. Author's edition. Original green cloth binding.

*254

PLINIUS SECUNDUS, C. *C. Plinii Secundi Naturae Historiarum libri XXXVII: e castigationibus Hermolai Barbari, quamemendatissime editi. Additus est admaiorem studiosorum commoditatem index Ioannis Camertis Minoritani, notis arithmeticis nunc dilucidior multo quamantea, quo Plynius ipse totus breui mora teporis edisci potest. Ad lectorem. Qui coelum, terras, aequor, genus omne animantum omne exors animae, quid ferat omnis ager inuentus rerum varios, arteisque, metalla marmora cum gemmis, quid iuuet, aut noceat denique naturae qui cuncta adoperta reuelat Plynion integrum, candide lector habes atque ita quod priscum praeseruat fronte nitorem lima viridocti praestitit Hermoleo cui fere te tantum (dicam) debere fatendum auctori quantum secula debuerunt.* Paris: P. Gaudoul, 1524.

[32] CCCCCXXXVI, [194]p. illus. 38 cm. Index has special title page. Includes manuscript notes. LSU copies 1 and 2 imperfect: c.1, p.123 incorrectly numbered 133, 171 incorrectly numbered 170, 334 duplicated in numbering only, 352 incorrectly numbered 357, 365 incorrectly numbered 345. Copy 2 also has p.48 incorrectly numbered 68, 172 incorrectly numbered 72, and 492 incorrectly numbered 493. Contemporary tooled leather over boards, spine rebacked in modern calf.
BM IV p.1586 (dif. eds.) Hunt 228 (1635 ed.) Pritzel 7207-8 (dif. eds.) Yale p.229 (dif. eds.)

255

PLOETZ, ALFRED J. (1860-   ). *Die tüchtigkeit unsrer rasse und der schutz der schwachen. Ein versuch über rassenhygiene und ihr verhältniss zu den humanen idealen, besonders zum socialismus.* Berlin: S. Fischer, 1895.

xi, 240 p. 24 cm. Bookplate: Library of Richard T. Ely, Ph.D. Stamped on title page: C.W. Gerritsen. Half red morocco, marbled boards, spine mended.

256

PORTER, CARLOS EMILIO (1870-1942). *Clasificaciones cientificas del Museo de Valparaiso.* [n.p., n.p., 1903?]

369 p. in various pagings. illus. 23 cm. Collection of publications authored by or associated with Carlos E. Porter, editor of *Revista Chilena de historia natural,* a series in which most of these publications were originally published. Contents:

1. *Breves instrucciones para la recolección de objetos de historia natural.* 2.ed., aumentada. Valparaiso: Gillet, 1903. (52, [3]p. illus. 23 cm.)
2. *Materiales para la fauna carcinolójica de Chile.* Valparaiso: Gillet, 1903. (pp.257-277. illus. 23 cm.)
3. *La yerba-mate, su cultivo, cosecho i preparacion.* Por Victor Ferreira do Amaral e Silva. Valparaiso: Gillet, 1902. (36 p. 23 cm.)
4. *Los congrios de Chile.* Por Federico T. Delfin. Valparaiso: Gillet, 1903. (40 p. illus. 23 cm.)
5. *Descripcion de un nuevo traquinido Chileno.* Por Federico T. Delfin. [Valparaiso: Gillet], 1899. (4 p. 23 cm.)
6. *El Museum de Historia Natural de Valparaiso durante el año de* 1902, memoria. Valparaiso: . Gillet, 1903. (80 p. illus. 23 cm.)
7. *Catálogo de los peces de Chile.* Por Federico T. Delfin. Valparaiso: Gillet, 1901. (133 p. 23 cm.)

Dark brown cloth bindings.

* 257

REDI, FRANCESCO (1626-1698). *Opere di Francesco Redi.* Seconda edizione Napoletana, corretta e migliorata. Napoli: A spese di Michele Stasi con licenza de' superiori e privilegio, 1778.

7 v. illus. (32 fold.), port. 22 cm. Stamped on title page of v.1, 2, 4, 5, 7: R. P. Manuscript notes on title page of v.1-4, 6-7 and on fly leaf and p.[3] of cover of v.1. Vol.1, pp.145-160 duplicated in numbering only; v.7, pp.255-264 incorrectly numbered 155-164. Vellum binding.
BM IV p.1660 (dif. eds.) Nissen ZBI 3316.

* 258

ROYAL SOCIETY OF LONDON. *Philosophical transactions.* v.1— 1665/66-19 . London, [1665]-19 .

v. illus., plates (part col.) maps, facsims., tables, diagrs. Split into series A & B with v.178, 1878. Some vols. on film. Various bindings.

259

ROYAL SOCIETY OF LONDON. *The philosophical transactions and collections, to the end of the year 1700.* Abridged and dispos'd under general heads. 2d edition. London: Printed by M. Matthews for R. Knaplock, R. Wilkin, and H. Clements, 1716-56.

10 v. in 13. fold. illus., fold. maps. 24 cm. Title varies. Imprint varies. Armorial bookplate of the Rev. Charles Lyttelton in v.1-3, c.1; 5, c.1; 6-10, pts.1-2. Armorial bookplate in v.3, c.2; 4; 5, c.2. Modern calf with gilt decoration.
BM III p.1754. Nissen ZBI 4742. Yale p.247.

260

ROYAL SOCIETY OF LONDON. *The philosophical transactions and collections.* Abridg'd and dispos'd under general heads... 3rd edition. London: Printed for J. Knapton, 1721-23.

6 v. in 7. tables, diagrs. 23.5 cm. V.1-3, to the end of 1700, compiled by John Lowthrop. V.4-5, from 1700 to 1720, compiled by Henry Jones. V.6, from 1720 to 1732, compiled by Andrew Reid and John Gray. Imprint varies. Contemporary calf binding.

* 261

SETON, ERNEST THOMPSON (1860-1946). *The worlds of Ernest Thompson Seton.* Edited with introduction & commentary by John G. Samson. 1st edition. New York: Knopf, distributed by Random House, 1976.

204 p. illus. 27 x 30 cm. Brown paper over boards, cloth spine, dustjacket.

* 262

SEWARD, ANNA (1742-1809). *Memoirs of the life of Dr. Darwin, chiefly during his residence in Lichfield, with anecdotes of his friends, and criticisms on his writings.* Philadelphia: At the Classic Press, for W. Poyntell, 1804.

xii, 313 p. 22 cm. Red imitation leather binding.

* 263

SHAW, GEORGE (1751-1813). *Cimelia physica. Figures of rare and curious quadrupeds, birds, &c. together with several of the most elegant plants.* Engraved and coloured from the subjects themselves by John Frederick Miller. With descriptions by George Shaw. London: Printed by T. Bensley for Benjamin and John White, and John Sewell, 1796.

106 p. 60 col. plates. 54 cm. Bookplate of John S. McIlhenny. Full calf with gilt decoration, marbled endpapers, spine rebacked.
BM IV p.1911. Yale p.263. Zimmer p.585.

264

SWAINSON, WILLIAM (1789-1855). *A preliminary discourse on the study of natural history.* London: Printed by A. Spottiswoode for Longman, Rees, Orme, Brown, Green & Longman, [1834].

viii, 462 p. 16.5 cm. (Lardner's cabinet cyclopaedia, 114). Half calf, marbled boards; mended spine.
BM V p.2054. Zimmer pp.369-71.

265

THOMAS, KEITH VIVIAN. *Man and the natural world: a history of the modern sensibility.* New York: Pantheon Books, 1983.

[426]p. illus. 25 cm. Quarter beige cloth/paper over boards, pictorial dustjacket.

* 266

Village science = or, The laws of nature explained. By the author of *"Peeps at nature," "Nature's wonders,"* etc. New York: Carlton & Porter, [1851?].

285 p. illus. 16 cm. In ms. on p.[2] of cover: No.76. Label mounted on fly leaves: Advent Church, Brier, Mass. LSU copy imperfect: various pages mutilated. Original cloth binding, faded.

267

VOSMAER, GUALTHERIUS CAREL JACOB (1854-1916). *The sponges of the Bay of Naples: Porifera incalcaria, with analyses of genera and studies in the variations of species.* Edited by C.S. Vosmaer-Röell and M. Burton. The Hague: Martinus Nijhoff, 1933-35.

3 v. 71 (i.e.69) double plates (part col.) tables, diagrs. 33 cm. Plates 43, 55 omitted in numbering. Brown cloth binding.

268

VRIES, HUGO DE (1848-1935). *Die mutationstheorie. Versuche und beobachtungen über die entstehung von arten im pflanzenreich.* Leipzig: Veit & company, 1901-03.

2 v. illus., 12 col. plates. 24.5 cm. Issued in parts. Tan cloth binding, covers retained on parts.
BM V p.2240. Plesch p.453.

269

WALLACE, ALFRED RUSSEL (1823-1913). *Contributions to the theory of natural selection. A series of essays.* London: Macmillan and co., 1870.

384 p. 19 cm. Original green cloth binding.
BM V p.2256. Yale p.304.

* 270

WALLACE, ALFRED RUSSEL (1823-1913). *The distribution of life, animal and vegetable, in space and time.* By Alfred Russel Wallace and W.T. Thiselton Dyer. New York: J. Fitzgerald, 1885.

48 p. 24 cm. (Humbolt library of popular science literature, no.64). Stamped on p.[2] of cover: Library of O.A. Johannsen, no.[ ]. In ms. on fly leaf: O. Johannsen, Dec. 19, 1892, Champaign, Ill. Quarter cloth, marbled boards.

271

WALLACE, ALFRED RUSSEL (1823-1913). *Natural selection and tropical nature; essays on descriptive and theoretical biology.* New edition with corrections and additions. London and New York: Macmillan and co., 1895.

xii, 492 p. illus. 20.5 cm. C.1, bookplate: John Dalton Shaffer, Ardoyne Plantation, Parish of Terrebonne, La. C.2 stamped on fly leaf: Gulf? Station Library, Cameron, La. Original green cloth binding.
BM V p.2256 (earlier eds.) Yale p.304 (later ed.)

272

WATERTON, CHARLES (1782-1865). *Essays on natural history.* Edited, with a life of the author, by Norman Moore. London: Frederick Warne and co.; New York: Scribner, Welford, & Armstrong, 1871, [c1870].

vii, [1], 631 p. front. (port.) plates. 19.5 cm. Original brown cloth, mended spine.
BM V p.2271 (earlier eds.) Yale p.306. Zimmer p.665 (earlier eds.)

* 273

WHEELER, CHARLES GILBERT (1836-1912). *Hand-colored illustrations of the beautiful and wonderful in animated nature, with full explanatory text.* Chicago: S.J. Wheeler, [188-?].

[25]leaves. 12 col. plates. 64 cm. LSU copy imperfect: 4 plates wanting (1 of birds, 1 of reptiles, and 2 of fishes). In portfolio, half red leather/original red cloth. Restored by William A. Payne.

274

WHITE, GILBERT (1720-1793). *The natural history of Selborne. With miscellaneous observations and explana-*

tory notes. Boston: Ticknor and Fields, 1866.

430 p. 13.5 cm. Original blue cloth binding, stamped; spine taped.

BM V p.2309 (various eds.) Nissen IVB 985. Zimmer p.672 (later ed.)

* 275

WHITEHEAD, PETER JAMES PALMER. *The British Museum (Natural History)*. Text by Peter Whitehead; photographs by Colin Keates. London: Philip Wilson in association with the British Museum (Natural History), 1981.

128 p. col. illus. 28 cm. Dark blue cloth binding, illustrated dustjacket.

* 276

WILSON, DAVID SCOFIELD. *In the presence of nature*. Amherst: University of Massachusetts Press, 1978.

xix, 234 p. illus. 24 cm. Black cloth binding, dustjacket.

277

WOLLE, FRANCIS (1817-1893). *Fresh-water algae of the United States (exclusive of the diatomaceae) complemental to Desmids of the United States*. With 2300 illustrations covering one hundred and fifty-one plates, a few colored, including nine additional of desmids. Bethlehem, Pa.: Comenius Press, 1887.

2 v. illus. plates. 24.5 cm. Tan library binding.

BM V p.2350.

278

WOOD, JOHN GEORGE (1827-1889). *The common objects of the sea shore; including hints for an aquarium*. With coloured illustrations from designs by Sowerby. London, New York: Routledge, Warne and Routledge, 1861.

viii, 204 p. illus., 12 col. plates (incl. front.) 17.5 cm. LSU copy imperfect: front. wanting. Original green cloth binding, spine mended.

BM V p.2352 (earlier ed.)

279

WOOD, JOHN GEORGE (1827-1889). *Wood's Bible animals: a description of the habits, structure, and uses of every living creature mentioned in the Scriptures, from the ape to the coral; and explaining all those passages in the Old and New Testaments in which reference is made to beast, bird, reptile, fish, or insect*. Illustrated with over one hundred new designs by Keyl, Wood, and E.A. Smith; engraved by G. Pearson... To which are added articles on evolution, by Rev. James McCosh, and research and travel in Bible lands, by Rev. Daniel March. Guelph, Ontario: J.W. Lyon & Company, publishers, 1877.

xxix, 800 p. illus. 25 cm. Original brown cloth binding, decorated spine.

BM V p.2352.

* 280

ZAHN, JOHANN. *Specula physico-mathematico-historica notabilium ac mirabilium sciendorum, in qua mundi mirabilis oeconomia, nec non mirificè amplus, et magnificus ejusdem abditè reconditus, nunc autem ad lucem protractus, ac ad varias perfacili methodo acquirendas scientias in epitomen collectus thesaurus curiosis omnibus cosmosophis inspectandus proponitur...* Norimbergae: sumptibus Joannis Christophori Lochner bibliopolae. Literis Knorzianis. Anno MDCXCVI.

3 v. in 1. plates (part fold.), ports., maps (part double), tables (part double). 42 cm. Each volume has added title page, illustrated and engraved. In double columns. LSU copy imperfect: v.1, p.327 incorrectly numbered 372; v.2, p.275 incorrectly numbered 285, p.364 incorrectly numbered 356; v.3, p.49 incorrectly numbered 41, p.51 incorrectly numbered 45, p.97 incorrectly numbered 91, p.102 incorrectly numbered 96, p.104 incorrectly numbered 98, p.238 incorrectly numbered 218. Vellum over boards, clasps.

PART V

Botany

Flower drawn by Redouté and named for him from Etienne Pierre Ventenat's *Description des Plantes Nouvelles et peu Connues Cultivées dan le Jardin de J. M. Cels.* Paris, circa 1799.

* 281

AGNEW, ANDREW D.Q. *Upland Kenya wild flowers; a flora of the ferns and herbaceous flowering plants of Upland Kenya.* London: Oxford University Press, 1974.

ix, 827 p. illus., map. 26 cm. Dark blue cloth binding.

282

ALDROVANDI, ULISSE (1552-1605?). *Dendrologia sive arboretum, Libris II de silva glandaria, acinosoque pomario comprehensum...opus theologis, jctis, medicis, philosophis, oratoribus, historicis, poetis adprime utile, multifariaque eruditione refertissimum, cum indice copioso & praefatione Georgii Franci...* Francofurti ad Moenum, 1690.

2 v. in 1 (4 p.l., 480, [30]p.) illus. 37 cm. At head of title: Ovidii Montalbani. Title page in red and black. Title page transcribed in ms. on fly leaf. Full calf, gilt decorated spine, red spine label.

BM I p.27 (earlier eds.) Hunt 306 (1st ed.) Jackson p.29. Nissen BBI 14. Pritzel 93.

* 283

ALDROVANDI, ULISSE (1522-1605?). *Ulyssis Aldrovandi patricii bononiensis Dendrologiae naturalis scilicet arborum historiae libri duo. Sylva glandaria, acinosumq. pomarium. Ubi eruditiones omnium generum una cum botanicis doctrinis ingenia quaecunque non parum ivvant, et oblectant. Ovidius Montalbanus opus summo labore collegit digessit, concinnavit...* Bononiae: Typis I.B. Ferronii, 1668.

2 p.l., [6], 660, [52]p. 36 cm. Colophon: Bononiae, Ex typographia Ferroniana 1667. Superiorum permissu. Errata slip mounted on p.11. LSU copy imperfect: p.200 incorrectly numbered 220. Vellum binding, marbled edges.

BM I p.27. Hunt 306. Jackson p.29. Nissen BBI 14. Pritzel 93.

* 284

ANASTASIA, GIUSEPPE EMILIO (d. 1934). *Araldica Nicotianae: nuove ricerche intorno alla filogenesi delle varieta di N. Tabacum.* L. Scafati: E. Fienga, 1914.

2 v. illus. (some col.) 27 cm. Stamped on cover and title page of each vol.: Dr. phil. Paul Graebner. At head of title: Direzione generale privative ministero finanze. R. Istituto sperimentale tabacchi. Original paper wrappers.

285

ARDE, WALKER R. *The Russulas.* Philadelphia: [Published by the author, 1957?].

13 p., 15 p. of hand col. plates. 29 cm. Printed only on recto of pages. Maroon cloth binding.

286

Arzneipflanzen-Tafeln. Stuttgart: Deutscher Apotheker-Verlag, [1957?].

76 mounted col. plates in box. 34 cm. Title from box.

287

AUBLET, FUSÉE (1720-1778). *Histoire des plantes de la Guiane Françoise, rangées suivant la méthode sexuelle, avec plusieurs mémoires sur différens objets intéressans, relatifs à la culture & au commerce de la Guiane Françoise, & une notice des plantes de l'Isle-de-France.* Londres & Paris: P.F. Didot jeune, 1775.

4 v. 393 plates. 26.5 x 22 cm. Vols.1-2 paged continuously. Vol.2 includes a Supplement (160 p.) and tables. Vols.3-4, plates. Bookplate of Hry. Pige Leschallas, with motto "De Tout Mon Couer." Full calf, gilt decorated spine, marbled endpapers. Binding by William A. Payne.

BM I p.70. Hunt 642. Nissen BBI 54. Plesch p.130. Pritzel 277.

* 288

BACKEBERG, CURT. *Cactus lexicon: enumeratio diagnostica Cactacearum...with descriptions of many newer species (1966-73) by Walther Haage.* [Translated from the German by Lois Glass]. Poole: Blandford Press, 1977.

828 p. illus. (some col.), maps. 25 cm. Translation of *Das Kakteenlexikon.* Brown cloth binding, dustjacket.

* 289

BAILEY, LIBERTY HYDE (1858-1954). *How plants get their names.* Reprint of the 1933 edition published by Macmillan, New York. Detroit: Gale Research Company, 1975.

vi, 209 p. illus., port. 23 cm. Green cloth binding.

290

Banks' florilegium: a publication in thirty-four parts of seven hundred and thirty-eight copperplate engravings of plants collected on Captain James Cook's first voyage round the world in H.M.S. Endeavour, 1768-1771. The specimens were gathered and classified by Sir Joseph Banks, Bart., and Daniel Solander and were accurately engraved between 1771 and 1784, after drawings taken from nature by Sydney Parkinson. London: Alecto

Historical Editions in association with the British Museum (Natural History), 1980- .

v. all col. illus. 79 cm. The plates from which this edition was printed were prepared under Banks' direction by various engravers, but were never used until 1973, when a selection of them was published under the auspices of the present owner, British Museum (Natural History), as: *Captain Cook's florilegium*. Limited edition of 100 copies. LSU owns copy no.43. Enclosed in green cloth-covered solander boxes.

*291

BARBEY, WILLIAM (1842-1914). *Epilobium: genus a cl. Ch. Cuisin illustratum*. Lausanne: G. Bridel, 1885.

[59]p., 24 leaves of plates, 24 illus. 32 cm. Title on half title and spine: *Genus Epilobium*. Quarter red cloth, printed paper over boards.
BM I p.96. Nissen BBI 75.

*292

BARTON, BENJAMIN SMITH (1766-1815). *Elements of botany; or, Outlines of the natural history of vegetables*. Illustrated by thirty plates. Philadelphia: Printed for the author, 1803.

xii, [2], 302, 168, 38 p., 1 l. XXX plates (part fold., incl. front.) 24.5 cm. LSU copy imperfect: pp.168, 38; 1 l., XXX plates missing. In ms. on title page: Presented to J.A. Read by her friend Dr. A. Thruston. In ms. on p.[iii]: Julia A. Hudson. In ms. on p.[v]: Alfred Thruston, 1806. Green imitation leather with hinge design.
BM I p.105. Jackson p.36. Nissen BBI 83. Plesch pp.133-34 (Brit. ed.) Pritzel 437.

293

BARTRAM, WILLIAM (1739-1823). *William Bartram, botanical and zoological drawings, 1756-1788, reproduced from the Fothergill album in the British Museum (Natural History)*. Edited with an introduction and commentary by Joseph Ewan. Philadelphia: American Philosophical Society, 1968.

x, 180 p. plates (part col.) 41 cm. (Memoirs of the American Philosophical Society, v.74). Printed at the Stinehour Press. "This volume presents a collection of drawings of plants and animals made between 1768 and 1776 for Dr. John Fothergill and for Robert Barclay by William Bartram." Green cloth binding.
Plesch p.135.

*294

BATEMAN, JAMES (1811-1897). *The Orchidaceae of Mexico and Guatemala*. Reprint of London edition: [For the author, J. Ridgway], 1837-42; New York: Johnson Reprint Corp., [1974].

5 p.l., 12, [5]p. illus., 40 col. plates. 50 cm. Engraved title within ornamental design. Each plate accompanied by leaf with descriptive letterpress. Limited to 1000 copies. LSU owns copy no.21. Green cloth binding.
BM I p.109 (orig. ed.) Jackson p.368 (orig. ed.) Nissen BBI 89 (orig. ed.) Pritzel 470 (orig. ed.)

*295

BAUER, FERDINAND LUCAS (1760-1826). *The Australian flower paintings of Ferdinand Bauer*. Text [by] William T. Stearn; introduction by Wilfrid Blunt. London: Basilisk Press, 1976.

30 p., [25] leaves of plates. col. ill., maps. 67 cm. "Five hundred and fifteen copies have been printed... This is copy number 170." Each plate accompanied by leaf with descriptive letterpress. Quarter green morocco/marbled boards, linen portfolio case.

296

BAUHIN, JOHANN (1541-1613). *Historia plantarvm vniversalis, nova et absolvtissima, cvm consensv et dissensv circa eas, auctoribus Ioh. Bavhino...et Ioh. Hen. Cherlero...quam recensuit et auxit Dominievs Chabraevs...Juris verò publici fecit Franciscvs Lvd. a Graffenried...* Ebrodvni, 1650-51.

3 v. illus. 39 cm. Vellum binding, restored.
BM I p.113. Hunt 251. Jackson p.28. Nissen BBI 103. Plesch p.136. Pritzel 504.

297

BAUHIN, KASPAR (1560-1624). Caspari Bauhini... Πινάξ *Theatri botanici sive Index in Theophrasti Dioscoridis, Plinii et botanicorvm qui à seculo scripserunt opera plantarvm circiter sex millivm ab ipsis exhibitarvm nomina cum earundem synonymijs & differentijs methodice secundum genera & species proponens. Opvs XL. annorvm summopere expetitum & ad autoris autographum recensitum*. Basileae: impensis Joannis regis, 1671.

12 p.l., 518, [20]p. 1 l. 28 cm. Bound with: Bauhin, Kaspar. *Caspari Bauhini* προ δρο ηοζ *Theatri botanici. in qvo plantae svpra sexcentae ab ipso primum descriptae cum plurimis figuris proponuntur*. Basileae, 1671. ([4], 160, [12]p.) Half brown morocco, green cloth binding.
BM I pp.113-14. Hunt 318, 319. Jackson p.28 (2nd title only) Nissen BBI 104 (2nd title only). Pritzel 507, 509.

298

BECK, LEWIS CALEB (1798-1853). *Botany of the northern and middle states; or, A description of the plants found in the United States, north of Virginia, arranged according to the natural system. With a synopsis of the genera according to the Linnaean system — a sketch of the rudiments of botany, and a glossary of terms.* Albany: Printed by Webster and Skinners, 1833.

2 p.l., iv, 471 p. 19 cm. Label on end sheet: John Wells. Green library binding.
BM I p.119. Jackson p.361. Pritzel 541.

* 299

BEDDOME, RICHARD HENRY (1830-). *The ferns of British India; being figures and descriptions of ferns from all parts of British India, including Bangla Desh, Burma, Ceylon, India and Pakistan.* New Delhi: Oxford & IBH Pub. Co., 1973.

2 v. illus. 29 cm. Reprint of the 1866 edition. V.1, red cloth binding; v.2, blue cloth binding.
BM I p.121 (orig. ed.) Jackson p.385 (orig. ed.) Nissen BBI 115 (orig. ed.) Pritzel 556 (orig. ed.)

300

BEDDOME, RICHARD HENRY (1830-). *The ferns of southern India; being descriptions and plates of the ferns of the Madras presidency.* 2d edition. Madras: Higginbotham and co., 1873.

1 p., xv, 88 p. 271 plates, xv p. illus. (plates). 32 cm. Green cloth binding.
BM I p.121. Jackson p.386. Nissen BBI 116. Pritzel 555 (1863 ed.)

301

BENTHAM, GEORGE (1800-1884). *Genera plantarum ad exemplaria imprimis in Herbariis kewensibus servata definita; auctoribus G. Bentham et J.D. Hooker.* London: Reeve & co. [etc.], 1862-83.

3 v. in 7. 25 cm. Green paper over boards.
BM I p.137. Jackson p.120. Pritzel 627.

302

BENTLEY, ROBERT (1821-1893). *Physiological botany.* An abridgement of *The students' guide to structural, morphological, and physiological botany...* Prepared as a sequel to *Descriptive botany,* by Eliza A. Youmans. New York: D. Appleton and company, 1886.

xiv, 292 p. illus. 19.5 cm. In ms. on fly leaf: W.R. Dodson. Original brown cloth binding.
BM I p.137 (earlier eds.)

303

BERGE, FR. *Giftpflanzen-buch; oder, Allgemeine und besondere naturgeschichte sämmtlicher inländischen sowie der wichtigsten ausländischen phanerogamischen und cryptogamischen giftgewächse...* Von Fr. Berge und Dr. V.A. Riecke. Stuttgart: Scheitlin & Krais, 1850.

xi, 329 p. 72 col. plates. 24 cm. Green library binding.
BM I p.139 (1855 ed.) Nissen BBI 143. Plesch p.139 (1855 ed.) Pritzel 654 (1845 ed.)

304

BERKELEY, MILES JOSEPH (1803-1889). *Outlines of British fungology; containing characters of above a thousand species of Fungi, and a complete list of all that have been described as natives of the British Isles.* London: Lovell Reeve, 1860.

1 p.l., xvii, 442 p. 24 plates (23 col.) 22 cm. Original black cloth binding.
——— *Supplement*, by Worthington G. Smith. London: L. Reeve, 1891 (xii, 386 p. 22 cm.) Red library binding.
BM I p.144. Jackson p.244. Nissen BBI 148. Pritzel 686.

305

BIGELOW, JACOB (1787-1879). *American medical botany, being a collection of the native medicinal plants of the United States, containing their botanical history and chemical analysis, and properties and uses in medicine, diet and the arts, with coloured engravings.* Boston: Cummings and Hilliard, 1817-1820.

3 v. in 6. plates (col.) 27 cm. In ms. on title page: J.L. Clarke. Original printed boards.
BM I p.162. Jackson p.360. Nissen BBI 164. Pritzel 773.

* 306

BIGELOW, JACOB (1787-1879). *Florula Bostoniensis. A collection of plants of Boston and its vicinity, with their generic and specific characters, principal synonyms, descriptions, places of growth, and time of flowering, and occasional remarks.* 2d edition greatly enlarged. To which is added a glossary of the botanical terms employed in the work. Boston: Cummings, Hilliard, 1824.

422 p. 25 cm. Original brown cloth binding.
BM II p.162. Jackson p.362. Pritzel 772.

307

BISCHOFF, GOTTLIEB WILHELM (1799-1854). *Allgemeine uebersicht organisation der phanerogamen und kryptogamen pflanzen.* 3911 lithographirte abbildungen auf 77 tafeln mit organologischem, systematischem und namen-register. Separat-abdruck aus dem handbuche der botanischen terminologie und systemkunde, von Prof. Dr. S.W. Bischoff. Neue wohlfeile ausgabe. Leipzig: T.L. Schrag's verlag, 1860.

2 v. illus., tables. 26 cm. LSU owns only v.2 (3 p.l., [27]-38 p., 1 l. plates XLVIII-LXVII). Half black calf, cloth binding.
Nissen BBI 165. Pritzel 798.

308

BLUNT, WILFRID (1901-). *The art of botanical illustration.* By Wilfrid Blunt, with the assistance of William T. Stearn. London: Collins, 1950.

xxxi, 304 p. illus. (part col.) 23 cm. (The New naturalist; a survey of British natural history [14]). Green cloth binding.
Plesch p.144.

*309

BLUNT, WILFRID (1901-). *The illustrated herbal.* By Wilfrid Blunt and Sandra Raphael. New York: Thames and Hudson, Inc., in association with the Metropolitan Museum of Art, 1979.

191 p. illus. (some col.), facsims. 31 cm. Profusely illustrated with numerous full page reproductions. First American edition; simultaneously published in London by Frances Lincoln Publishers Ltd. Beige cloth binding, dustjacket.

310

BOJER, WENZEL (1800-1856). *Hortus Mauritianus ou Énumération des plantes exotiques et indigènes, qui croissent à l'Ile Maurice, disposées d'après la méthode naturelle.* Maurice: Impr. d'A. Mamarot et compagnie, 1837.

vi p., 1 l., 456 p. 22.5 cm. Quarter green morocco, marbled boards, marbled endpapers. Original paper covers retained.
BM I p.189. Jackson p.353. Pritzel 940.

311

Botanical Miscellany; containing figures and descriptions of such plants as recommend themselves by their novelty, rarity, or history, or by the uses to which they are applied in the arts, in medicine, and in domestic oeconomy; together with occasional botanical notices and information. By William Jackson Hooker. v.1-3. London: John Murray, 1830-33.

3 v. plates (part col.) 24 cm. First of a series of four periodicals edited by Sir William Jackson Hooker. Succeeded by the *Journal of Botany* (1834-42), in turn succeeded by the *London Journal of Botany* (1842-48) and *Hooker's Journal of Botany* (1849-57). Half green morocco marbled boards.
BM II p.873. Jackson p.471. Nissen BBI 2356. Pritzel 10810.

312

BOULGER, GEORGE SIMONDS (1853-1922). *British flowering plants.* Illustrated by three hundred full-page coloured plates reproduced from drawings by Mrs. Henry Perrin, with detailed descriptive notes and an introduction by Professor Bougler. London: B. Quaritch, 1914.

4 v. 300 col. plates. 32 cm. Limited to 1000 copies. LSU owns copy no.856. In ms. on fly leaf of v.1: To Mary from Wallie, 1944. In ms. on fly leaf of vols. 2 & 3: Presented to Mary with love from Pop, July 1944. In ms. on fly leaf of v.4: Presented to Mary with love from Pop, July 1944 with the injunction that she will read, mark & digest. W.V.B. Ivory cloth binding, gilt lettering.
BM VI p.115, VII p.1000. Nissen BBI 1509 (Perrin).

313

BOVIUS, HYACINTHUS. *Novi flores medicinales, sive obseruationes, sententiae, dicta, historiae, & medicamenta morbis probata, quae per multa probatissimorum authorum inueniuntur volumina.* Venetiis: Typis Catanei, 1675.

5 pts. in 1 v. (554 p.) 16 cm. Ms. notes in Latin on fly leaf. Vellum binding.

314

BRAITHWAITE, ROBERT (1824-1917). *The British moss-flora.* London: L. Reeve, 1887-1905.

3 v. CXXVIII plates. 27 cm. Issued in 23 parts. Label on p.[2] of cover of v.1: F.J. Chittenden. Half black morocco/cloth, marbled endpapers.
BM I p.221. Jackson p.241. Nissen BBI 222.

315

BRIOSI, GIOVANNI (1846-1919). *Intorno alla anatomia della canapa (Cannabis sativa L.) ricerche de Giovanni Briosi e Filippo Tognini.* Milano: Tip. Bernardoni di C. Rebesahini etc., 1894-1896.

2 v. illus. 29 cm. Half red morocco/cloth, marbled endpapers; original printed wrappers retained.

316

BRITISH MUSEUM (NATURAL HISTORY). DEPT. OF BOTANY. *A monograph of the British lichens; a descriptive catalogue of the species in the Department of botany, British Museum.* 2d edition. By Annie Lorrain Smith. London: Printed by order of the Trustees, 1918-26.

2 v. illus., plates. 22 cm. LSU owns only pt.2. Green cloth binding.
BM VI p.135 (1st ed.)

317

BRITTEN, JAMES (1846-1924). *European ferns...with coloured illustrations from nature by D. Blair.* London, Paris & New York: Cassell, Petter, Galpin & co., [1879-81].

vii, [1], xliv, 196 p. col. front., illus., col. plates. 28.5 cm. Issued in 30 parts, 1879-81. On verso of fly leaf a pressed New Zealand fern, 1888. Half red morocco/cloth, marbled endpapers, marbled edges.
BM I p.250. Jackson p.227. Nissen BBI 235.

318

BRITTON, NATHANIEL LORD (1859-1934). *The Cactaceae, descriptions and illustrations of plants of the cactus family.* By N.L. Britton and J.N. Rose. Washington: The Carnegie Institution of Washington, 1919-23.

4 v. fronts., illus., plates (part col.) 29.5 cm. (Carnegie Institution of Washington. Publication no.248, vol.I-IV). Green library binding.
Nissen BBI 236. Plesch p.157.

*319

BRITTON, NATHANIEL LORD (1859-1934). *An illustrated flora of the northern United States, Canada and the British possessions, from Newfoundland to the parallel of the southern boundary of Virginia, and from the Atlantic Ocean westward to the 102d meridian.* By Nathaniel Lord Britton and Hon. Addison Brown. 2d edition, revised and enlarged. New York: New York Botanical Garden, 1936, [c1923].

3 v. illus. 27 cm. Original dark red cloth binding.

320

BROWN, ROBERT (1773-1858). *The miscellaneous botanical works of Robert Brown.* Edited by John J. Bennett. London: Published for the Ray Society by Robert Hardwicke, 1866-68.

2 v. 23 cm. and atlas of 38 plates. 35 cm. Green cloth, gilt decoration; atlas, blue library binding.
BM I p.260. Hunt 118. Nissen BBI 253. Pritzel 1213.

*321

BRUNET, OVIDE (1826-1876). *Énumération des genres de plantes de la flore du Canada, précédée des tableaux analytiques des familles, et destinée aux élèves qui suivent le cours de botanique descriptive donné à l'Université Laval par l'abbé Ovide Brunet.* Québec: G. & G.E. Desbarats, 1864.

45 p. 20.5 cm. In ms. on cover: L.C. Choquette, etudiant (?) Universite Laval, Quebec, 1871-72. Embossed on cover, title page, p.[3]-7, 25-27: O.F.M. Quebec. Original printed paper wrappers.

*322

BUDD, ARCHIBALD C. *Wild plants of the Canadian prairies.* Ottawa: Experimental Farms Service, Canada Department of Agriculture, 1957.

348 p. illus. 26 cm. (Canada. Dept. of Agriculture Publication, 983). Paperbound.

323

BULLIARD, PIERRE (1742-1793). *Dictionnaire élémentaire de botanique ou Exposition par ordre alphabétique, des préceptes de la botanique, & de tous les termes, tant françois que latins, consacrés à l'étude de cette science.* Paris: Auteur, 1783.

viii, 242 p., [8]l. X plates (9 col.) 36 cm. Paper over boards.
BM I p.285. Nissen BBI 296. Plesch p.163. Pritzel 1355.

324

BULLIARD, PIERRE (1742-1793). *Herbier de la France.* Paris: Auteur; etc., etc., 1780-1812.

9 v. 606 col. plates. 36 cm. Imperfect: only first volume of plates has correct title page. Plates bound in order of publication. Publisher varies. Original red marbled boards.
BM I p.285. Hunt 432. Jackson p.273. Nissen BBI 296. Plesch p.163. Pritzel 1356.

*325

BULLIARD, PIERRE (1742-1793). *Histoire des plantes vénéneuses et suspectes de la France.* 2d édition. Paris: A.J. Dugour, 1798.

2 p.l., xvii, 398 p. 21 cm. Full contemporary calf, restored, gilt decorated spine.
BM I p.285. Hunt 432. Jackson p.206. Nissen BBI 293. Plesch p.163. Pritzel 1354.

326

BURNETT, GILBERT THOMAS (1800-1835). *An encyclopaedia of useful and ornamental plants, consisting of beautiful and accurate coloured figures of plants used in the arts, in medicine, and for ornament, with copious scientific and popular descriptions of each, accounts of their uses, and mode of culture, and numerous interesting anecdotes.* New edition. Edited by M.A. Burnett. London: George Willis, 1852.

2 v. 260 col. plates. 29 cm. Half green morocco, marbled boards, gilt edges.

BM I p.291 (1st ed.) Nissen BBI 305 (1st ed.) Plesch p.167 (1st ed.) Pritzel 1400 (1st ed.)

*327

Cacti. [By] Walter Kupper [and] Pia Roshardt. Translated and edited by Vera Higgins. [Edinburgh]: Nelson, 1960.

127 p. col. illus. 30 cm. Translation of *Kakteen.* Blue cloth binding.

Nissen BBI: 3 1114na (orig. ed.) Plesch p.293 (French trans.)

328

CAMUS, EDMOND GUSTAVE (1852-1915). *Iconographie des orchidées d'Europe et du bassin Méditerranéen.* Par E.G. Camus...avec la collaboration, pour l'anatomie et la biologie, de A. Camus... Paris: P. Lechevalier, 1921-29.

2 v. illus., port. 28.5 cm. and atlas of 133 plates (part col.) 40 cm. Texte published 1928-29. Atlas (1921) accompanied by "Explication des planches" of 72 p. (1921). Atlas, 2 pte., planches nos.123 à 133, issued with t.II of Texte. Texte consecutively paged. Vols.1 & 2, black library binding. Atlas, original printed paper boards.

Nissen BBI 316.

329

CAMUS, EDMOND GUSTAVE (1852-1915). *Monographie des orchidées de l'Europe, de l'Afrique septentrionale, de l'Asie Mineure et des Provinces Russes transcaspiennes.* Par E.G. Camus...P. Bergon... Mlle. A. Camus... Paris: Jacques Lechevalier, 1908.

484 p. 32 plates. 28 cm. Mimeographed. Black library binding.

Nissen BBI 316nb.

330

CANDOLLE, ALPHONSE LOUIS PIERRE PYRAMUS DE (1806-1893). *Monographiae phanerogamarum prodromi nunc continuatio, nunc revisio auctoribus Alphonso et Casimir de Candolle aliisque botanicis ultra memoratis.* Parisiis sumptibus G. Masson, 1878-1896.

9 v. VIII plates. 25.5 cm. A supplementary work to Augustin Pyramus de Candolle's *Prodromus systematis naturalis regni vegetabilis.* Quarter black morocco, marbled boards.

BM I p.309. Jackson p.120. Nissen BBI 318.

331

CANDOLLE, ALPHONSE LOUIS PIERRE PYRAMUS DE (1806-1893). *La phytographie; ou, L'art de décrire les végétaux considérés sous différents points de vue.* Paris: G. Masson, 1880.

xxiv, 484 p. 22 cm. Green library binding.

BM I p.309. Plesch p.171.

*332

CANDOLLE, AUGUSTIN PYRAMUS DE (1778-1841). Augustini-Pyrami Decandolle, *Astragalogia, nempe astragali, biserulae et oxytropidis; nec non phacae, colutae et lessertiae historia, iconibus illustrata.* Parisiis: sumptibus Joann. Bapt. Garnery, Typis Didot Junioris, an XI.—1802.

viii, 218 p. 50 plates. 50.5 cm. Full brown morocco.

BM I p.310. Jackson p.124. Nissen BBI 319. Plesch p.170. Pritzel 1464.

333

CANDOLLE, AUGUSTIN PYRAMUS DE (1778-1841). *Prodromus systematis naturalis regni vegetabilis, sive Enumeratio contracta ordinum generum specierumque plantarum huc usque cognitarum, juxta methodi naturalis normas digesta.* Parisiis: sumptibus sociorum Treuttel et Wurtz, 1824-73.

17 v. in 18. 21 cm. Quarter black morocco, marbled boards.

______ *Genera, species et synonyma Candolleana alphabetico ordine disposita, seu Index generalis et specialis ad A.P. Decandolle Prodromum systematis naturalis regni vegetabilis...* Auctore H.W. Buek, M.D. Berolini: sumptibus librariae Nauckianae, 1840- . (4 v. in 2. 20.5 cm.) Quarter black morocco, marbled boards.

BM I p.310. Jackson p.119. Pritzel 1485.

*334

CARLSON, RAYMOND, ed. *The flowering cactus; an informative guide, illustrated in full-color photography, to one of the miracles of America's Southwest.* Photographs and technical data by R.C. and Claire Meyer Proctor; sketches and design by George M. Avey. New York: McGraw-Hill, [c1954].

96 p. illus. (part col.), map. 30 cm. Quarter coral cloth/tan paper over boards.

Plesch p.177.

335

CARSON, JOSEPH (1808-1876). *Illustrations of medical botany: consisting of coloured figures of the plants affording the important articles of the materia medica.* And descriptive letterpress, by Joseph Carson. The drawings on stone by J.H. Colen. Philadelphia: Lloyd P. Smith, 1847.

2 v. C col. plates. 34 x 27.5 cm. Vol.2 published by Robert P. Smith, Philadelphia. Quarter black cloth, grey cloth, marbled endpapers.
Nissen BBI 333. Pritzel 1545.

336

CAVANILLES, ANTONIO JOSÉ (1745-1804). *Icones et descriptiones plantarum.* With bibliographical notes by F.A. Stafleu. Lehre: J. Cramer; New York: Stechert-Hafner Service Agency, 1965.

6 v. in 1. 600 plates. 28 cm. (Historiae naturalis classica, tomus 42). Reproduced from the 1791-1801 edition published in Madrid. Green cloth binding.
BM I p.329 (orig. ed.) Jackson p.339 (orig. ed.) Nissen BBI 341 (orig. ed.) Pritzel 1616 (orig. ed.)

337

CHAUMETON, FRANÇOIS PIERRE (1775-1819). *Flore Médicale, décrite par F.P. Chaumeton...* Peinte par Mme. E.P..., et par P.J.F. Turpin. Paris: C.L.F. Panckoucke, 1815-1820.

7 v. in 9. illus. 21 cm. T.3-6 par F.P. Chaumeton, Chamberet et Poiret. T.7 Partie élémentaire par J.L.M. Poiret. Iconographie végétale par P.J.F. Turpin. T.7 in 3 parts, with separate titles and pagination. Red library binding.
BM I p.339. Jackson p.201. Nissen BBI 349. Plesch p.176. Pritzel 1679.

338

CONARD, HENRY SHOEMAKER (1874-). *The waterlilies: a monograph of the genus Nymphaea.* [Washington]: Published by the Carnegie Institution of Washington, 1905.

1p.l., xiii, 279 p. illus., 30 plates (11 col., incl. front.) 31.5 x 24.5 cm. (Carnegie Institution of Washington. Publication no.4). Black library binding.
BM VI p.222. Nissen BBI 392.

339

COOKE, MORDECAI CUBITT (1825-1914). *Handbook of British Fungi, with descriptions of all the species.* 2d and revised edition. London, 1883.

398 p. 22 cm. Issued as an appendix to *Grevillea.* Planned as the first volume of reprint and revision of the *"Handbook of British Fungi"* (2 vols., 1871). No more published. Green cloth binding.
BM I p.378. Jackson p.244.

340

COOKE, MORDECAI CUBITT (1825-1914). *Illustrations of British Fungi (Hymenomycetes), to serve as an atlas to the "Handbook of British Fungi."* London: Williams and Norgate, 1881-91.

8 v. in 7. col. plates. 23 cm. Quarter brown morocco, marbled boards, marbled endpapers.
BM I p.379. Nissen BBI 395.

341

COOKE, MORDECAI CUBITT (1825-1914). *A plain and easy account of British Fungi: with descriptions of the esculent and poisonous species...and a tabular arrangement of orders and genera.* London: R. Hardwicke, 1866.

vi, [2], 166 p. illus., 24 col. plates (incl. front.) 16 cm. Added title page, engraved, with colored vignette. Green library binding.
BM I p.378. Pritzel 1855 (1862 ed.)

342

COOKE, MORDECAI CUBITT (1825-1914). *Rust, smut, mildew, & mould. An introduction to the study of microscopic Fungi.* 5th edition, revised and enlarged. London: W.H. Allen & co., 1886.

4 p.l., 262 p. illus., XVI col. plates (incl. front.) 17.5 cm. Original brown stamped cloth binding.
BM I p.378 (earlier eds.) Jackson p.167 (earlier eds.) Nissen BBI 398 (earlier eds.) Pritzel 1857 (earlier eds.)

343

COOPER, CHARLES SAMUEL. *Trees & shrubs of the British Isles, native & acclimatised.* By C.S. Cooper and W. Percival Westell. Sixteen full-page coloured plates and 70 full-page black and white plates drawn direct from nature by C.F. Newall. London: J.M. Dent & co.; New York: E.P. Dutton & co., 1909.

2 v. col. fronts., plates (part col.) 31 cm. Original green cloth binding, gilt decoration.
BM VI p.230. Nissen BBI 400.

344

CORDIER, FRANÇOIS SIMON (1797-1874). *Les Champignons de la France: histoire, description, culture,* *usages des espèces comestibles, vénéneuses, suspectes, employées dans les arts, l'industrie, l'économie domestique et la médecine.* Paris: J. Rothschild, 1870.

2 pts. in 1 v. illus., 60 col. plates. 28 cm. Quarter red morocco/cloth, marbled endpapers.

BM I p.384. Jackson p.164. Nissen BBI 404. Pritzel 1882.

345

CROSS, CHARLES FREDERICK (1855-). *Cellulose: an outline of the chemistry of the structural elements of plants, with reference to their natural history and industrial uses.* By Cross & Bevan (C.F. Cross, E.J. Bevan, and C. Beadle). London, New York: Longmans, Green, 1895.

vi, 320 p. 14 plates. 21 cm. In. ms. on title page: Charles E. Coates. Stamped on title page: Louisiana State University Chemical Laboratory. Green library binding.

BM I p.402.

346

CRUZ, MARTÍN DE LA. *The Badianus manuscript (Codex Barberini, Latin 241) Vatican library; an Aztec herbal of 1552.* Introduction, translation, and annotations by Emily Walcott Emmart. With a foreword by Henry E. Sigerist. Baltimore: The Johns Hopkins Press, 1940.

xxiv, 341 p., incl. illus., plates (part col.) 31 cm. "Facsimile of the Badianus manuscript *[Libellus de medicinalibus Indorum herbis]*: pp.[83]-202. Tan cloth binding.

* 347

CURTIS, WILLIAM (1746-1799). *Flora londinensis: or, Plates and descriptions of such plants as grow wild in the environs of London: with their places of growth, and times of flowering; their several names according to Linnaeus and other authors: with a particular description of each plant in Latin and English.* London: Printed for and sold by the author, 1777-98.

6 fasc. 433 col. plates. 51 cm. Originally issued in 6 fascicles, to be bound as 2 v. Title page for v.1 in fasc.1; title page for v.2 in fasc.6. Plates are numbered in ms. in each fascicle. In ms. on p.[2] of cover of fasc.1-3: W. Darby, 1783. "A catalogue of certain plants growing wild, chiefly in the environs of Settle, in Yorkshire, observed by W. Curtis, in a six weeks botanical excursion from London, made at the request of J.C. Lettsom, M.D.F.R.S. &c. in the months of July and August, 1782": [6]p. at end of fasc.4. Half brown morocco, marbled boards.

BM I p.407. Hunt 650. Jackson p.256. Nissen BBI 439. Plesch p.186. Pritzel 2004.

348

Curtis's Botanical Magazine. v.1-; 1787- . London: 1787- .

v. illus., col. plates. 26 cm. Title varies: v.1-14, *Botanical magazine.* Vols.54-70 also numbered n.s., v.17; v.71-130 also numbered ser.3, v.1-60; v.131-146 also numbered ser.4, v.1-16. Publication suspended 1921-Sept.1922. "Beginning April, 1948 called new series to mark a change from hand-colouring to mechanical processing of the plates. The numbering of the plates in the new series will start from one, although for continuity of reference the volume number has been retained." LSU lacks vols.90-92, 94-95, 103-104, 108-110, 146. Various bindings.

BM I p.408. Hunt 473. Nissen BBI 2350. Plesch pp.187-89. Pritzel 2007.

* 349

DANIELS, GILBERT S. *Artists from the Royal Botanic Gardens, Kew.* Pittsburgh, Pa.: Hunt Institute for Botanical Documentation, Carnegie-Mellon University, 1974.

73 p. illus. 23 cm. Paperbound, illustrated cover.

350

DARWIN, CHARLES ROBERT (1809-1882). *The different forms of flowers on plants of the same species.* New York: D. Appleton and company, 1886.

xxiv, 352 p. illus. 21 cm. Label on p.[2] of cover: Agricultural Museum, A&M College, La. Original brown cloth binding, stamped in black.

Jackson p.92.

351

DARWIN, CHARLES ROBERT (1809-1882). *The different forms of flowers on plants of the same species.* New York: Appleton, 1897.

viii, 352 p. illus. 21 cm. "Authorized edition." Quarter red morocco, marbled boards.

BM I p.423 (1877 ed.) Jackson p.92 (1877, 1880 eds.) Plesch p.192 (1892 ed.)

Garland-Bullard Memorial Collection. Gift of Rietta Garland Albritton and Florence Garland Brookes.

352

DARWIN, CHARLES ROBERT (1809-1882). *The effects of cross and self fertilisation in the vegetable kingdom.* London: J. Murray, 1876.

viii, 482 p. incl. tables. 19.5 cm. Original green cloth binding.

BM I p.423. Plesch p.192 (1900 ed.)

353

DARWIN, CHARLES ROBERT (1809-1882). *La faculté motrice dans les plantes.* Par Ch. Darwin avec la collaboration de Fr. Darwin. Ouvrage traduit de l'Anglais, annoté et augmenté d'une préface, par Edouard Heckel. Paris: C. Reinwald, 1882.

xxvi, 599 p. illus. 23 cm. Translation of *The power of movement in plants.* Original dark green cloth binding.

*354

DARWIN, CHARLES ROBERT (1809-1882). *Insectivorous plants.* [Authorized ed.]. New York: D. Appleton, [pref. 1888].

xiv, 376 p. illus., port. 22 cm. (His Selected works. Westminster ed.) "Limited to one thousand copies of which this is no.54." Red cloth binding.

355

DARWIN, CHARLES ROBERT (1809-1882). *Insectivorous plants.* New York: D. Appleton, 1899.

x, 462 p. illus. 21 cm. "Authorized edition." Quarter red morocco, marbled boards.

BM I p.423 (1875 ed.) Jackson p.96. Plesch p.192 (1893 ed.)
Garland-Bullard Memorial Collection. Gift of Rietta Garland Albritton and Florence Garland Brookes.

356

DARWIN, CHARLES ROBERT (1809-1882). *The movements and habits of climbing plants.* 2d edition, revised... New York: D. Appleton and company, 1884.

viii, 208 p. illus. 20 cm. "This essay first appeared in the ninth volume of the *Journal of the Linnean Society*, published in 1865" — Preface. Original brown cloth binding, stamped in black.

BM I p.423 (1875 ed.) Plesch p.192 (1891 ed.) Pritzel 2059 (1865 ed.)

357

DARWIN, CHARLES ROBERT (1809-1882). *The various contrivances by which orchids are fertilised by insects.* 2d edition, revised. New York: D. Appleton and company, 1886.

xvi, 300 p. illus. 20 cm. Label on p.[2] of cover: Agricultural Museum, A & M College, La. Original brown cloth binding, stamped in black.

BM I p.422 (1862, 1877 eds.) Pritzel 2058 (1862 ed.)

358

DARWIN, ERASMUS (1731-1802). *Phytologia; or, the philosophy of agriculture and gardening. With the theory of draining morasses, and with an improved construction of the drill plough.* London: J. Johnson, 1800.

viii, 612, [12]p. 12 plates (part fold.) 28 cm. Armorial bookplate of John Hutton, Esq., Marske. Full mottled calf, gilt decorated spine.

BM I p.423. Nissen BBI 452. Plesch p.191. Pritzel 2062.

359

DESCOLE, HORACIO RAÚL (1910-). *Genera et species plantarum argentinarum, opus quod in ordinem redegit et direxit Horatius R. Descole, adiuvante personali Institutionis Michaelis Lillo Universitati nationali tucumanensi annexae, et cooperantibus botanicis argentinis et extraneis in capite totius familiae citatis...* Bonis Auris: in aedibus Guillermo Kraft ltda., 1943-56.

5v. in 7. plates (part col.), maps. 51.5 cm. LSU owns copy no.333. Ivory cloth binding.

Nissen BBI 3: 470n.

*360

DESFONTAINES, RENÉ LOUICHE (1750-1833). *Flora atlantica, sive Historia plantarum quae in Atlante, agro tunetano et algeriensi crescunt.* Parisiis: apud editorem L.G. Desgranges, anno sexto Reipublicae gallicae, 1798?.

3v. 261 (i.e.263) plates. 31 cm. In ms. on title page of v.1-2: John Subboch, Algiers, 15 Oct. 1878. Manuscript notes in margin throughout the text of v.1-2. Various pages in each vol. numbered in manuscript. Plate 190 duplicated. Vol.3 consists of plates. Half green morocco, marbled boards, marbled endpapers.

BM I p.444. Jackson p.347. Nissen BBI 475. Plesch p.197. Pritzel 2176.

361

DESPORTES, JEAN BAPTISTE POUPPE (1704-1748). *Traite ou abrege des plantes usuelles de S. Domingue.* Paris: Chez Lejay, 1770.

v. 17.5 cm. LSU owns only v.3. (2 p.l., 455, [7]p.) Full contemporary calf, marbled endpapers.

*362

DICKSON, JAMES (1738-1822). *Jacobi Dickson Fasciculus plantarum cryptogamicarum Britanniae Lusitanorum botanicorum in usum, celsissimi ac potentissimi*

Lusitaniae principis regentis domini nostri, et jussu, et auspiciis denou typis mandatus curante Josepho Mariano Veloso. Ulysipone: Typographia Domus Chalcographicae, ac Litterariae ad Arcum Caeci, 1800.

2p.l., 94 p. 18 plates. 20 cm. Quarter calf, marbled boards.
BM I p.451. Jackson p.239. Nissen BBI 477. Pritzel 2224.

363

DIETRICH, DAVID NATHANAEL FRIEDRICH (1798-1888). *Deutschlands kryptogamische gewächse.* 2d auflage. Jena: Adolph Suckow, 1858-1860.

4 v. in . plates (col.) 21, 28 cm. LSU owns v.2, pts. 1 & 2. V.1, tan library binding; v.2, brown cloth, marbled endpapers, marbled edges.
BM I p.461. Jackson p.296. Pritzel 2267 (1864 ed.)

* 364

DIOSCORIDES, PEDANIUS, OF ANAZARBOS. ΔΙΟΣΚΟΡΙΔΗΣ. *Dioscorides.* [Venetiis, in aedibvs Aldi at Andreae soceri mense ivnio, 1518].

235 (i.e. 243) leaves. 22 cm. On verso of title page: Pedacij Dioscoridis De materia medica libri sex. Eiusdem de uenenatis animalibus libri duo, quibus canis rabidi signa, et curatio eorum continetur, quibus uenenata animalia morsum defixerint. Index omnium plantarum, animalium metallorum quorum utilitatem author Dioscorides praesenti in libro docet. Carmina de uirtute, siue facultate quarundam plantarum in antiquis reperta exemplaribus. Edited by Franciscus Asulanus, with annotations also by Hieronymus Roscius. "Omnes quaterniones praeter, sexternionem & ultimum GH duernionem." Leaves 59-62 incorrectly numbered 51-54; leaves 89-90 incorrectly numbered 97-98; leaves 95-96 incorrectly numbered 103-104; leaf 126 incorrectly numbered 129; leaf 154 incorrectly numbered 146; leaf 160 incorrectly numbered 152; leaf 164 incorrectly numbered 174; leaf 166 incorrectly numbered 176; leaves 203-206 incorrectly numbered 201-204; leaf 208 incorrectly numbered 200; leaves 223-243 incorrectly numbered 224, 223, 217-223, 225-229, 233, 231-233, 227, 235. Manuscript notes on fly leaves, leaves 33, 88-[89], 124, [243] and colophon. Vellum binding, spine label.
BM I p.463. Hunt 50 (1543 ed.) Nissen BBI 496 (later eds.) Plesch p.202 (1549 ed.) Pritzel 2292.

* 365

DORMON, CAROLINE (1889-1971). *Wild flowers of Louisiana. Including most of the herbaceous wild flowers of the Gulf states, with the exception of mountainous regions, and the sub-tropical portions of Florida and Texas.* Garden City, N.Y.: Doubleday, Doran & company, inc., 1934.

xviii p., 1 l., 172 p. illus., XXIV col. plates (incl. front.) 24 cm. With author's autograph and bookplate of Ellen Shipman. Inscription to "Lady Ellen" by Edith A. Stern on dedication page [v]. Original green cloth binding.

366

DOUGLAS, DAVID (1798-1834). *Journal kept by David Douglas during his travels in North America 1823-1827, together with a particular description of thirty-three species of American oaks and eighteen species of Pinus, with appendices containing a list of the plants introduced by Douglas and an account of his death in* 1834. Edited, with "memoir" and various notes by W. Wilks, assisted by H.R. Hutchinson. Published under the direction of the Royal Horticultural Society. London: W. Wesley & son, 1914.

4 p.l., 364 p. front. (port.) 1 illus. 25.5 cm. Brown cloth binding.
BM VI p.277.

* 367

DRAKE DEL CASTILLO, EMMANUEL (1855-1904). *Illustrationes florae insularum Maris Pacifici.* Vaduz: J. Cramer, 1977. Reprint of original edition, Paris: G. Masson, 1886.

458, [25] leaves of plates. illus. 29 cm. Historiae Naturalis Classica, CII. Originally issued in 7 fascicules. Dark green cloth binding.
BM I p.477 (orig. ed.) Nissen BBI: 527 (orig. ed.)

368

DUHAMEL DU MONCEAU, HENRI LOUIS (1700-1782). *Traité des arbres et arbustes que l'on cultive en France en pleine terre...* Paris: Chez Didot aîné [etc., 1800-19].

7 v. 488 (i.e.498, incl. 2 not col.) col. plates. 49 cm. Vol.1 has engraved title page. Vol.2 published by E. Michel; v.3-7 by Librairie Encyclopedique de Roret. Issued in 83 pts., 1800-1819. Half red morocco, marbled boards, marbled endpapers.
BM I p.487. Nissen BBI 549. Plesch p.212. Pritzel 2467.
Gift of the LSU Foundation in honor of The Land Grant Centennial.

369

EATON, AMOS (1776-1842). *Manual of botany, for North America: containing generic and specific descriptions of the indigenous plants and common cultivated exotics, growing north of the Gulf of Mexico.* 5th edition, revised, corrected, and much extended. Albany: Printed by Websters and Skinners, 1829.

451, [1], 12 p., 1 l., [13]-63, [71]p. 19 cm. In ms. on title page: Rd Butler, Jackson, La.? College, June 2, 1830. On

LSU Library bookplate: The Thomas Butler Library, Cottage Plantation. Full calf, taped spine.
BM II p.502 (various eds.) Jackson p.354 (1st ed.) Pritzel 2593.

370

EATON, AMOS (1776-1842). *Manual of botany for North America: containing generic and specific descriptions of the indigenous plants and common cultivated exotics, growing north of the Gulf of Mexico.* 7th edition. Albany: O. Steele, 1836.

672 p. 19 cm. With this is bound: His *Botanical grammar and dictionary*, 4th ed., 1836 (125 p.) Albany: Published by Oliver Steele. LSU owns two copies. C.2, gift of Henry L. Fuqua. Both copies, full contemporary calf, taped spines.
BM II p.502 (various eds., both titles). Jackson p.355(1st ed., 1st title), p.38 (2nd title). Pritzel 2593 (1st title).

371

EATON, AMOS (1776-1842). *A manual of botany for the northern and middle states. Part I. Containing generic descriptions of the plants to the north of Virginia, with references to the natural orders of Linnaeus and Jussieu. Part II. Containing specific descriptions of the indigenous plants, which are well defined and established; and of the cultivated exotics.* 2nd edition, corrected and enlarged. Albany: Printed and published by Websters and Skinners, 1818.

524 p. 18.5 cm. In ms. on title page: Joseph Jones, M.D. Black cloth library binding.
BM II p.502. Jackson p.354.

372

EATON, AMOS (1776-1842). *North American botany; comprising the native and common cultivated plants, north of Mexico. Genera arranged according to the artificial and natural methods.* In the present edition the author is associated with John Wright. 8th edition. With the very valuable additions of the properties of plants, from *Lindley's New Medical Flora.* Troy, N.Y.: Elias Gates, 1840.

vii, [1], 625 p. 22.5 cm. Stamped on fly leaves front and back: Herbarium Louisiana State University. Full calf binding.
BM II p.502.

*373

EATON, DANIEL CADY (1834-1895). *Beautiful ferns.* From original water-color drawings after nature, by C.E. Faxon and J.H. Emerton. Boston: D. Lothrop, 1882, [c1881].

158 p. 14 col. plates. 32 cm. In ms. on fly leaf: From Samuel Perrins Jr. To Addie B. Clark, Dec. 25th 1891-Present. Gold cloth decorated binding.
Nissen BBI 596.

374

EATON, DANIEL CADY (1834-1895). *The ferns of North America. Colored figures and descriptions, with synonymy and geographical distribution, of the ferns (including the Ophioglossaceae) of the United States of America and the British North American possessions.* The drawings by J.H. Emerton and C.E. Faxon. Salem: S.E. Cassino, 1879-80.

2v. LXXXI col. plates. 33.5 cm. Vol.2 published in Boston. Issued in parts, 1877-1880. Original green cloth binding.
BM II p.502. Jackson p.359. Nissen BBI 575.

375

ELLIOTT, STEPHEN (1771-1830). *A sketch of the botany of South Carolina and Georgia.* Charleston, S.C.: J.R. Schenck, 1821-24.

2 v. XII plates. 22.5 cm. Black library binding.
BM II p.523. Jackson p.363. Pritzel 2664.

376

ENGELMANN, GEORGE (1809-1884). *The botanical works of the late George Engelmann, collected for Henry Shaw, Esq.* Edited by William Trelease and Asa Gray. Cambridge, Mass.: John Wilson and son, 1887.

1 p.l., ix, [11]-548, [4]p., 1 l. front. (port.) illus., 103 plates (1 fold.) 31.5 cm. Compiler's autographed presentation copy to Louisiana State University. Dark red cloth, marbled endpapers.
BM II p.531. Nissen BBI 603.

377

ESSER, PETER, *i.e.* JOHANNES PETER HEINRICH (1859-). *Die giftpflanzen Deutschlands...mit 660 einzeldarstellungen auf 113 zum text gehörenden farbentafeln.* Braunschweig: Friedrich Vieweg und sohn, 1910.

xxii, 212 p. 112 (i.e. 113) col. plates. 24.5 cm. Grey cloth binding.
BM VI p.308. Nissen BBI 607.

378

FARLOW, WILLIAM GILSON (1844-1919). *Icones Farlowianae; illustrations of the larger Fungi of eastern*

North America...with descriptive text by Edward Angus Burt. Cambridge, Mass.: The Farlow Library and Herbarium of Harvard University, c1929.

x, 120 p. 103 plates (1 fold.) 36.5 cm. Green cloth binding.

379

Flora medica: containing coloured delineations of the various medicinal plants admitted into the London, Edinburgh, and Dublin pharmacopoeias; with their natural history, botanical descriptions, medical and chemical properties, &c. &c.; together with a concise introduction to botany; a copious glossary of botanical terms; and a list of poisonous plants, &c. &c. Edited by a member of the Royal college of physicians, and fellow of the Linnaean society [i.e. George Spratt]; with the assistance of several eminent botanists... London: Callow and Wilson, 1829-30.

2 v. hand col. plates (part fold.) 22 cm. In ms. on verso of marbled fly leaf: Presented by David Rice, Stratford upon Avon to David Rice Jr. upon his return to Charlbury, Oxon, Sept. 16, 1860. Vol.1 inscription is dated Aug. 16. Vol.2 inscription is dated Sept. 16, 1860. Quarter linen, grey boards.

BM II p.584. Jackson p.201. Nissen BBI 1882. Plesch p.224.

*380

The Floral Cabinet, and Magazine of Exotic Botany. Conducted by G.B. Knowles and Frederic Westcott. London: W. Smith, 1837-40.

3 v. 137 col. plates. 28 cm. Half green morocco, green cloth.

BM II p.999. Jackson p.472. Nissen BBI 2229. Plesch p.290. Pritzel 4759.

*381

FORSTER, JOHANN REINHOLD (1729-1798). *Flora Americae Septentrionalis; or, a Catalogue of the plants of North America, containing an enumerative of the known herbs, shrubs, and trees, many of which are but lately discovered; together with their English names, the places where they grow, their different uses, and the authors who have described and figured them.* London: B. White, 1771.

viii, 51 p. 22 cm. Bookplate of the Essex Museum Library, County Borough of West Ham, Essex Field Club mounted on p.[2] of cover. Stamp of the Essex Museum Library embossed on fly leaf. In ms. on fly leaf: Edward Farberger, 1803, and Presented by Dr. M.C. Cook, Feby, 1895. Paper over boards, taped spine.

BM II p.595. Hunt 619. Jackson p.354. Plesch p.228. Pritzel 2979.

382

FOURNIER, EUGÈNE (1834-1884). *Mexicanas plantas nuper a collectoribus expeditionis scientificae allatas aut longis ab annis in herbario Musei Parisiensis depositas, praedise J. Decaisne. Enumerandas curavit Eug. Fournier.* Parisiis: ex typographeo reipublicae [Imprimerie Nationale], 1872-86.

2 v. 12 plates. 26 x 28 cm. Mission scientifique au Mexique et dans l'Amerique centrale. Recherches botaniques. LSU owns only v.1. Unbound.

BM II p.604.

383

FRANCE. CENTRE NATIONAL DE LA RECHERCHE SCIENTIFIQUE. *Les botanistes français en Amérique du Nord avant 1850.* Paris 11-14 septembre 1956. Paris, 1957.

360 p. plates (1 col.), ports., maps (1 fold.), facsims. 24 cm. Black cloth binding, patterned endpapers.

384

FUCHS, LEONHART (1501-1566). *New Kreüterbüch/in welchem nit allein die gantz histori/das ist/namen/gastalt/statt und zeit der wachsung/natur/krafft und würckung/des meysten theyls der kreüter so in teütschen vnnd andern landen wachsen/mit dem besten vleiss beschriben/sonder auch aller derselben wurtzel/stengel/bletter/ blümen/ samen/frücht/und in summa die gantze gestalt/ allso artlich vnd kunstlich abgebildet vnd contrafayt ist/ das dessgleichen vormals nie geschen/noch an tag kom̃en...* Basle: Durch M. Isingrin, 1543. Facsimile of the first German edition, [Leipzig: K.F. Köhlers antiquarium, 1938].

facsim. ([888]p. 38.5 cm. incl. 516 full-page illus., ports.) Supplement bound in: *Leonhart Fuchs und sein New kreüterbuch* (1543) von prof. dr. Heinrich Marzell. Leipzig: F. Koehlers antiquarium, 1938. (80 p., 1 l., with special title page). Tan paper over boards.

BM II p.629 (orig. ed.) Jackson p.25 (orig. ed.) Nissen BBI 659. Pritzel 3139 (orig. ed.)

385

FÜNFSTÜCK, MORITZ (1856-1925). *Naturgeschichte des Pflanzenreichs. Grosser Pflanzenatlas mit Text für Schule und Haus. 80 Grossfoliotafeln mit mehr als 2000 fein kolorierten Abbildungen und 40 Bogen erläuterndem Text nebst zahlreichen Holzschmitten.* 4. Auflage. Stuttgart: Süddeutsches Verlags-Institut, [1891].

172 p. illus., 80 col. plates. 33 cm. Quarter leather, marbled boards.

*386

FYSON, PHILIP FURLEY (1877-). *The flora of the South Indian hill stations, Ootacamund, Coonoor, Kotagiri, Kodaikanal, Yercaud, and the country round...with 611 full page plates by Mrs. Fyson, Lady Bourne, R. Natesan, the author and others.* Madras: Printed by the superintendent, Government press, 1932.

2 v. plates. 23 cm. In ms. on p.[2] of cover and on fly leaf of each vol.: A.D.G. Raphavan, Presidency College. Stamped on title page: Taru, no.3643/111, Mori Gate, Delhi-110006, India. Original red cloth binding.

387

GALLØE, OLAF (1881-). *Natural history of the Danish lichens, original investigations based upon new principles.* Copenhagen: H. Aschehoug & co., 1927-1939.

6 v. plates (part col.) 30.5 cm. Green library binding.
Nissen BBI 3: 683n.

388

GAMBLE, JAMES SYKES (1847-1925). *The Bambuseae of British India.* Calcutta: Bengal Secretariat Press, 1896.

2 p.l., xvii, [3] 133, 3, 4 p. 119 plates. 35 cm. (Annals of the Royal Botanic Garden, Calcutta, vol. VII) Black library binding.
BM II p.637. Jackson p.387. Nissen BBI 684.

*389

GERARD, JOHN (1545-1612). *Gerard's Herball; the essence thereof distilled by Marcus Woodward from the edition of Th. Johnson, 1636.* London: Spring Books, 1964.

xix, 303 p. illus. 26 cm. Green cloth binding, dustjacket.
Nissen BBI 698 (1927 ed.)

390

GERARD, JOHN (1545-1612). *The herball or generall historie of plantes.* London: Edm. Bollifant, for Bonham and Iohn Norton, 1597.

1 p.l., [18], 1392, [71]p. illus. 31 cm. Title, within elaborate engraved border, coat of arms of Lord Burghley on verso. Full page portrait of author facing p.1, dated 1598. LSU copy imperfect: pages of index mutilated, lacks original title page, photostat title page supplied. Bookplate of Otto Vernon Darbyshire. Full mottled calf, marbled endpapers.
BM II p.660. Hunt 175. Nissen BBI 698. Plesch p.236 (1633 ed.) Pritzel 3282.

*391

GESNER, KONRAD (1516-1565). *Conradi Gesneri Historia plantarum.* Herausgegeben von Heinrich Zoller, Martin Steinmann und Karl Schmid. Faksimileausgabe. Dietikon-Zürich: Urs Graf Verlag, [c1972-].

v. illus. (part col.) 47 cm. Published in an edition of 550 numbered copies; LSU owns copy no.427. Paper over boards.
BM II p.667 (orig. ed.) Pritzel 3297 (orig. ed.)

*392

GISSING, THOMAS WALLER (1829-1870). *The ferns and fern allies of Wakefield and its neighbourhood.* Illustrated by J.E. Sowerby. Wakefield, [England]: Published for the author by R. Micklethwaite, 1862.

x, 54 p. 26 leaves of plates, col. illus. 22 cm. George Gissing's autograph presentation copy to Mrs. Pound (?) Spool. Original green cloth binding.
Jackson p.261.

393

GMELIN, JOHANN GEORG (1709-1755). *Flora sibirica, sive Historia plantarvm Sibiriae...* Avctore d. Joanne Georgio Gmelin... Petropoli: ex typographia Academiae scientiarvm, 1747-69.

4 v. 286 plates (part fold.) 26 cm. Vols.3-4 edited by Samuel Gottlieb Gmelin. Full brown morocco, gilt decoration. Binding by William A. Payne.
BM II p.685. Hunt 531. Jackson p.393. Nissen BBI 721. Pritzel 3384.

394

GRAY, ASA (1810-1888). *Genera floræ Americae boreali-orientalis illustrata.* The genera of the plants of the United States illustrated by figures and analyses from nature, by Isaac Sprague. Superintended and with descriptions, &c, by Asa Gray. New York: G.P. Putman; London: Putnam's American agency, 1849.

2v. 186 plates. 24 cm. Black cloth binding with "Burgmeier Book Bindery, Chicago" in design on endpapers.
BM II p.710. Jackson p.355. Nissen BBI 749. Pritzel 3526.

395

GRAY, ASA (1810-1888). *Synoptical flora of North America.* By Asa Gray and Sereno Watson. Continued and edited by Benjamin Lincoln Robinson. New York: American book company, 1878- .

[v.1, pt.1, 1895].
v. 27 cm. LSU owns only v.1, pt.1, Fascicules I & II. Polypetalae from the Ranunculaceae to the Polygalaceae. (xv, 506 p.) Half calf, cloth, patterned endpapers.
BM II p.710. Jackson p.358.

396

GREENE, EDWARD LEE (1843-1915). *Landmarks of botanical history.* Edited by Frank N. Egerton; with contributions by Robert P. McIntosh and Rogers McVaugh. Stanford, Calif: Stanford University Press, 1983.
2 v. (x,1139 p.) illus. 25 cm. "A publication of the Hunt Institute for Botanical Documentation, Carnegie-Mellon University." Text of v.1 first published in 1909; text of v.2 never before published. Dark blue cloth binding, dust-jacket.

397

GREVILLE, ROBERT KAYE (1794-1866). *Scottish cryptogamic flora, or coloured figures and descriptions of cryptogamic plants, belonging chiefly to the order Fungi; and intended to serve as a continuation of "English botany."* Edinburgh: Printed for Maclachlan and Stewart; [etc., etc.], 1823-27.
6 v. 360 col. plates. 25 cm. Half green morocco, marbled boards.
BM II p.733. Jackson p.246. Nissen BBI 757. Plesch p.243. Pritzel 3550.

398

GREW, NEHEMIAH (1644-1712). *The anatomy of plants. With an idea of a philosophical history of plants, and several other lectures, read before the Royal society.* [London: Printed by W. Rawlins, for the author, 1682].
10p.l., 24, [10], 304 (i.e. 300), [13]p., 1 l. 83 plates (part fold.) 32 cm. Full calf, gilt decorated spine.
BM II p.733. Hunt 362. Jackson p.33. Nissen BBI 758. Plesch p.243. Pritzel 3557.

*399

GUBERNATIS, ANGELO DE, COMTE (1840-1913). *La mythologie des plantes; ou, Les légendes du règne végétal.* Paris: C. Reinwald, 1878-82.
2v. 24 cm. Red cloth binding.
BM II p.745. Jackson p.214. Plesch p.244.

400

HALLE, JOHANN SAMUEL (1727-1810). *Die Deutsche Giftplanzen, zur Verhütung der tragischen Vorfälle in den Haushaltungen, nach ihren botanischen Kennzeichen, nebst den Heilungsmittein.* Berlin: J. Pauli, 1784-1793.
2 v. col. illus. (part fold.) 20 cm. LSU owns only v.2. (1 p.l., 126 p. 8 plates). Half calf, paper boards, marbled endpapers.
BM II p.774. Plesch p.248. Pritzel 3712.

401

HARVEY, NORMAN BRUCE. *New Zealand botanical paintings* [by] Norman B. Harvey, with text by E.J. Godley. [Christchurch]: Whitcombe & Tombs, [c1969].
86 p. 40 col. plates. 37 cm. Green cloth illustrated binding, illustrated paper board slipcase.

402

HARVEY, WILLIAM HENRY (1811-1866). *A manual of the British marine Algae: containing generic and specific descriptions of all the known British species of sea-weeds with plates to illustrate all the genera.* London: John Van Voorst, 1849.
lii, 252 p. 27 plates (part col.), port. 23 cm. Manuscript notes throughout text. Green cloth binding.
BM II p.795. Jackson p.242. Pritzel 3828.

*403

HEATH, FRANCIS GEORGE (1843-1913). *The fern world.* 6th edition. London: Sampson Low, Marston, Searle & Rivington, 1879.
xi, 459 p. 1 leaf of plates, illus. (some col.) 21 cm. Label mounted on p.[3] of cover: Bound by Burn & Co. Bookplate of F. Ernest Haworth. Gray cloth decorated binding.
BM II p.806 (1877 ed.) Jackson p.241 (1877 ed.)

404

HEATHCOTE, EVELYN DAWSONNE. *Flowers of the Engadine drawn from nature.* Winchester: Warren and son, 1891.
4 p.l., 22 p. [5]l. 224 plates (col.) 28 cm. In ms. on fly leaf: To Dr. Fuller England in affectionate remembrance of October 5-11 from Helen E. Puckle. Red cloth binding, marbled edges.
BM II p.806. Nissen BBI p.823.

405

HEDWIG, JOHANN (1730-1799). *Theoria generationis et frvctificationis plantarvm cryptogamicarvm Linnaei, retracta et avcta...* Lipsiae ex officina Breitkopfio-Haerteliana, 1798.

2 v. 42 plates. 26 cm. Vol.2 composed entirely of plates. V.2 has bookplate of W. Sowerby. In ms. on fly leaf of both vols.: Sowerby's Museum. Manuscript notes throughout v.2. V.1, half calf, marbled boards; v.2, full calf.
BM II p.810. Jackson p.148. Nissen BBI 832. Pritzel 3879.

406

HEIM, ROGER (1900-). *Les champignons hallucinogènes du Mexique; études ethnologiques, taxinomiques, biologiques, physiologiques et chimiques.* Par Roger Heim et R. Gordon Wasson, avec la collaboration de Albert Hofmann [et. al.]. Paris: Muséum National d'histoire Naturelle, [1958, 1959].

322 p. illus. (part col.), 36 plates (part col.) 33 cm. Black library binding, original illustrated paper wrappers bound in.
Plesch p.254.

*407

HEIM, ROGER (1900-). *Nouvelles investigations sur les champignons hallucinogènes.* Par Roger Heim, avec la collaboration de Roger Cailleux, R. Gordon Wasson, Pierre Thévenard. [Paris: Muséum National d'Histoire Naturelle, 1967].

116-218 p. illus., plates (part col.) 32 cm. Original illustrated paper wrappers.

408

HELMCKE, JOHANN GERHARD (1908-). *Diatomeenschalen im elektronenmikroskopischen Bild.* [Von] J.G. Helmcke [und] W. Krieger. 2. unveränderte. Auflage. Weinheim: J. Cramer, 1961- . [v.1, 1962].

v. plates. 22 cm. In portfolios. Vol.3- without edition statement; vol. 6- has imprint: Lehre, J. Cramer; vol.10- has imprint: Vaduz, J. Cramer; vol.10- bearbeitet von J. Gerloff und J.G. Helmcke.

409

HEPP, PHILIPP (d. 1867), ed. *Abbildungen und beschreibung der sporen zum I. II. III. und IV. band der Flechten Europas in getrockneten mikroskopisch untersuchten exemplaren.* Zürich, 1853-67.

4 v. in 1. 110 plates. 29 cm. Hepp's *Flechten Europa* is in part a continuation of *Lichenes Helvetiae* exsiicate by L.E. Schaerer. Half brown morocco, decorated endpapers.

——— *Synonymen-register zu Dr. Philipp Hepp's Flechten Europa's.* Band I-XVI und zu dessen abbildungen der flechten-sporen, heft I-IV. Zürich, 1867 (22 p., 28.5 cm.) Pamphlet binder.
BM II p.826. Nissen BBI 856.

410

HERBERT, WILLIAM, DEAN OF MANCHESTER (1778-1847). *Amaryllidaceae; preceded by an attempt to arrange the monocotyledonous orders, and followed by a treatise on cross-bred vegetables, and supplement.* London: James Ridgway and sons, 1837.

vi, [2], 428 p. 48 plates (incl. front.) 26 cm. Original dark green cloth binding.
BM II p.827. Jackson p.123. Nissen BBI 857. Plesch p.255. Pritzel 3984.

411

Herbier général de l'amateur, contenant la description, l'histoire, les propriétés et la culture des végétaux utiles et agreables...par Mordant de Launay: continué par m. Loiseleur Deslongchamps, avec figures peintes d'après nature par m. P. Bessa. Paris: Audot, 1816-27.

8 v. 572 col. plates. 29 cm. Continued as *Nouvel Herbier de l'amateur, and Herbier général de l'amateur,* 2d sér. Stamped on title pages: State Library, Louisiana. Paper over boards, marbled endpapers. Vols. I, VI black library binding.
BM II p.828. Nissen BBI 2323. Plesch p.256. Pritzel 5586.

412

Herbier général de l'amateur, deuxième série, contenant les figures coloriées des plantes nouvelles, rares et intéressantes, des jardins de l'Europe, avec leurs description, histoire, propriétés et culture. Par m. Loiseleur Deslongchamps. Paris: H. Cousin, 1839-44.

4 v. col. plates. 29 cm. LSU owns vols.1-3. Black paper over boards, spines taped, marbled endpapers.
BM II p.828. Nissen BBI 2326. Plesch pp.256/57. Pritzel 5586.

*413

HERTODT, JOHANN FERDINAND (1645-1714). *Tartaro-mastix Moraviae: per quem rariora & admiranda â natura in faecundo hujus regionis gremio effusa, comprimis tartarus, illiusque effectus morbosi curiosè examinantur, & cura tam therapeutica quam prophylactica proponitur à Ioanne Ferdinando Hertodât.* Viennae: Typis S. Rickesin viduae, 1669.

2 p.l., [14], 263, [12]p. 16 cm. Added title page, engraved. Errata (1 leaf) inserted. Stamped on fly leaf and engraved title page: Bibl. Lainc. Vellum over boards.

*414

HOEHNE, FREDERICO CARLOS (1882-). *Iconografia de orchidaceas do Brasil; gêneros e principais espécies em texto e em pranchas; resumo e complemento da monografia das orchidaceas na "Flora Brasilica."* S[ão] Paulo: [Secretaria de Agricultura, 1949].

301 p. illus., 316 plates (part col.) 33 cm. Half brown morocco, cloth binding.

*415

HOFFMANN, GEORG FRANZ (1761-1826). *Genera plantarum umbelliferarum eorumque characteres naturales secundum numerum, figuram, situm et proportionem omnium fructificationis partium. Accedunt icones et analyses aeri incisae.* [Vol. I]. Mosquae: Typis N.S. Vsevolozskianis, 1814.

xxix, 182, 16 p. 3 fold. illus. 23 cm. No more published. Manuscript notes on fly leaf. Paper over boards.
BM II p.857. Jackson p.146. Nissen BBI 894. Pritzel 4140.

*416

HOFFMANN, JULIUS (d. 1904). *Jul. Hoffmann's Alpenflora für Alpenwanderer und Pflanzenfreunde; mit 283 farbigen Abbildungen auf 43 Tafeln meist nach Aquarellen von Hermann Friese.* 2. Auflage, mit neuem text hrsg. von K. Giesenhagen. Stuttgart: E. Schweizerbartsche Verlagsbuchhandlung, 1914.

147 p. 43 leaves of plates, col. illus. 21 cm. Beige cloth with color illustration inset on cover.
BM VI p.473 (Engl. trans.)

417

HOLLÓS, LÁSZLO (1859-). *Gasteromycetes Hungariae. Die gasteromyceten Ungarns.* Im auftrage der Ungarischen Akademie der Wissenschaften, bearb. von dr. Ladislaus Hollós. Leipzig: O. Weigel, 1904.

278 p. XXIX (i.e.31) plates (1 double, 23 col.) 42.5 cm. Printed paper over boards.
BM VI p.476.

*418

HOOKER, JOSEPH DALTON, SIR (1817-1911). *Illustrations of Himalayan plants, chiefly selected from drawings made for the late J.F. Cathcart, esquire.* The descriptions and analyses by J. D. Hooker. The plates executed by W.H. Fitch. London: L. Reeve, 1855.

3 p.l., ix-x, iv p., 1 l. 24 col. plates. 51 cm. Each plate accompanied by leaf with descriptive letterpress (plate XV: 2 l; plate XVI: 3 l.) Quarter green morocco, green cloth binding.
BM II p.869. Jackson p.388. Nissen BBI 910. Plesch p.266. Pritzel 4201.

419

HOOKER, WILLIAM JACKSON, SIR (1785-1865). *Musci exotici; containing figures and descriptions of new or little known foreign mosses and other cryptogamic subjects.* London: Longman, Hurst, Rees, Orme, and Brown, 1818-20.

2 v. CLXXVI col. plates (incl. front.) 22 cm. V.1, green library binding; v.2, original green cloth binding.
BM II p.870. Jackson p.153. Nissen BBI 925. Pritzel 4210.

420

HOOKER, WILLIAM JACKSON, SIR (1785-1865). *Synopsis filicum; or, A synopsis of all known ferns, including the Osmundaceae, Schizaeaceae, Marattiaceae, and Ophioglossaceae (chiefly derived from the Kew herbarium). Accompanied by figures representing the essential characters of each genus.* By the late Sir William Jackson Hooker...and John Gilbert Baker... London: Robert Hardwicke, 1868.

482 p. ix plates (col.) 22.5 cm. Brown library binding.
BM II p.872. Jackson p.151.

421

Hooker's Journal of Botany and Kew Garden Miscellany. Edited by Sir William Jackson Hooker. v.1-9, 1849-57. London: Reeve and Co., 1849-57.

9 v. 22 cm. LSU owns only v.4. Half green morocco, marbled boards.
BM II p.873. Jackson p.472. Nissen BBI 2347. Pritzel 10787.

422

HOUSE, HOMER DOLIVER (1878-1949). *Wild flowers of New York.* Albany: University of the State of New York, 1923.

2 v. illus., 264 col. plates. 30 cm. (State Museum. Memoir 15). In ms. on fly leaf: Clair A. Brown, Port Allegany, Pa. Original green cloth binding.
BM I sup. p.486.
Gift of Clair A. Brown.

423

HULME, FREDERICK EDWARD (1841-1909). *Familiar wild flowers, figured and described by F. Edward*

Hulme. 1st-3d? ser. London, Paris, New York: Cassell, Petter & Galpin, [1878-85?].

5v. in 1. fronts., illus., col. plates. 20 cm. LSU copy imperfect: title page for 3d? ser. wanting. Half black morocco, cloth binding.

BM II p.888. Jackson p.238. Nissen BBI 951.

424

HUMBOLDT, ALEXANDER, FREIHERR VON (1769-1859). *Ideen zu einer Physiogomik der Gewächse.* Tubingen: J.G. Cotta, 1806.

28 p. 21 cm. Paper wrappers.

BM II p.889. Pritzel 4326.

425

HUNT BOTANICAL LIBRARY . *A catalogue of Redoutéana exhibited at the Hunt Botanical Library, 21 April to 1 August* 1963. Pittsburgh, 1963.

vii, 117 p. plates (part col.) 26 cm. Dark orange library binding.

*426

HUNT, PETER FRANCIS. *Orchidaceae...plates drawn by Mary A. Grierson.* Bourton, England: Bourton Press, 1973.

144 p., 40 leaves of plates, illus. (some col.), map. 49 cm. "Of this work 600 copies numbered 1-600 are for public sale... This copy is number 193." With author's and illustrator's autographs. Vellum binding by Zaehnsdorf, London, England. Issued in slipcase.

427

Icones Plantarum; or, Figures, with brief descriptive characters and remarks, of new or rare plants, selected from the author's herbarium. By Sir William Jackson Hooker. v.1.- ; 1837-19 . London: Longman, Rees, Orme, Brown, Green, & Longman, 1837-19 .

v. plates. 22 or 25 cm. Continuous volume numbering. Also numbered v.1-4; NS., v.1-6; Ser.3, v.1-10; Ser.4, v.1-10; Ser.5, v.1- . Vols.1-32, bookplate of Gerald Walter Erskine Loder. V.34, bookplate of William Jocelyn Lewis Palmer. Vols.1-34, half dark green morocco, marbled boards. V.35, black library binding.

BM II p.873. Jackson p.120. Nissen BBI 2341. Plesch pp.267/68. Pritzel 4224.

428

IINUMA, YOKUSAI. *Somoku-dzusetsu. [Iconography of plants indigenous to, cultivated in, or introduced into Nippon.]* 2d edition. Ogaki, 1874.

20 v. illus. 27 cm. Preface in French by Tanaka Yosiwo. Text in Japanese. Blue library binding.

429

IRVING, CHRISTOPHER (d. 1856). *Irving's catechism of botany: containing a description of the most familiar and interesting plants, arranged according to the Linnaean system. With an appendix on the formation of an herbarium.* 4th American edition, revised and improved, by M.J. Kerney. Adapted to the use of schools in the United States. Baltimore: J. Murphy & co., 1853.

94 p. front. 14 cm. Printed paper boards.

430

ISTVANFFI, GYULA (1860-). *Études et commentaires sur le Code de L'Escluse [sic] augmentés de quelques notices biographiques.* Budapest: L'auteur, 1900.

5 p.l., 287 p. illus. (incl. ports., facsims.) 86 (i.e.89), col. plates (2 double). 40.5 cm. Added title page in Hungarian; text in Hungarian and French; L'Écluse's work in Latin. Pages 1-38, and the plates, are facsimiles of the *Fungorum in Panoniis Observatorium Brevis Historia a Carolo Clusio,* originally published in 1601 as appendix to his *Rariorum Plantarum Historia.* Brown cloth binding.

BM II p.912.

431

JACQUIN, NIKOLAUS JOSEPH, FREIHERR VON (1727-1817). *Collectanea ad botanicam, chemiam, et historiam naturalem spectantia, cum figuris.* Vindobonae: Ex officina Wappleriana, 1786-90.

4 v. illus. 29 cm. Paper over boards.

BM II p.918. Hunt 681. Nissen BBI 970. Plesch p.278. Pritzel 4370.

*432

JACQUIN, NIKOLAUS JOSEPH, FREIHERR VON (1727-1817). *Stapeliarum in hortis Vindobonensibus cultarum.* Sandton, South Africa: Constantia Classics, 1982.

[149]p., [64] leaves of plates. col. illus., port. 47 cm. (Botanica classica series) Facsimile. Original imprint: Vindobonae: Prostant apud Wappler et Beck, Londini apud White, 1806. Edition limited to 150 numbered copies. LSU owns copy no.44. Quarter green morocco/tan cloth binding. Tan cloth slipcase.

BM II p.918. Nissen BBI 981. Pritzel 4373.

433

JENNINGS, OTTO EMERY (1877-). *Wild flowers of western Pennsylvania and the upper Ohio basin.* Watercolors by Andrey Avinoff. Pittsburgh: University of Pittsburgh Press, c1953.

2v. 200 col. plates, maps. 37 cm. Each plate accompanied by descriptive letterpress. 3,000 copy edition. Original green cloth binding.
Nissen BBI 3: 991n.

*434

JESSEN, KARL FRIEDRICH WILHELM (1821-1889). *Deutsche Excursions-Flora: die Pflanzen des deutschen Reichs und Deutsch-Oesterreichs nördlich der Alpen mit Einschluss der Nutzpflanzen und Zierhölzer tabellarisch un geographisch bearbeitet...mit 34 Original-Holzschnitten, 320 verschiedene Zeichnungen enthaltend geschnitten von Ad. Closs, Stuttgart.* Hannover: P. Cohen, 1879.

vii, 32, 711 p. illus. 17 cm. Quarter brown morocco, marbled boards, marbled edges.
Jackson p.295.

*435

JOHNSTON, ELIZA GRIFFIN. *Texas wild flowers.* With a biography of Mrs. Johnston by Mildred Pickle Mayhall. Collector's edition. Austin, Texas: Shoal Creek Publishers, 1972.

xl, 205 p. illus. (part col.), map (on lining paper), ports. 33 cm. "Collector's edition...consisting of ninety-seven lithographs...with the author's descriptions, was limited to three hundred copies, this copy being number 73." "Preserved for future generations by the Daughters of the Republic of Texas and the William Barret Travis Chapter of the Daughters of the Republic of Texas." With autograph of Mrs. M.M. O'Dowd, president general, Daughters of the Republic of Texas. Seal of the Daughters of the Republic of Texas on p.[iii]. Full brown calf binding with monogram in gilt on front cover.

*436

JOHNSTON, GEORGE (1797-1855). *The botany of the eastern borders, with the popular names and uses of the plants, and of the customs and beliefs which have been associated with them.* London: J. Van Voorst, 1853.

xii, 336 p. illus., 15 plates (incl. front.) 22 cm. (Terra Lindisfarnensis). Bookplate of Robert H. Elliot mounted on p.[2] of cover. "Our wild flowers in their relations to our pastoral life. A lecture read to the Mechanics Institute of Berwick-upon Tweed in...1851," pp.221-246. "The fossil flora of the mountain limestone formation of the eastern borders, in connection with the natural history of coal. By George Tate," pp.289-317. Original green cloth binding.
BM II p.938 (Berwick-upon-Tweed not incl.) Jackson pp.246, 249. Nissen BBI 1001. Pritzel 4458, 4459.

*437

JONAS, FRITZ. *Entwicklung und Besiedlung Ostfrieslands.* Dahlem bei Berlin: Im Selbstverlag, 1942.

2 v. (181 p.) illus., maps. 26 cm. (Repertorium specierum novarum regni vegetabilis. Beihefte; Bd. 125). Issued in 2 parts. In brown cloth case.

438

The Journal of Botany, being a second series of the Botanical Miscellany; containing figures and descriptions of such plants as recommend themselves by their novelty, rarity, or history, or by the uses to which they are applied in the arts, in medicine, and in domestic oeconomy; together with occasional botanical notices and information. By William Jackson Hooker. v.1-4; [1834]-Jan. 1842. London: Longman, Rees, Orme, Brown, Green & Longman [etc]; Edinburgh: A. & C. Black [etc.], 1834-42.

4 v. plates (part col.) ports. 23 cm. Title varies. Second of a series of four periodicals edited by Sir William Jackson Hooker. Succeeded by the *London Journal of Botany* (1842-48), in turn succeeded by *Hooker's Journal of Botany* (1849-57). V.1, black library binding; v.2-4, original brown cloth binding.
BM II p.873. Jackson p.472. Nissen BBI 2347. Pritzel 10785.

439

JUSSIEU, ADRIEN DE (1797-1853). *Botanique. Ouvrage adopté par le Conseil royal de l'instruction publique pour l'enseignement dans les colléges; et approuve par Monseigneur l'Archevèque de Paris pour l'enseignement dans les établissements religieux.* Paris: V. Masson, [1844].

viii, 552 p. illus. 19 cm. (Le cours élémentaire d'histoire naturelle). Full green morocco, gilt decorated spine.
BM III p.952 (various eds.) Jackson p.45. Pritzel 4544.

*440

JUSSIEU, ANTOINE LAURENT DE (1748-1836). Antonii Laurentii de Jussieu. *Genera plantarum, secundum ordines naturales disposita, juxta methodum in Horto regio Parisiensi exaratam, anno M.DCC.LXXIV.* Parisiis: apud viduam Herissant...Chez Theophile Barrois, 1789.

24, lxxii, 498 p. 1 l. 21 cm. Full tan calf, gilt decorated spine with label. Binding by William A. Payne.
BM II p.959. Hunt 703. Jackson p.112. Plesch pp.285-86. Pritzel 4549.

*441

KARSCH, ANTON (1822-1892). *Vademecum botanicum: Handbuch zum Bestimmen der in Deutschland wildwachsenden, sowie im Feld und Garten, im Park, Zimmer und Gewächshaus kultivierten Pflanzen.* Leipzig: O. Lenz, 1894.

lv, 1094, 1 p. illus. 21 cm. Stamped on title page: Hermann Jaenichen Dr. agr., Diplomgärtner, Dozent für Botanik und Pflanzenschutz, Berlin-Dahlem, Königin-Luise-Str. 22. Paper over boards, black cloth spine.
BM II p.959.

442

KARTESZ, JOHN T. *A synonymized checklist of the vascular flora of the United States, Canada, and Greenland.* By John T. Kartesz and Rosemarie Kartesz in confederation with Anne H. Lindsey and C. Ritchie Bell. Chapel Hill: University of North Carolina Press, c1980.

xlviii, 498 p. 29 cm. (Vol. II, The biota of North America.) Beige cloth binding.

443

KERCHOVE DE DENTERGHEM, OSWALD, COMTE DE (1844-1906). *Les palmiers; histoire iconographique: géographie, paléontologie, botanique, description, culture, emploi, etc.* Avec index général des noms et synonymes des espèces connues. Dessines d'après nature par P. De Pannemaker. Paris: J. Rothschild, 1878.

viii, 348 p. illus., 40 col. plates. 28 cm. Tan library binding.
BM II p.970. Jackson p.139. Nissen BBI 1032. Plesch p.287.

444

KONRAD, PAUL. *Icones selectae fungorum.* Par P. Konrad et A. Maublanc. Préface de René Maire. Paris: P. Lechevalier, 1924-37.

6 v. 500 col. plates, ports. 28 cm. Quarter red morocco, marbled boards, marbled endpapers.

445

KRÄUSEL, RICHARD (1890-), ed. *Flore d'Europe.* Ouvrage publié sous la direction de Richard Kräusel avec la collaboration de H. Merxmüller et de H. Nothdurft et avec l'appui de divers instituts et sociétés savantes. Adaptation française avec la collaboration de C. Radt sous le contrôle de Jean F. Leroy. Paris: Société Française du Livre, 196-?—.

v. col. plates. 28 cm. (Collection de documents d'histoire naturelle). Issued in portfolio. LSU owns vols.1-2. Green cloth portfolio.

446

KRÄUSEL, RICHARD (1890-), ed. *Mitteleuropäische Pflanzenwelt. Kräuter und Stauden, eine Auwahl unter Berücksichtigung der in Deutschland geschützen Arten. Ausgewählt und bearb. von Richard Kräusel, unter Mitarbeit von Hermann Merxmüller und Heinrich Nothdurft.* Hamburg: Kronen-Verlag E. Cramer, 1954-1955, c1954.

[16]p., 168 col. plates. 27 cm. (Sammlung naturkundlicher Tafeln). Green cloth portfolio.
Nissen BBI 3: 1096n.

447

KRÄUSEL, RICHARD (1890-), ed. *Mitteleuropäische Pflanzenwelt; Sträucher und Bäume.* Ausgewählt und bearb. von Richard Kräusel, unter Mitarbeit von Hermann Merxmüller und Heinrich Nothdurft. Hamburg: Kronen-Verlag E. Cramer, c1960.

11 p. 144 col. plates. 27 cm. (Sammlung naturkundlicher Tafeln). Issued in parts, in green cloth portfolio.
Nissen BBI 3: 1096n.

448

LABILLARDIÈRE, JACQUES JULIEN HOUTON DE (1755-1834). *Novae Hollandiae plantarum specimen.* With an introduction by Frans A. Stafleu. Lehre: J. Cramer; New York: Stechert-Hafner Service Agency, 1966.

xxxxi, 112, 130 p. 265 plates. 28 cm. (Historiae naturalis classica, v.45). Reproduction, with added new introduction in English, of the work published in Paris in 2 v. (27 pts.) in 1804-07 (original title pages dated 1804-06). Dark green cloth binding.
BM III p.1040 (orig. ed.) Jackson p.398 (orig. ed.) Nissen BBI 1116 (orig. ed.) Pritzel 4963 (orig. ed.)

449

LAMARCK, JEAN BAPTISTE PIERRE ANTOINE DE MONET DE (1744-1829). *Encyclopédie méthodique. Botanique.* Paris: Panckoucke; Liege: Chez Plomteux, 1783-1808.

8v. 27 x 20.5 cm. Marbled boards, spines taped.

——— *Supplément.* Continuée par J.L.M. Poiret. Paris: Agasse, 1810-17. (5 v. 27 cm.) Marbled boards, spines taped.

——— *Tableau encyclopédique et méthodique. Des trois règnes de la nature. Botanique.* Par M. le Chemalier de la Mark. Continuée par J.L.M. Poiret. Paris: Panckoucke, 1791-1823. (3 v. 28 cm.) Marbled boards, spines taped.

——— *Recueil de planches de botanique de l'encyclopédie.* Paris: Agasse, 1823. (4 v. 1000 plates. 29 cm.) Marbled boards, spines taped. V.4, brown library binding.

BM II p.528. Jackson pp.12, 114. Nissen BBI 2244. Pritzel 5004.

*450

LAMARCK, JEAN BAPTISTE PIERRE ANTOINE DE MONET DE (1744-1829). *Histoire naturelle des végétaux, classés par familles: avec la citation de la classe et de l'ordre de Linné, et l'indication de l'usage que l'on peut faire des plantes dans les arts, le commerce, l'agriculture, le jardinage, la médecine, etc. des figures dessinées d'après nature, et un genera complet, selon le systême de Linné; avec des renvois aux familles naturelles de A.L. de Jussieu.* Par J.B. Lamarck et par B. Mirbel. Paris: Deterville, 1803.

15 v. illus. (some col.) 15 cm. Full brown morocco, gilt decorated spine, marbled edges.

BM I p.282, III p.1049. Jackson p.115. Pritzel 5006.

451

LAMBERT, AYLMER BOURKE (1761-1842), comp. *An illustration of the genus Cinchona; comprising descriptions of all the officinal Peruvian barks including several new species. Baron de Humboldt's Account of the Cinchona forests of South America, and Laubert's Memoir on the different species of Quinquina. To which are added several dissertations of Don Hippolito Ruiz on various medicinal plants of South America... And a short account of the spikenard of the ancients...* London: Printed for J. Searle [etc.], 1821.

ix, [1], 181 p. 5 fold. plates. 32.5 cm. Black cloth binding.

BM III p.1050. Jackson p.128. Nissen BBI 1128. Pritzel 5012.

452

LANGE, JAKOB EMANUEL (1864-). *Flora agaricina Danica.* Published under the auspices of the Society for the advancement of mycology in Denmark and the Danish Botanical Society. Collaborators: N.F. Buchwald, M.P. Christiansen, C. Ferdinandsen, Poul Larsen, F.H. Møller, Ø. Winge, et al. Copenhagen: Printed by Recato a/s, 1935-40.

5 v. in 2. 200 col. plates. 32.5 cm. Green cloth binding.

Nissen BBI 1132.

*453

LANKESTER, PHEBE (POPE) (1825-1900), ed. *A plain and easy account of the British ferns; together with their classification, arrangement of genera, structure, and functions; and a glossary of technical and other terms.* London: R. Hardwicke, [1860].

xv, 108 p. illus. 17 cm. Original green cloth binding, gilt edges.

BM III p.1058. Jackson p.240.

*454

LARBER, GIOVANNI (1785-1845). *Sui funghi saggio generale di Giovanni Larber, con tavole in rame ed una descrizione e tavola sinottica de'funghi mangerecci più comuni d'Italia.* Bassano: Dalla tipografia Baseggio, 1829.

2v. 21 plates. 29 cm. Cinnamon-colored cloth binding.

BM III p.1061. Plesch p.298. Pritzel 5076.

455

L'ÉCLUSE, CHARLES DE (1526-1609). *Caroli Clvsi Atrebatis... Rariorvm plantarvm historia...* Antverpiae: ex officina Plantiniana apud Ioannem Moretum, 1601.

[14], 364, cccxlviii, [10]p., 1 l. illus. 34 cm. Vellum binding.

BM III p.1075. Jackson p.27. Nissen BBI 372. Plesch p.302. Pritzel 1759.

*456

L'ÉCLUSE, CHARLES DE (1526-1609). *Rariorum aliquot stirpium historia...* Reprint of the original edition, Antverpiae: C. Plantini, 1583. Graz, Austria: Akademische Druck, 1965.

766, [47]p. illus. 19 cm. Includes facsimile reproduction of the title page of the original edition of 1583: *Caroli Clvsii Atrebatis Rariorum aliquot stirpium, per Pannoniam, Austriam & vicinas quasdam prouincias obseruatarum historia, qvatvor libris expressa.* Green cloth binding.

BM III p.1074 (orig. ed.) Hunt 144 (orig. ed.) Nissen BBI 371 (orig. ed.) Pritzel 1758 (orig. ed.)

457

LÉGER, CHARLES (1880-). *Redouté et son temps.* [Paris]: Editions de la Galerie Charpentier, 1945.

169 p. col. front., illus. (incl. ports., facsims.) mounted col. plates and portfolio of 12 col. plates. 33 cm. In an edition of 980 copies: LSU owns copy no.533. Half red morocco/cloth, marbled endpapers.

Plesch p.378.

458

LEIGHTON, WILLIAM ALLPORT (1805-1889). *The British species of angiocarpous lichens, elucidated by their sporidia.* London: Printed for the Ray Society, 1851.

3p.l., 101, 18, 3 p. XXX col. plates. 22 cm. (The Ray Society. [Publications, v.20, 1851]). Original black cloth binding, spine restored.

BM III p.1084 (Ray Society). Jackson p.243. Nissen BBI 1168. Pritzel 5188.

459

LEVIN, KARL SEMENOVICH. *Risunki sanktpeterburgskoĭ flory, s kratkīm opisanīem eia rasteniĭ... Izdavaemye K. Levinym...* Sanktpeterburg: V Tip. A. Smirdina, 1836-43.

2 v. 259 hand col. plates. 21-24.5 cm. Colored hand-drawing of *Viola umbrosa* inscribed. LSU owns only vol.1.

460

L'HÉRITIER DE BRUTELLE, CHARLES LOUIS (1746-1800). *Sertum Anglicum,* 1788. Facsimile with critical studies and a translation. Pittsburgh: Hunt Botanical Library, 1963.

xcviii p., facsim. ([4], 36 p. 35 plates). illus. (part col.) 32 cm. (The Hunt facsimile series, no.1). Green marbled boards, ivory paper spine.

*461

L'HÉRITIER DE BRUTELLE, CHARLES LOUIS (1746-1800). *Stirpes, novæ, aut minus cognitæ, quas descriptionibus et iconibus illustravit Carolus-Ludovicus L'Héritier, dom. de Brutelle...* Parisiis: ex typographia Philippi-Dionysii Pierres, 1784-85.

6 pts. in 2 v. LXXXIV plates (part col. and fold.) 62 cm. Plates 47 and 48 duplicated. Half blue morocco, blue cloth binding.

BM III p.1108. Hunt 673. Jackson p.111. Nissen BBI 1190. Plesch p.307. Pritzel 5268.

*462

LIGHTFOOT, JOHN (1735-1788). *Flora scotica: or, A systematic arrangement, in the Linnæan method, of the native plants of Scotland and the Hebrides.* London: Printed for B. White, 1777.

2 v. XXXV plates. 21.5 cm. Paged continuously: v.1: 1 p.l., xii, 530 p.; v.2: 2 p.l., 531-1151 p., [24]p. Slip mounted on lining paper of each vol.: Auchincruive. Full contemporary calf, gilt decorated spine.

BM III p.1113. Hunt 651. Jackson p.246. Nissen BBI 1193. Plesch p.308.

463

LINDLEY, JOHN (1799-1865). *Flora medica; a botanical account of all the more important plants used in medicine, in different parts of the world.* London: Longman, Orme, Brown, Green, and Longmans, 1838.

xiii p., 1 l., 655, [1]p. 23 cm. In ms. on title page: H. Pope Noble. Green library binding.

BM III p.1120. Jackson p.201. Pritzel 5359.

464

LINDLEY, JOHN (1799-1865). *Rosarum monographia; or, A botanical history of roses. To which is added, an appendix for the use of cultivators, in which the most remarkable garden varieties are systematically arranged.* London: Printed for J. Ridgway, 1820.

xxxix, 156 p. 19 col. plates (incl. front.) 24.5 cm. Plate 19 not colored. No plate of 19 included bound in frontispiece position. Half black morocco, marbled boards, marbled endpapers, marbled edges.

BM III p.1119. Jackson p.142. Nissen BBI 1204. Plesch p.309. Pritzel 5343.

*465

LINDLEY, JOHN (1799-1865). *Sertum orchidaceum; a wreath of the most beautiful orchidaceous flowers, selected by John Lindley.* Reprint of the London edition of 1838. Amsterdam: Theatrum Orbis Terrarum, Ltd; New York: Johnson Reprint Corp., [1974].

8 p.l., [76]p. 49 col. plates. 50 cm. Added title pages, engraved in colors. London, 1840. Originally issued in 10 fasc., 1837-42. Plates accompanied by descriptive letterpress. Edition limited to 1000 copies. LSU owns copy no.63. Green cloth binding.

BM III p.1120 (orig. ed.) Jackson p.138 (orig. ed.) Nissen BBI 1205 (orig. ed.) Pritzel 5360 (orig. ed.)

466

LINDLEY, JOHN (1799-1865). *The vegetable kingdom; or, The structure, classification, and uses of plants, illustrated upon the natural system.* 3d edition, with corrections and additional genera. London: Bradbury & Evans, 1853.

lxviii, 908 (i.e. 984)p. front., illus. 22.5 cm. Extra pages to the number of 76 are intercalated throughout the text, as 11 a-11 b, 50 a-50 d, etc. Original brown cloth binding.

BM III p.1120. Jackson p.117. Pritzel 5367.

467

LINNÉ, CARL VON (1707-1778). Caroli Linnaei... *Philosophia botanica in qva explicantvr fvndamenta bo-*

tanica cvm definitionibvs partivm, exemplis terminorvm, observationibvs rariorum, adiectis figuris æneis. Viennae Avstriae: typis Ioannis Thomae Trattner, 1755.

2 p.l., 364 p. IX plates. 20.5 cm. Full contemporary calf, gilt decorated spine.

BM III p.1130 (1751 ed.) Hunt 541 (1751 ed.) Jackson p.16 (1st ed.) Plesch p.311 (1783 ed.) Pritzel 5426.

468

LINNÉ, CARL VON (1707-1778). Caroli Linnaei... *Species plantarum, exhibentes plantas rite cognitas, ad genera relatas, cum differentiis specificis, nominibus trivialibus, synonymis selectis, locis natalibus, secundum systema sexuale digestas.* Holmiae: impensis L. Salvii, 1753.

2 v. 21.5 cm. Paged continuously: v.1: 6 p.l., 560 p; v.2: 1 p.l., 561-1200, [29]p. Quarter red morocco, marbled boards, decorated endpapers, marbled edges.

BM III p.1132. Hunt 548. Jackson p.110. Pritzel 5427.

469

LINNÉ, CARL VON (1707-1778). *A system of vegetables, according to their classes, orders, genera, species, with their characters and differences...* Translated from the 13th edition (as published by Dr. Murray) of the *Systema vegetabilium* of the late Professor Linneus; and from the *Supplementum plantarum* of the present Professor Linneus... By a botanical society at Lichfield. Lichfield: Printed by J. Jackson for Leigh and Sotheby, 1783.

2 v. XI plates. 22 cm. LSU owns only v.2. Marbled boards, spine taped.

BM III p.1133. Hunt 665. Jackson p.111. Pritzel 5431.

470

LOUDON, JOHN CLAUDIUS (1783-1843), ed. *Loudon's encyclopædia of plants; comprising the specific character, description, culture, history, application in the arts, and every other desirable particular respecting all the plants indigenous to, cultivated in, or introduced into Britain.* Edited by Mrs. Loudon, assisted by George Don, F.L.S., and David Wooster. New impression. London: Longmans, Green and co., 1866.

xxii, 1574 p. illus. 22 cm. In ms. on fly leaf: J. Griffiths from his affectionate mother, Nov. 8th, 1870. Full calf with gilt decoration.

BM III p.1182. Jackson pp.12, 118. Nissen BBI 1241. Plesch p.317 (1st ed.) Pritzel 5625.

*471

LOWE, EDWARD JOSEPH (1825-1900). *Ferns, British and exotic.* London: Groombridge, 1858-64.

8 v. 479 col. plates. 26 cm. Full green morocco, gilt decoration, gilt edges, marbled endpapers.

BM III p.1184 (1856-60 ed.) Jackson p.240. Nissen BBI 1243. Plesch p.318. Pritzel 5640.

*472

LOWE, EDWARD JOSEPH (1825-1900). *A natural history of British grasses.* With coloured illustrations. London: Groombridge, 1865.

vi, 245 p. 74 col. plates. 26 cm. Stamped on lining papers: Library of Willard M.L. Robinson, New York. Original green cloth stamped binding.

BM III p.1184 (1858 ed.) Jackson p.239. Nissen BBI 1246. Plesch pp.318/19. Pritzel 5641.

*473

LOWE, EDWARD JOSEPH (1825-1900). *A natural history of British grasses.* London: Groombridge, [186-?].

vii, 12 p., [4] leaves of plates. col. illus. 26 cm. On cover: part 1. Original printed wrappers.

BM III p.1184 (1858 ed.) Jackson p.239. Nissen BBI 1246. Plesch pp.318/19. Pritzel 5641.

*474

LOWE, EDWARD JOSEPH (1825-1900). *A natural history of new and rare ferns; containing species and varieties, none of which are included in any of the eight volumes of "Ferns, British and exotic," amongst which are the new hymenophyllums and Trichomanes.* With coloured illustrations and wood-cuts. London: Groombridge, 1865.

viii, 192 p. 72 col. plates. 26 cm. Stamped on title page and lining papers: Clinton Hall Association, N.Y., For the use of the Mercantile Library. Modern full green morocco, gilt decorated spine, original binding preserved on front cover. Binding by William A. Payne.

BM III p.1184 (1862 ed.) Jackson p.150. Nissen BBI 1245. Plesch p.319 (1864 ed.) Pritzel 5644.

*475

LOWE, EDWARD JOSEPH (1825-1900). *Our native ferns; or, A history of the British species and their varieties.* London: Groombridge, 1867-69.

2 v. illus., 76[i.e. 79] col. plates. 26 cm. Half brown morocco, marbled boards, marbled endpapers.

BM III p.1184 (1865-67 ed.) Jackson p.240. Nissen BBI 1244 (1865-67 ed.) Pritzel 5643.

* 476

MANGIN, ARTHUR (1824-1887). *Les plantes utiles.* Illustration par Yan 'Dargent et W. Freeman. Tours: A. Mame et Fils, 1870.

402, 4 p. illus. 22 cm. Quarter green morocco, gilt decorated spine, green cloth boards, marbled endpapers, gilt edges.
BM III p.1233. Jackson p.193. Plesch p.324 (2nd ed.)

* 477

MANICKAM, V.S. *Enumeration of ferns of the Palni Hills, South India.* By V.S. Manickam and C.A. Ninan. Lucknow: Scholar Publishing House, 1976.

xii, 52, [1]p. [1] leaf of plates, illus., maps. 25 cm. (Botanical records and monographs, no.1, 1976). Published under the auspices of the New Botanist Association. Printed paper wrappers.

478

MARTIUS, KARL FRIEDRICH PHILIPP VON (1794-1868). *Icones plantarum cryptogamicarum quas in itinere annis MDCCCXVII-MDCCCXX per Brasiliam...* Monachii: impensis auctori, 1828-1834.

138 p. 76 col. plates. 35 cm. Grey cloth binding.
BM III p.1255. Jackson p.373. Nissen BBI 1287. Plesch p.327. Pritzel 5892.

479

MARTYN, THOMAS (1735-1825). *Thirty-eight plates, with explanations; intended to illustrate Linnaeus's System of vegetables, and particularly adapted to the Letters on the elements of botany.* London: Printed for J. White, 1799.

vi, 72 p. 38 plates. 24 cm. Manuscript note on p.[2] of cover. Quarter brown calf, plain boards.
BM III p.1259. Hunt 732 (1794 ed.) Jackson p.16. Nissen BBI 1292. Plesch p.328. Pritzel 5928.

* 480

MASON, HILDA. *Western Cape Sandveld flowers. Illustrated by Hilda Mason.* Text by Enid du Plessis and collaborators. 1st edition. Cape Town: C. Struik, 1972.

203 p. illus. (part col.) 29 cm. "Deluxe collector's edition." "Limited to three hundred copies. This copy is no.216." With author's autograph. Green imitation leather.

* 481

MEE, MARGARET. *Flowers of the Brazilian forests; collected and painted by Margaret Mee.* Foreword on the Brazilian forests by Robert Burle Marx, with a preface by Sir George Taylor. London: Tryon Gallery, 1969.

[79]p. 32 plates, col. illus., 5 col. maps. 54 cm. Limited edition of 500 copies. LSU owns copy no.438. In slipcase. Each plate accompanied by descriptive letterpress. Quarter green morocco, marbled boards, pictorial endpapers. Bound by Zaehnsdorf of London.

* 482

MEEHAN, THOMAS (1826-1901). *The native flowers and ferns of the United States in their botanical, horticultural, and popular aspects.* Illustrated by chromolithographs. Series I-II. Boston: L. Prang, 1878-80.

4 v. col. plates. 27 cm. LSU owns only v.1-2. Half black morocco, black cloth, marbled endpapers.
BM III p.1278. Jackson p.358. Nissen BBI 1331.

* 483

MEYER, CARL ANTON (1795-1855). *Über einige Cornus-Arten: aus der Abtheilung Thelycrania...* St. Petersburg: Gedruckt bei der Kaiserlichen Academie der Wissenschaften, 1845.

1 p.l., 33 p. 26 cm. (Aus den Mémoires de l'Académie Impériale des Sciences, Sc. Natur. T.V. besonders abgedruckt). Brown cloth binding.
BM III p.1299. Pritzel 6179.

484

MICHAUX, ANDRÉ (1746-1802). *Flora boreali-Americana, sistens caracteres plantarum quas in America septentrionali collegit et detexit, Andre Michaux...* Aeditio nova. Paris: Bibliopola journaux junior, 1820.

2 v. plates. 22 cm. Marbled boards, taped spine, marbled endpapers.
BM III p.1305. Jackson p.508. Nissen BBI 1357. Pritzel 7611 (under Richard).

485

MICHAUX, ANDRÉ (1746-1802). *Histoire des chênes de l'Amérique; ou, Descriptions et figures de toutes les espèces et variétés de chênes de l'Amérique Septentrionale.* Paris: Levrault Frères, an IX-1801.

2, [8, 36]p. 36 plates. 43 cm. Half red morocco, marbled boards.
BM III p.1305. Jackson p.360. Nissen BBI 1358. Plesch pp.334/35. Pritzel 6194.

486

MICHAUX, FRANÇOIS ANDRÉ (1770-1855). *The North American sylva; or, A description of the forest trees of the United States, Canada, and Nova Scotia...to which is added a description of the most useful of the European forest trees.* Illustrated by 156 coloured copperplate engravings, by Redouté, Bessa, etc. Translated from the French of F. Andrew Michaux... With notes by J. Jay Smith. Philadelphia: Robert P. Smith, 1852.

6 v. col. plates. 28 cm. Vols.1, 5, 6 half green morocco, marbled boards, gilt edges. Vols.2, 3 green library binding. Vol.4 modern half light green morocco, original marbled boards, gilt edges.

BM III p.1305. Jackson p.360 (1859 ed.) Nissen BBI 1361 (1859 ed.) Plesch p.335. Pritzel 6196 (1859 ed.)

* 487

MICHAUX, FRANÇOIS ANDRÉ (1770-1855). *The North American sylva; or, A description of the forest of the United States, Canada, and Nova Scotia, considered particularly with respect to their use in the arts and their introduction into commerce. To which is added a description of the most useful of the European forest trees.* Illustrated by 156 colored engravings. Translated from the French of F. Andrew Michaux, with notes by J. Jay Smith. Philadelphia: Published by D. Rice & A.N. Hart, 1859, [c1857].

5 v. 277 col. plates. 28 cm. Vols.1-3 translated by Augustus L. Hillhouse. Vols.4-5 have title: *The North American sylva; or, A description of the forest trees of the United States, Canada, and Nova Scotia, not described in the work of F. Andrew Michaux...* By Thomas Nuttall. Full brown morocco, stamped, gilt edges.

BM III p.1305. Jackson p.360. Nissen BBI 1361. Plesch p.335. Pritzel 6196.

488

MICHELI, PIERANTONIO (1679-1737). *Nova plantarvm genera iuxta Tovrnefortii methodvm disposita quibus plantae MDCCCC recensentur, scilicet fere MCCCC nondum observatæ, reliquæ suis sedibus restitutæ; quarum vero figuram exhibere visum fuit, eæ ad DL æneis tabulis CVIII. graphice expressæ sunt; adnotationibus, atque observationibus, præcipue fungorum, mucorum, affiniumque plantarum sationem, ortum, & incrementum spectantibus, interdum adiectis...* Florentiae: typis B. Paperinii, 1729.

[24], 234 p. 108 plates. 31 cm. Vellum, spine taped.

BM III p.1306. Hunt 480. Jackson p.31. Nissen BBI 1363. Plesch pp.335-36. Pritzel 6202.

489

MILLAIS, JOHN GUILLE (1865-1931). *Magnolias.* With illustrations by R. Millais and from photographs. London, New York [etc.]: Longmans, Green and co. ltd., 1927.

v., [3], 250, [2]p. front., plates. 23.5 cm. Blue library binding.

BM VII p.842.

490

MILLAIS, JOHN GUILLE (1865-1931). *Rhododendrons and the various hybrids.* Second series...with 17 coloured plates by Miss Beatrice Parsons, Miss Winifred Walker and Miss Lilian Snelling. 14 collotype plates, and numerous illustrations from photographs. London, New York [etc]: Longmans, Green and co., c1924.

xii, 263 p., 1 l. col. front., plates (part col.) 41.5 cm. "Only five hundred and fifty copies of this book have been printed. This copy is no.135." Dark red cloth binding.

491

MILLAIS, JOHN GUILLE (1865-1931). *Rhododendrons, in which is set forth an account of all species of the genus Rhododendron (including azaleas) and the various hybrids...* With 17 coloured plates by Miss Beatrice Parsons, Miss Winifred Walker, Miss E.F. Brennand, and Archibald Thorburn; 14 collotype plates, and numerous illustrations from photographs. London, New York [etc.]: Longmans, Green & co., c1917.

xi, 267, [1]p. col. front., illus., plates (part col.) 42 cm. "Only five hundred and fifty copies of this book have been printed, this copy is no.185." Dark red cloth binding.

Nissen BBI 1369.

492

MILLSPAUGH, CHARLES FREDERICK (1854-1923). *Medicinal plants; an illustrated and descriptive guide to plants indigenous to and naturalized in the United States which are used in medicine, their description, origin, history, preparation, chemistry and physiological effects fully described.* Illustrated with 180 full-page plates...colored from life embodying over 1000 drawings by the author. Philadelphia: John C. Yorston & co., 1892.

2 v. 180 col. plates (1 fold.) 29.5 x 22 cm. Pages numbered to correspond to plate numbers. Half red morocco with original boards, marbled edges and endpapers.

Nissen BBI 1381. Plesch p.338.

*493

MOORE, THOMAS (1821-1887). *British ferns and their allies; an abridgment of the Popular history of British ferns, and comprising the ferns, club-mosses, pepperworts, & horsetails.* Illustrated by W.S. Coleman. London, New York: Routledge, Warne, and Routledge, 1859.

124 p. 12 leaves of plates, illus. 17 cm. Issued with Thomson, S., *Wild flowers.* London, New York, 1858. In ms. on fly leaf: E.C. Wyme. Half green calf, marbled boards. Jackson p.240.

494

MOORE, THOMAS (1821-1887). *The ferns of Great Britain and Ireland.* Edited by John Lindley. Nature-printed by Henry Bradbury. London: Bradbury and Evans, 1855.

5 p.l., [154]p. illus., LI col. plates (incl. front.) 57 cm. Frontispiece lacking. Bookplate of Mowbray Morris. Half dark green morocco/cloth.
BM III p.1345. Nissen BBI 1400. Pritzel 6405.

*495

MOORE, THOMAS (1821-1887). *The octavo nature-printed British ferns: being figures and descriptions of the species and varieties of ferns found in the United Kingdom.* Nature-printed by Henry Bradbury. London: Bradbury and Evans, 1859-60.

2 v. illus., col. plates. 25 cm. Each vol. has added engraved title page, dated 1859, with title: *The nature printed British ferns...* Octavo edition. Vol.1-2: Ex libris Charles Bauman. Stamped on fly leaf of v.1-2: Estes & Lauriat, Boston. Half green morocco, marbled boards, marbled endpapers, gilt edges.
BM III p.1345. Jackson p.240. Nissen BBI 1401.

*496

MOORE, THOMAS (1821-1887). *A popular history of the British ferns: and the allied plants comprising the club-mosses, pepperworts, and horsetails.* New edition. London, New York: G. Routledge, [186-?].

xvi, 394 p. 22 leaves of plates, illus. (some col.) 16 cm. Label mounted on p.[3] of cover: Bound by Westleys & Co. Red pictorial cloth binding.
BM III p.1345 (1851 ed.) Jackson p.240 (1851 ed.) Nissen BBI 1402 (1st ed.) Pritzel 6404 (1855 ed.)

*497

MOORE, THOMAS (1821-1887). *A popular history of the British ferns, and the allied plants, comprising the club-mosses, pepperworts, and horsetails.* A new and revised edition. London, New York: Routledge, Warne & Routledge, 1862.

xvi, 394 p. xxii col. plates. 17 cm. Red stamped cloth binding.
BM III p.1345 (1851 ed.) Jackson p.240 (1851 ed.) Nissen BBI 1402 (1st ed.) Pritzel 6404 (1855 ed.)

498

MUÑOZ PIZARRO, CARLOS. *Flores silvestres de Chile.* Prólogo de Sir George Taylor. 51 láminas originales de Eugenio Sierra Ráfols. Santiago: Ediciones de la Universidad de Chile, 1966.

245 p. col. plates. 35 cm. Grey cloth binding.

499

NEES VON ESENBECK, CHRISTIAN GOTTFRIED DANIEL (1776-1858). *Das system der pilze und schwämme.* Würzburg: In der Stahelschen buchhandlung, 1816-17.

xxxvi p., 1 l., 329, [2]p. 2 fold. tab. 26.5 x 21 cm. and atlas of XLIV (i.e. 46) col. plates. 26.5 x 23 cm. Quarter red morocco, marbled boards, marbled endpapers.
BM III p.1407. Nissen BBI 1444 (1837-58 ed.) Pritzel 6643.

*500

NEWMAN, JOHN B. *The 'Ladies' flora: containing a dictionary and glossary of botanical names.* New York: T.L. Magagnos, [185-?].

352 p. col. plates. 24 cm. Added title page in color has imprint: New York: H.S. Samuels & co. Earlier editions have title: *The Illustrated Botany.* Original black morocco boards with gilt decoration; new blue synthetic spine.

501

NICHOLLS, WILLIAM HENRY (1885-1951). *Orchids of Australia.* The complete edition drawn in natural colour by W.H. Nicholls with descriptive text. Edited by D.L. Jones and T.B. Muir. [Melbourne]: Nelson, [c1969].

xxii, 141 p. 476 col. plates. 34 cm. Red cloth binding.

502

NOOTEN, MME. BERTHE HOOLA VAN. *Fleurs, fruits et feuillages choisis de la flore et de la pomone de l'île de Java; peints d'après nature par Madame Berthe Hoola van Nooten.* Bruxelles: Émile Tarlier, 1863.

3 p.l., 43 leaves. 40 col. plates. 60.5 cm. Issued in 10 pts., 1863-64. In French and English. Each plate accompanied by

descriptive letterpress. Half dark green morocco, decorated endpapers.
BM III p.1442 (1880 ed.) Jackson p.396. Nissen BBI 931. Plesch p.349 (1880 ed.) Pritzel 6739.

* 503

NUTTALL, THOMAS (1786-1859). *An introduction to systematic and physiological botany*. Cambridge, [Mass.]: Hilliard and Brown; Boston: Hilliard, Gray, Little, Wilkins, and Richardson and Lord, 1827.
xi, 360 p. 12 plates (incl. front.) 21 cm. Full brown morocco, gilt decorated spine.
BM III p.1456. Jackson p.68. Pritzel 6773.

504

OBERMEYER, WILHELM. *Unsere wichtigsten Pilze in Wort und Bild*. Mit 43 Abbildungen auf 24 Tafeln in feinstem Farbdruck. Stuttgart: K.G. Lutz, [19-?].
36 p. 23 col. plates. 32 cm. Grey cloth binding.

* 505

Orchids from Curtis's Botanical Magazine. Edited by David R. Hunt. London: Curwen Books, 1981.
125 p., [31] leaves of plates. illus. (some col.) 26 cm. Tan cloth binding, dustjacket.

506

PABST, GUSTAV (d. 1911), ed. *Cryptogamen-flora, enthaltend die Abbildung und Beschreibung der vorzüglichsten Cryptogamen deutschlands und der angrenzenden Länder*... Gera: C.B. Griesbach, 1876-77.
3 pts. in 1 v. illus., plates (part col.) 35 cm. Contents: T.1, *Flechten*, O. Muller und G. Pabst. T.2, *Pilze*, hrsg. von G. Pabst. T.3, *Die Moose*. 1 abth. *Lebermoose...mit Zeichnungen* von W.O. Müller und G. Pabst. Red library binding.
BM IV p.1495. Jackson p.149.

507

PARKINSON, JOHN (1567-1650). *Theatrum botanicvm: the theater of plants. Or, an herball of a large extent: containing therein a more ample and exact history and declaration of the physicall herbs and plants that are in other authors, encreased by the accesse of many hundreds of new, rare, and strange plants from all the parts of the world...shewing withall the many errors, differences, and oversights of sundry authors that have formerly written of them...distributed into sundry classes or tribes...collected by the many years travaile, industry, and experience in this subject*. By John Parkinson apothecary of London, and the Kings herbarist, and published by the Kings Majestyes especiall priviledge. London: Printed by Tho. Cotes, 1640.
[18], 1745 p. 34.5 cm. LSU copy imperfect: missing pages photographically reproduced at end of book. Added title page, engraved. In ms. on title page and p.[9]: Benj. M. Everhart, Westchester, 1874; on title page only: B.J. Lutmans, Burlington, Vt., 1937. In ms. on engraved title page: John Torr, Doncaster, 1812. In ms. on p.[5]: John Torr, Doncaster. Tan cloth binding with red spine label.
BM IV p.1523. Hunt 235. Jackson p.28. Nissen BBI p.1490. Plesch p.356. Pritzel 6934.

508

Paxton's magazine of botany, and register of flowering plants. v.1-16. London: Orr and Smith [etc.], 1834-49.
16 v. illus., col. plates. 24 cm. Half red morocco, cloth, gilt edges.
BM IV p.1534. Jackson p.472. Nissen BBI 2351. Plesch p.358. Pritzel 7003.

509

PEARLESS, ANNE (PRATT), MRS. (1806-1893). *Haunts of the wild flowers*. London: Routledge, Warne and Routledge, 1863.
320 p. front. (col.) plates (col.) 17.5 cm. Blue cloth binding.
Jackson p.236.

510

PECK, CHARLES HORTON (1833-1917). *Report of the state botanist [on fungi]*. Albany: University of the State of New York, [187?-1913].
5 v. illus. 24 cm. Some reports issued as *New York State Museum Bulletins*. Green cloth binding.

511

PENZIG, OTTO ALBERT JULIUS (1856-). *Icones fungorum javanicorum*. Von O. Penzig und P.A. Saccardo. Mit 80, z. th. colorirten tafeln in photozincotypie. Leiden: Brill, 1904.
3 p.l., 124 p. and atlas of 1 p.l., LXXX plates (part col.) 25.5 cm. LSU copy bound in two vols: v.1 Text, v.2. Plates. Text forms revised and enlarged edition of the authors' "Diagnoses fungorum novorum in insula Java collectorum," which appeared in 3 articles in *Malpighia*, vols. XI, XV, 1897-1902. Green library binding.
BM IV p.1545. Nissen BBI 1711.

512

PFEFFER, WILHELM FRIEDRICH PHILIPP (1845-1920). *The physiology of plants; a treatise upon the metabolism and sources of energy in plants.* 2d. fully revised edition, translated and edited by Alfred J. Ewart. Oxford: Clarendon Press, 1900-06.

3 v. illus. 25 cm. Stamped on p.[2] of cover of v.2-3 and fly leaf of v.2: Dept of Bacteriology, La. Sugar Experiment Station, Audubon Park, New Orleans, La. Quarter black morocco/cloth.
BM IV p.1561.

*513

Plants of the gods: origins of hallucinogenic use. By Richard Evans Schultes and Albert Hofmann. New York: McGraw-Hill, c1979.

192 p. illus. 27 cm. Green cloth binding with gilt cover design, dustjacket.

*514

PLUES, MARGARET. *British ferns; an introduction to the study of the ferns, lycopods, and equiseta indigenous to the British Isles...with chapters on the structure, propagation, cultivation, diseases, uses, preservation, and distribution of ferns.* London: L. Reeve, [1914].

viii, 281 p. illus. 19 cm. Stamped on title page: Hermann Jaenichen Dr. agr., Diplomgärtner, Dozent für Botanik und Pflanzenschutz, Berlin-Dahlem, Königin-Luise-Str. 22. In ms. on fly leaf: This book is the personal property of P.O.W. Nr. B 817 Jaenichen Hermann. E.D. Galloway, Capt., Accounts officer. Stamped on fly leaf: P.O.W. Camp no.21, Great Britain. Red cloth binding.
BM IV p.1587 (1866 ed.) Jackson p.240 (1866 ed.) Nissen BBI 1537 (1866 ed.)

515

POELT, JOSEF (1924-). *Mitteleuropäische Pilze.* 180 Tafeln in 6-8 farbigem Offsetdruck nach Originalen von Claus Caspari. Ausgewählt und bearb. von Josef Poelt und Hermann Jahn mit Unterstützung der Botanischen Staatssammlung, München und der Deutschen Gesellschaft für Pilzkunde. Hamburg: Kronen-Verlag, [1963-64].

3 pts. illus., plates. 30 cm. (Sammlung naturkundlicher Tafeln 6. Bd.) 180 loose plates in green portfolio.

*516

PREST, JOHN M. *The garden of Eden: the botanic garden and the re-creation of paradise.* New Haven: Yale University Press, 1981.

121 p. illus. (some col.) 27 cm. Red cloth binding, dustjacket.

517

PRITCHARD, ANDREW (1804-1882). *A history of Infusoria, including the Dismidiaceæ and Diatomaceæ, British and foreign.* 4th edition. Enlarged and revised by J.T. Arlidge, W. Archer, J. Ralfs, W.C. Williamson, and the author. London: Whittaker and co., 1861.

xii, 968 p. XL plates (part col.) 23 cm. Original brown cloth binding.
Jackson p.158.

518

PURSH, FREDERICK (1774-1820). *Flora Americae Septentrionalis; or, A systematic arrangement and description of the plants of North America. Containing, besides what have been described by preceding authors, many new and rare species, collected during twelve years travels and residence in that country.* London: Printed for White, Cochrane, and co., 1814.

2 v. 24 hand col. plates. 21.5 cm. Plate 8 lacking. Black library binding.
BM IV p.1624. Jackson p.354. Nissen BBI 1570. Pritzel 7370.

519

QUÉLET, LUCIEN (1832-1899). *Champignons trouvés dans la Forêt de Fontainebleau dessinés par Delthil et dénommés par le Dr. Quélet; avec une notice du Cdt Hermary.* [n.p., n.d.].

108 (i.e.109) col. plates on 56 leaves. 44 cm. Title from handwritten label mounted on preliminary leaf. "Note synonymique et critique relative aux espèces de champignons représentées sur des planches appartenant ou Commandant Renard [?] et determinées par le Docteur Quélet," signed by Hermary: [15]p. at end. Preliminary leaf bears handwritten note: "109 aquarelles numérotées de 1 à 108 (il y a deux numéros 41). Elles furent exécutées entre 1857 et 1862. La notice du Commandant Hermary a été rédigée en 1888. Original watercolor drawings; believed to be the only copy made. Half red morocco, marbled boards, marbled endpapers.

520

RABENHORST, LUDWIG (1806-1881). *Kryptogamenflora, von Deutschland, Oesterreich und der Schweiz.* 2te auflage, vollständig neu bearbeitet von A. Grunow, Dr. F. Hauck, G. Limpricht...[und anders]. Leipzig: Edward Kummer, 1884-1910.

v. in. illus. 24 cm. LSU owns Band I, 9 pts. Some pts. lacking. Half brown calf/cloth, decorated endpapers.
BM IV p.1635. Jackson p.296. Pritzel 7383 (1st ed.)

521

RADDI, GIUSEPPE (1770-1829). *Plantarum brasiliensium nova genera et species novae, vel minus cognitae.* Florentiae: ex typ. A. Pezzati, 1825.

1 p.l., 87, [1], 88-101 p., 1 l. 84 (i.e. 97) plates. 37 cm. No more published. Quarter green morocco, marbled boards, marbled endpapers.
BM IV p.1637. Nissen BBI 1576. Pritzel 7391.

522

RAFINESQUE, CONSTANTINE SAMUEL (1783-1840). *Autikon botanikon. Icones plantarum select. nov. vel rariorum, plerumque americana, interdum african. europ. asiat. oceanic. &c. centur. XXV. — Botanical illustrations by select specimens or self-figures in 25 centuries of 2500 plants, trees, shrubs, vines, lilies, grasses, ferns &c., chiefly new or rare, doubtful or interesting, from North America and some other regions; with accounts of the undescribed, notes, synonyms, localities &c.* In 5 parts of 5 centuries each of text with 25 volumes folio of self-figures. Philadelphia: Collected, ascertained and described between 1815 & 1840. [Ann Arbor, Mich.: Edwards Bros., 1942].

3 plates. 23 cm. Paged continuously: 1st pt. (cent. I-V): [2], 3-72 p. — 2d pt. (cent. VI-X): p.[73]-140. — 3d pt. (cent. XI-XV): p.[141]-200. Green library binding, original paper covers bound in.
Jackson p.116 (orig. ed.) Nissen BBI 1577 (orig. ed.)

523

RALFS, JOHN (1807-1890). *The British desmidieae...the drawings by Edward Jenner.* London: Reeve, Benham, and Reeve, 1848.

xxii, 226 p. 35 plates. 25 cm. Original dark green stamped cloth binding.
BM IV p.1640. Jackson p.243. Nissen BBI 1582.

524

RASPAIL, FRANÇOIS VINCENT (1794-1878). *Nouveau système de physiologie végétale et de botanique, fondé sur les méthodes d'observation qui ont été développées dans le nouveau système de chimie organique, accompagné d'un atlas de soixante planches d'analyses.* Bruxelles: Société Typographique Belge, 1837.

xii, 450, xlvii p. 2 fold. tables and atlas of 60 plates. 27 cm. Brown library binding.
BM IV p.1646. Plesch p.372. Pritzel 7418.

525

RAVENSCROFT, EDWARD JAMES (1816-1890). *The pinetum britannicum. A descriptive account of hardy coniferous trees cultivated in Great Britain.* Edinburgh and London: W. Blackwood & sons [etc.], 1884.

3 v. col. fronts., 45 col. plates, 4 photographs, map. 56 cm. Published in 52 parts, 1863-84. Botanical descriptions for pts.1-3 by J. Lindley; pts.4-37, by Andrew Murray; pts.38-52, by M.T. Masters. Pts.1-33 issued from the private press of P.& C. Lawson; pts.34-52 published by Ravenscroft in London and by W. Blackwood & sons in Edinburgh. Half dark green morocco, green boards, gilt decorated spine, marbled endpapers, gilt edges.
BM IV p.1652. Nissen BBI 1588.

* 526

RAY, JOHN (1627-1705). *Historia plantarum generalis: species hactenus editas aliasque insuper multas noviter inventas & descriptas complectens; in qua agitur primò de plantis in genere, earúmque partibus, accidentibus & differentiis; deinde genera omnia tum summa; tum subalterna ad species usque infimas, notis suis certis & characteristicis definita, methodo naturae vestigiis insistente disponuntur; species singulae accurate describuntur, obscura illustrantur, omissa supplentur superflua resecantur, synonyma necessaria adjiciuntur; vires denique & usus recepti compendio traduntur; accesserunt lexicon botanicum et nomenclator botanicus totum opus in duobus tomis cum indicibus necessariis nominum, morborum, & remediorum.* Londini: Samuelis Smith & Benjamini Walford, 1693-1704.

3 v. 37 cm. Vols.1-2 paged continuously; v.1: 13 p.l., 983 p.; v.2: 4 p.l., 985-1944, [35]p. V.3, paging irregular: various repetitions in numbering. Stamped on title page of v.1-2: Royal Microscopical Society. Full contemporary calf binding, new calf spines.
BM IV p.1653. Jackson p.30. Plesch pp.373-74. Pritzel 7436.

527

REAL EXPEDICIÓN BOTÁNICA DEL NUEVO REINO DE GRANADA (1783-1816). *Flora de la Real Expedición Botánica del Nuevo Reino de Granada. Publicada bajo los auspicios de los Gobiernos de España y Colombia y merced a la colaboración entre los Institutos de Cultura Hispánica de Madrid y Bogotá.* Madrid: Ediciones Cultura Hispánica, 1954- .

v. illus., plates (part col.), ports. (part col.) 55 cm. LSU owns vols.1, 7, 8, 27, 44. Full calf, decorated endpapers. V.8, unbound.

528

REDOUTÉ, PIERRE JOSEPH (1759-1840). *The best of Redouté's Roses.* Selected and introduced by Eva

Mannering. New York: Viking Press, [1960, c1959].

xvi p., 29 col. plates, illus. 41 cm. Descriptive notes in French. Grey cloth binding.
Nissen BBI 3: 1599.

*529

REDOUTÉ, PIERRE JOSEPH (1759-1840). *Les roses.* Avec le texte, par Cl. Ant. Thory. Reprint of Paris edition: Imprimerie de Firmin Didot, 1817-24. [Antwerp, Belgium: De Schutter, 1974-78].

4 v. col. plates, port. 56 cm. "A complete and facsimile reprint of the three folio volumes of the original 1817-1824 edition. This edition is limited to 510 copies." LSU owns copy no.52. Half black morocco, brown cloth binding.
BM IV p.1661 (orig. ed.) Jackson p.42 (orig. ed.) Nissen BBI 1599 (orig. ed.) Plesch p.375 (orig. ed.) Pritzel 7455 (orig. ed.)

*530

REDOUTÉ, PIERRE JOSEPH (1759-1840). *Les roses, peintes par P.J. Redouté, décrites et classées selon leur ordre naturel, par C.A. Thory.* 3d. édition., publiée sous la direction de M. Pirolle. Paris: P. Dufart, 1828-29.

3 v. 184 col. plates, 2 ports. 26 cm. Originally issued in 30 livraisons. In ms. on fly leaf of vols. 2 & 3: CWD Lynch, 1906/7. Half red morocco, marbled boards, marbled endpapers.
BM IV p.1661. Nissen BBI 1599. Plesch p.376.

531

REDOUTÉ, PIERRE JOSEPH (1759-1840). *Roses.* Selected and introduced by Eva Mannering. London: Ariel Press, [1954-56].

2 v. illus., col. plates. 41 cm. Descriptive notes in French. V.1, beige cloth binding; v.2, red library binding.
Nissen BBI 3: 1599.

*532

REDOUTÉ, PIERRE JOSEPH (1759-1840). *Roses and lilies; selected reproductions from Les roses and Les liliacées.* n.p., [1954?].

24 col. plates in portfolio. 55 x 37 cm. Names of flowers in Latin and French. On 2 plates: Printed in the Netherlands. Laid in pamphlet binder, unbound.

533

REDOUTÉ, PIERRE JOSEPH (1759-1840). *Die schönsten Rosen, von Redouté.* Ausgewählt und eingeleitet von Eva Mannering. Stuttgart: Schuler, [196-].

xviii p. 29 col. plates, illus. 41 cm. Blue cloth binding.
Nissen BBI 3: 1599.

534

REICHENBACH, HEINRICH GOTTLIEB LUDWIG (1793-1879). *Agrostographia germanica, sistens Icones Graminearum et Cyperoidearum...die Gräser und Cyperoideen der deutschen Flora, in getreuen Abbildungen auf Kupfertafeln.* Leipzig: F. Hofmeister, 1834.

2 p.l., 50 p. 110 col. plates. 26 cm. Binding error: plates LXXX and LXXIII laid in respective place of each; plates LXXXIII to LXXXVI laid in reverse order. Text in Latin and German. Quarter black morocco, marbled boards, marbled endpapers.
BM IV pp.1668, 1669 (Iconographia Botanica & Icones Florae Germanicae).

*535

REICHENBACH, HEINRICH GOTTLIEB LUDWIG (1793-1879). *Flora germanica excursoria ex affinitate regni vegetabilis naturali disposita, sive Principia synopseos plantarum in Germania terrisque in Europa media adjacentibus sponte nascentium cultarumque frequentius...* Lipsiae: apud C. Cnobloch, 1830-1832.

2 p.l., viii, 878 p. 2 fold. maps. 16 cm. Stamped on title page: Institut für Staatsarzneikunde Fürst (?) Universität Graz. Label mounted on cover: Institut für Staatsarzneikunde Graz. Stamped on half title page and label mounted on fly leaf: Institut für gerichtliche Medizin der Universität Graz. Brown cloth binding.
BM IV p.1668. Pritzel 7505.

536

REICHENBACH, HEINRICH GOTTLIEB LUDWIG (1793-1879). *Icones florae Germanicae et Helveticae simul Pedemontanae, Tirolensis, Istriacae, Dalmaticae, Austriacae, Hungaricae, Transylvanicae, Moravicae, Borussicae, Holsaticae, Belgicae, Hollandicae, ergo Mediae Europae. Inconographia et supplementum ad opera willdenowii, schkuhrii, persoonii, decandolli, gaudini, kochii aliorumque, exhibens nuperrime detectis novitiis additis collectionem compendiosam imaginum characteristicarum omnium generum atque specierum quas in sua flora Germanica excursoria recensuit.* Leipzig: F. Hofmeister, 1837-1912.

25 v. in 26. col. plates. 30 cm. LSU vol.25, title page wanting. Vol.1: 2d ed., 1850. Vols.15-19, pts.1 and 20-21: published by Ambrosii Abel, Leipzig; Vol.19, pts.2-3 and 22-25: published by F. de Zezschwitz, Leipzig. Quarter brown morocco, marbled boards, marbled endpapers.

Vol.25, quarter red morocco, marbled boards, marbled endpapers.
BM IV p.1669. Jackson p.295. Nissen BBI 1604. Pritzel 7511.

537

RENNIE, JAMES (1787-1867). *Alphabet of medical botany, for the use of beginners*. A new edition. London: Orr and Smith, 1834.
vi, 152 p. illus. 16 cm. Green cloth binding, cover and spine labels.
Jackson p.38.

538

RICHARDS, PAUL. *A book of mosses;* with 16 plates from Johannes Hedwig's *Descriptio muscorum*. London: Penguin Books, [1950].
39 p. illus. 16 col. plates. 19 cm. Decorative paper over boards.

539

RICKEN, ADALBERT. *Die blätterpilze (Agaricaceae) Deutschlands und der angrenzenden länder, besonders Oesterreichs und der Schweiz*. Leipzig: T.O. Weigel, 1915.
2 p.l., xxiv, 480 p, and atlas of 2 p.l., vii, [1]p., 112 col. plates. 24.5 cm. Published in 15 lieferungen. Half black cloth, marbled boards.
BM VIII p.1076. Nissen BBI 1637.

*540

RICKETT, HAROLD WILLIAM (1896-). *Wild flowers of the United States*. General editor, William C. Steere. Collaborators: Rogers McVaugh and others. 1st edition. New York: McGraw-Hill, c1966.
v. inpts. illus., col. map (on lining papers), col. plates. 33 cm. "Publication of the New York Botanical Garden." Cloth bindings in slipcases, various colors.

*541

RIX, MARTYN. *The art of the botanist*. Guildford, [England]: Lutterworth Press, c1981.
224 p. illus. (some col.) 35 cm. "Published simultaneously in the United States as *The art of the plant world*. Dark red cloth binding, dustjacket.

542

ROHDE, ELEANOUR SINCLAIR. *The old English herbals*. With coloured frontispiece and 17 illustrations. London, New York, [etc.]: Longmans, Green and co., 1922.
xii, 243, [1]p. col. front., plates, ports., facsims. 26 cm. Tan cloth binding with gilt decoration.
BM VIII p.1088. Plesch p.384.

543

ROLLAND, LÉON LOUIS (1831-). *Atlas des champignons de France, Suisse, et Belgique*. Paris: P. Klincksieck, 1910.
126 p. 120 plates (col.) 25 cm. Published in 15 parts, 1906-10, consisting of 8 plates each, accompanied by a "texte résumé et provisoire" which was replaced by a "texte définitif" at the completion of the series. Quarter tan cloth, marbled boards.
BM IV p.1724. Nissen BBI 1670.

*544

RÖMER, JOHANN JAKOB (1763-1819). *Scriptores de plantis hispanicis, lustanicis, brasiliensibus, adornavit et recudi curavit I.I. Römer*. Norimbergae: in officina Raspeana, 1796.
1 p.l., 184 p. VII (i.e. 8) fold. plates. 22 cm. In ms. on fly leaf: H. Bassler. Bookplate of the Library of Lehigh University mounted on p.[2] of cover. Label mounted on p.[2] of cover: R. Friedländer & Sohn, Buchhandlung, Berlin N.W. 6, 11. Charlstrasse 11. LSU copy imperfect: engraved title page wanting. Quarter calf, marbled boards, decorated endpapers.
BM IV p.1720. Hunt 745. Pritzel 7709.

*545

ROSSER, CELIA. *The Banksias*. By Celia E. Rosser and Alexander S. George. London; New York: Academic Press, 1981- .
vol. col. illus. maps. 78 cm. "Designed by Academic Press and the Curwen Press. Printed at the Curwen Press." Bound by A.W. Lumsden of Edinburgh. "This edition comprising three volumes is limited to 730 copies of which 720 are for sale." LSU owns copy no.151. Dark green half leather/green cloth binding, gilt lettering.

546

ROUSSEAU, JEAN JACQUES (1712-1778). *La botanique de J.J. Rousseau, ornée de soixante-cinq planches, imprimées en couleurs d'après les peintures de P.J. Redouté*. Paris: Delachaussée, 1805.
x, 124 p. 65 col. plates. 55 cm. Half green morocco, marbled boards.
BM IV p.1742. Jackson p.35. Nissen BBI 1688. Plesch p.388. Pritzel 7824.

*547

RUMPF, GEORG EBERHARD (1627-1702). *Herbarium amboinense, plurimas conplectens arbores, frutices, herbas, plantas terrestres & aquaticas, quae in Amboina, et adjacentibus reperiuntur insulis, adcuratissime descriptas juxta earum formas, cum diversis denominationibus, cultura, usu, ac virtutibus. Quod & insuper exhibet varia insectorum animaliumque genera, plurima cum naturalibus eorum figuris depicta...* Amstelaedami: Apud F. Changuion, J. Catuffe, H. Uytwerf, 1741-50.

6 v. in 4. illus. (part fold.) 42 cm. Latin and Dutch in parallel columns. Vols.5-6 published by F. Changuion and H. Uytwerf. Vol.3, pp.129-130 omitted in numbering only. Vol.5, p.257 incorrectly numbered 275. "Ex libris Andreae, equit belli med et chir doct," on slip in each vol.

———— *Herbarii amboinensis auctuarium reliquas complectens arbores, frutices, ac plantas, quae in Amboina, et adjacentibus demum repertae sunt insulis, omnes accuratissime descriptae, & delineatae juxta earum formas, cum diversis indicis denominationibus, cultura, usu, ac viribus; nunc primum in lucem editum, & in latinum sermonem versum, cura & studio Joannis Burmanni. Qui varia adjecit synonyma, suasque observationes.* Amstelaedami: Apud, Mynardum Uytwerf & viduam ac filium S. Schouten, 1755.

74, [20]p. 30 plates (part fold.) 42 cm. Latin and Dutch in parallel columns. Bound as v.7 of the author's *Herbarium amboinense.* Half tan modern calf, marbled boards. Rebound by William A. Payne.
BM V p.1766. Hunt 518. Jackson p.378. Nissen BBI 1700. Pritzel 7908.

*548

SACCARDO, PIER ANDREA (1845-1920). *Cronologia della flora italiana.* Bologna: Edagricole, 1971.

vi, iv-xxxvii, 390 p. 23.5 cm. (Opera botanica, 4). Reproduction of the Padua 1909 edition. Half white leather, decorated boards.

549

SACCARDO, PIER ANDREA (1845-1920). *Sylloge fungorum omnium hucusque cognitorum.* Patavii: sumptibus auctoris, 1882-1931.

v. port. 24.5 cm. LSU owns 26 v. in 23. Imprint varies. V.26 has imprint: New York, Johnson Reprint Corp., 1972. Half brown morocco/cloth, marbled endpapers. Some vols. rebound in black library binding.
BM IV p.1777.

*550

SAINT-PIERRE, JACQUES HENRI BERNARDIN DE (1737-1814). *Botanical harmony delineated: or, Applications of some general laws of nature to plants.* Translated by Henry Hunter. 1st American edition. Worcester: Printed for J. Nancrede, No.49, Marlborough Street, Boston, 1797.

179 p. 4 plates. 21 cm. A translation of the eleventh of the author's *Études de la nature.* In ms. on recto of front.: Thos G. Wright. Full contemporary mottled calf, spine new. Plesch p.395 (French ed.)

551

SÃO PAULO, BRAZIL (STATE). DEPARTAMENTO DE BOTANICA. *Flora brasílica, planejada e iniciada por F.C. Hoehne.* São Paulo, Brazil: Instituto de Botanica, 1940.

v. plates (part col.) 31 cm. Scattered holdings. Black library binding.

552

SAPORTA, GASTON, MARQUIS DE (1823-1895). *L'évolution du règne végétal.* Les cryptogames par G. de Saporta... [et] A.F. Marion... Avec 85 figures dans le texte. Paris: G. Baillière et cie,1881.

xii, 238 p. illus. 22 cm. Bookplate of J.C. Mansel-Pleydell. Half black calf/marbled boards, marbled endpapers.
BM IV p.1804.

553

SARGENT, CHARLES SPRAGUE (1841-1927). *The silva of North America; a description of the trees, which grow naturally in North America exclusive of Mexico...* Illustrated with figures and analyses drawn from nature by Charles Edward Faxon and engraved by Philibert and Eugène Picart. Boston and New York: Houghton, Mifflin and company, 1890-1902.

14 v. 740 plates. 37 x 28.5 cm. Brown cloth binding.
BM IV p.1805. Nissen BBI 1728.

*554

SAVONIUS, MOIRA. *All color book of flowers: 100 color photographs of spectacular flowers of the world.* London: Octopus Books, 1974.

[72]p. chiefly col. illus. 30 cm. Illustrated paper over boards, illustrated endpapers, dustjacket.

555

SCHEUCHZER, JOHANN (1684-1738). *Agrostographia sive Graminum, juncorum, cyperorum, cyperoidum, iisque affinium historia.* Tiguri: Typis & Sumptibus Bodmerianis, 1719.

[38], 512, [24]p. fold. plates. 21 cm. Title page in red and black. Full contemporary calf, gilt decorated spine, marbled endpapers, marbled edges.
BM IV p.1830. Jackson p.132. Nissen BBI 1751. Pritzel 8170.

556

SCHIMPER, WILHELM PHILIPP (1808-1880). *Bryologia europaea; seu, Genera muscorum europaeorum monographice illustrata auctoribus Ph. Bruch, W. Ph. Schimper & Th. Gümbel, editore W. Ph. Schimper...* Stuttgartiae: sumptibus librariae E. Schweizerbart, 1836-55.

6 v. 641 plates. 30.5 cm. Various pagings. "Corollarium" has special title page. Descriptions in Latin, notes in German and French. Vols.1 & 6, half black morocco, marbled edges. Vols.2-5, black library binding.

——— *Musci europaei novi; vel, Bryologiae europaeae supplementum auctore W. Ph. Schimper...* Stuttgartiae: sumptibus librariae E. Schweizerbart, 1864-[66]. (Cover title, [54]p. 40 plates. 30.5 cm. With his *Bryologia europaea*. v.6).
BM IV p.1833. Jackson p.228. Nissen BBI 1759. Pritzel 8191.

*557

SCHLECHTENDAL, DIEDRICH FRANZ LEONHARD VON (1794-1866). *Abbildung und Beschreibung aller in der Pharmacopoea Borussica aufgeführten Gewaechse.* Hrsg. von. Friedrich Guimpel. Text von D.F.L. v. Schlechtendal. Berlin: L. Oehmigke, 1830-37.

3 v. 308 hand col. plates. 26 cm. In ms. on fly leaf: Benoit 1840. Quarter mottled calf, marbled boards.
BM IV p.1837. Nissen BBI 1769. Plesch p.399. Pritzel 3653.

*558

SCHOEPF, JOHANN DAVID (1752-1800). *Materia medica americana potissimum regni vegetabilis.* Erlangae: Sumtibus Io. Iac. Palmii, MDCCLXXXVII. 1974 reprint of the 1787 edition. Leipzig: Zentralantiquariat der Deutschen Demokratischen Republik, 1974.

xviii, 170 p. 21 cm. Contains reproduction of original title page. Dark orange cloth binding.
BM IV p.1855. Pritzel 8346 (orig. ed.)

*559

SCHREBER, JOHANN CHRISTIAN DANIEL (1739-1810). *Beschreibung der Graser nebst ihren Abbildungen nach der Natur.* Leipzig: S.L. Crusius, 1769-1810.

3 v. in 2. 54 plates (part col.) 41 cm. LSU owns parts 1 and 2 only, in one volume; title page of pt.2 wanting. A third part containing 14 plates was issued in 1810. Paper over boards.
BM IV p.1861. Jackson p.132. Nissen BBI 1807. Plesch p.402 (1st ed.) Pritzel 8395.

*560

SCHUMANN, KARL MORITZ (1851-1904), ed. *Blühende Kakteen (Iconographia Cactacearum) Im Auftrage der Deutschen Kakteen-Gesellschaft.* Begründet und herausgegeben von Karl Schumann. Fortgesetzt von Max Gürke. Neudamm: J. Neumann, [1900-21].

3 v. 180 col. plates. 34 cm. Originally issued in 45 pts. Vol.2 edited by Max Gürke; v.3 edited by F. Vaupel. Each plate is accompanied by leaf with descriptive letterpress. Green cloth binding.
Nissen BBI 1818.

*561

SCHWEINFURTH, CHARLES. *Orchids of Peru.* [Chicago]: Chicago Natural History Museum, 1958-61.

4 v. illus., map. 24 cm. (Fieldiana: botany, v.30, no.1-4), ([Chicago Natural History Museum], Publication 837, 868, 885, 913). Vols.1-4, coral cloth binding, decorated endpapers.

——— *Supplement.* 1- . [Chicago]: Field Museum of Natural History, 1970. (no. 24 cm.) (Fieldiana: botany), (Field Museum of Natural History. Publication). LSU lacks supplement.
Nissen BBI 3: 1819n.

562

SCHWEINITZ, LEWIS DAVID VON (1780-1834). *Synopsis fungorum in America. Boreali media degentium.* Secundum observationes Ludovic Davidig de Schweinitz. Communicated to the American Philosophical Society, Philadelphia, 15 April 1831... [Philadelphia, 1834].

caption title, pp.141-316. plates. 29 cm. Detached from American Philosophical Society, Philadelphia. *Transactions*, v.4, 1834. In ms. on fly leaf: Hollis Webster, Cambridge, Mass. Index in manuscript, [13]p. Half brown calf, marbled boards, marbled endpapers.

563

SHARPE, RICHARD BOWDLER (1847-1909). *A review of recent attempts to classify birds; an address delivered before the Second International Ornithological Congress on the 18th of May, 1891.* Budapest: Published at the office of the congress, 1891.

1 p.l., 90 p. incl. 9 diagr. 3 fold. diagr. 25.5 cm. At head of title: ...Hungarian committee of II. International Orni-

thological Congress. (Budapest, Hungarian National Museum). Black library binding.
BM IV p.1910.

564

SIBTHORP, JOHN (1758-1796). *Floræ graecæ prodromus: sive Plantarum omnium enumeratio, quas in provinciis aut insulis Graeciæ invenit Johannes Sibthorp, M.D... Characteres et synonyma omnium cum annotationibus elaboravit Jacobus Edvardus Smith...* Londini: typis Richardi Taylor et socii etc, 1806-13.
2 v. 23.5 cm. Original green cloth binding.
BM IV p.1920. Jackson p.312. Nissen BBI 1840. Plesch p.406/7. Pritzel 8660.

565

SIEBOLD, PHILIPP FRANZ VON (1796-1866). *Flora Japonica; sive, Plantae, quas in imperio japonico collegit, descripsit, ex parte in ipsis locis pingendas... Sectio prima continens plantas ornatui vel usui inservientes.* Digessit dr. J.G. Zuccarini. Lugduni Batavorum, apud auctorem, 1835-70.
2 v. in 1. illus. 27 cm. "Volumen secundum, ab auctoribus inchoatum relictum ad finem perduxit F.A. Guil. Miquel." Vol.2 has imprint: Lugduni Batavorum: in horto sieboldiano acclimatationis dictor, 1870. Half brown morocco/red cloth binding.
BM IV p.1923. Jackson p.382. Nissen BBI 1842. Plesch p.407. Pritzel 8674.

566

SOWERBY, JAMES (1757-1822). *Coloured figures of English fungi or mushrooms.* London: Printed by J. Davis, 1797-[1815?].
3 v. 440 (i.e.437) col. plates. 33.5 cm. Published in 32 parts, 1795-1815. Supplement bound with vol.1. Four notebooks with manuscript notes bound in front of v.1. Manuscript notes throughout text. Manuscript index to plates laid in. Half brown calf, marbled boards.
BM V p.1982. Jackson p.244. Nissen BBI 1874. Plesch p.413. Pritzel 8788.

567

SOWERBY, JAMES (1757-1822). *English botany; or, Coloured figures of British plants, with their essential characters, synonyms, and places of growth.* By Sir James Edward Smith, the figures by James Sowerby. 2d edition arranged according to the Linnaean method, with the descriptions shortened, and occasional remarks added. London: Taylor, [1832]-1846.
12 v. 2756 col. plates. 22 cm. Armorial bookplate of T.L. Bolitho. Full calf binding, marbled endpapers, marbled edges.
BM V p.1982.

*568

SOWERBY, JOHN EDWARD (1825-1870). *The ferns of Great Britain: illustrated by John E. Sowerby.* The descriptions, synonyms, &c., by Charles Johnson. London: J.E. Sowerby, 1855.
87 p. 49 col. plates. 24 cm. Issued with the author's *The fern allies.* London, 1856 (1 p.l., 52 p.) Original green cloth, gilt decoration.
BM II p.935. Jackson p.240. Nissen BBI 994. Plesch p.414 (1859 ed.) Pritzel 8793.

*569

STEPHENSON, JOHN (1790-1864). *Medical botany; or, Illustrations and descriptions of the medicinal plants of the London, Edinburgh, and Dublin pharmacopoeias; including a popular and scientific description of poisonous plants.* By John Stephenson and James Morss Churchill. London: J. Churchill, 1828-31.
4 v. 185 hand col. plates (1 fold.) 24 cm. Signed in ms. on title page of each vol: William Pritchard. Half green morocco, cloth.
BM V p.2016. Jackson p.201. Nissen BBI 1891. Plesch p.416. Pritzel 8946.

570

STEPHENSON, JOHN (1790-1864). *Medical botany; or, Illustrations and descriptions of the medicinal plants of the London, Edinburgh, and Dublin pharmacopoeias, comprising a popular and scientific account of poisonous vegetables indigenous to Great Britain.* By John Stephenson and James Morss Churchill. New edition. Edited by Gilbert T. Burnett. London: Printed for J. Churchill, 1834-36.
3 v. 185 col. plates (part fold.) 24 cm. V.1, tan library binding; vols.2-3, green library binding.
BM V p.2016. Jackson p.201. Nissen BBI 1891. Plesch p.416 (1st ed.) Pritzel 8946.

571

STONES, MARGARET. *The endemic flora of Tasmania.* Painted by Margaret Stones; botanical and ecological text by Winifred Curtis. London: Ariel Press, c1967-1978.
6 v. col. plates. 41 cm. Col. map on endpapers. Coral cloth binding.

572

STRABO, WALAHFRID (809-849). *Hortulus.* Translated by Raef Payne. Commentary by Wilfrid Blunt. Pittsburgh: Hunt Botanical Library, 1966.

[xii], 20 p., 20 l., [21]-91 p. illus. 27 cm. (The Hunt facsimile series, no.2). Edition limited to 1500 copies. Ivory paper over boards, marbled endpapers.

*573

SUMNER, GEORGE (1793-1855). *A compendium of physiological and systematic botany with plates.* Hartford: O.D. Cooke, 1820.

xii, 300 p. 8 plates. 19 cm. Bookplate of Gray Herbarium of Harvard University laid in. Full contemporary calf; new calf spine gilt decorated.

574

TAKATORI, JISUKE. *Saishiki shasei zu, Nihon no yakuyo shokubutsu.* Tokyo: Hirokawa Publishing Company, 1966.

7, 80 p. 80 mounted col. illus. 31 cm. Text also in English, with added title page: Color atlas, medicinal plants of Japan; 80 colored plants, 113 species. Red cloth binding.

575

THOMSON, ANTHONY TODD (1778-1849). *The London dispensatory, containing I. The elements of pharmacy. II. The botanical description, natural history, chymical analysis, and medicinal properties, of the substances of the materia medica. III. The pharmaceutical preparations and compositions of the pharmacopoeias of London, of Edinburgh, and of Dublin. The whole forming a practical synopsis of materia medica, pharmacy, and therapeutics: illustrated with many useful tables, and wood-cuts of the pharmaceutical apparatus.* 9th edition. London: Printed for Longman, Orme, Brown, Green, & Longmans, 1837.

1164 p. illus. Original brown cloth binding, spine missing.

576

THORNTON, ROBERT JOHN (1768?-1837). *Temple of flora; with plates faithfully reproduced from the original engravings and the work described by Geoffrey Grigson.* With bibliographical notes by Handasyde Buchanan [and botanical notes by William T. Stern]. London: Collins, 1951.

viii, 20 p., 2 l. 36 plates (incl. port., facsims., part col.) 44 cm. Includes a reproduction of the title page of the original *The temple of flora, or garden of nature* which was first published as the concluding part of the author's *New illustration of the sexual system of Carolus von Linnaeus,* completed in 1807. "All the plates, with the exception of plates VII and VIII and the alternative plates IX A and XXIV A, are reproduced from the original volume in the Library of Eton College." Half linen, boards with label on front cover, monogram on endpapers.

BM V p.2103 (orig. ed.) Nissen BBI 1955 (orig. ed.) Plesch p.433 (orig. ed.)

577

THUNBERG, KARL PETER (1743-1828). *Flora Japonica; sistens plantas insularum Japonicarum.* Reprint of the 1784 edition (Lipsiae, In Bibliopolio I.G. Mülleriano). New York: Oriole Editions, 1975.

lii, 418 p. 39 leaves of plates (5 fold.) illus. 24 cm. Orange cloth binding.

BM V p.2107 (orig. ed.) Jackson p.381 (orig. ed.) Nissen BBI 1959 (orig. ed.) Plesch p.434 (orig. ed.) Pritzel 9257 (orig. ed.)

578

TITFORD, WILLIAM JOWIT. *Sketches towards a Hortus Botanicus Americanus; or Coloured plates (with a catalogue and concise and familiar descriptions of many species) of new and valuable plants of the West Indies and North and South America. Also of several others, natives of Africa and the East Indies: Arranged after the Linneaean System with a concise and comprehensive glossary of terms, prefixed, and a general index.* London: Printed for the author by C. Stower, published by Sherwood, Neely, and Jones [etc.], 1811—[-12].

2 p.l., [199]p. col. front., 17 col. plates. 30 x 24 cm. Paging irregular. In ms. on title page: John Redman Coxe. Black library binding.

BM V p.2117. Jackson p.354. Nissen BBI 1968. Plesch p.436. Pritzel 9370.

579

TORREY, JOHN (1796-1873). *Plantæ frémontianæ; or, Descriptions of plants collected by Col. J.C. Frémont in California.* Washington, D.C.: Smithsonian Institution, 1853.

24 p. X plates. 33 cm. (Smithsonian contributions to knowledge vol. VI, art. 2). Incorrectly numbered vol. V, art. 1. (Smithsonian Institution publication, 46). Pamphlet binding.

BM V p.2125. Pritzel 9413.

580

TOURNEFORT, JOSEPH PITTON DE (1656-1708). *Elemens de botanique ou Methode pour connoître les*

plantes. Paris: De l'Imprimerie royale, 1694.
3 v. illus. plates. 21 cm. V.3 has title: *Institutiones rei herbariae*. Each volume has added title page engraved. Full brown morocco, gilt decoration.
BM V p.2128. Hunt 392. Jackson p.30. Nissen BBI 1698. Plesch p.438 (2nd ed.) Pritzel 9423.

581

The Treasury of botany: a popular dictionary of the vegetable kingdom; with which is incorporated a glossary of botanical terms. Edited by John Lindley and Thomas Moore. Assisted by numerous contributors. London: Longmans, Green, and co., 1866.
2 v. illus., 20 plates (incl. front.) 17 cm. Half calf, marbled boards, marbled endpapers.
BM III p.1121. Jackson p.12. Nissen BBI 1405. Pritzel 5373.

582

UNDERWOOD, LUCIEN MARCUS (1853-1907). *A century of illustrative fungi, with generic synopses of the basidiomycetes and myxomycetes*. Prepared by L.M. Underwood and O.F. Cook. Syracuse, N.Y.: n.p., 1889.
21 p. 24 cm. 100 mounted specimens. On p.[2] of cover: A.J. Wallon & Son, Bookbinders, Syracuse, N.Y. Brown cloth binding, in portfolio.

583

VACHEROT, MAURICE. *Charme et diversité des orchidées*. Paris: Éditions J.B. Baillière, [1957].
67 p. illus. (part col.) 31 cm. "Les illustrations en couleurs sont l'oeuvre de Guy Richard." Pictorial paper over boards.
Plesch p.445.

584

VALLENTIN, E.F., MRS. *Illustrations of the flowering plants and ferns of the Falkland Islands with descriptions by Mrs. E.M. Cotton*. London: L. Reeve & co., ltd., 1921.
xii p., 65 l. 64 col. plates (part fold.) 25.5 cm. Maroon cloth binding.
BM VIII p.1354. Nissen BBI 2038.

*585

VANCE, FENTON R. (1907-   ). *Wildflowers across the prairies*. By F.R. Vance, J.R. Jowsey, and J.S. McLean. Saskatoon: Western Producer Prairie Books, 1977.
5 p.l., 214 p. col. illus., map. 22 cm. Red cloth binding, dustjacket.

586

VENTENAT, ETIENNE PIERRE (1757-1808). *Description des plantes nouvelles et peu connues cultivées dans le jardin de J.M. Cels*. Avec figures par E.P. Ventenat. Paris: de l'Imprimerie de Crapelet, An VIII [1799?].
6 p.l., 100 leaves, [1]l. 100 plates. 35 cm. Half green morocco, marbled boards, gilt and red decorated spine.
BM V p.2203. Nissen BBI 2048. Plesch pp.448-49. Pritzel 9731.

587

VICKERS, ANNA (d. 1907). *Phycologia barbadensis. Iconographie des algues marines récoltées à l'Ile Barbade (Antilles) (chlorophycées et phéophycées) par Anna Vickers*. Avec texte explicatif, par Mary Helen Shaw. 93 planches coloriées dessinées par Mlles Trottet d'après les analyses de l'auteur. Paris: P. Klincksieck, 1908.
44 p. 93 col. plates. 34 cm. Green printed paper over boards.
BM V p.2211. Nissen BBI 2060. Plesch p.450.

588

VORONIN, MĪKHAĪL STEPANOVICH (1883-1903). *Über die sclerotienkrankheit der vaccinieen-beeren. Entwickelungsgeschichte der diese krankheit verursachenden sclerotinien*. Mit 10 tafeln. St. Petersbourg: Commissionnaires de l'Académie Impériale des sciences, 1888.
2 p.l., 49 p. 10 col. plates. 31 cm. (Mémoires de l'Académie Impériale des sciences de St. Petersbourg, 7e. serie, tome 36, no.6). Maroon cloth binding.
BM V p.2237.

589

WAKEFIELD, PRISCILLA (BELL), MRS. (1751-1832). *An introduction to botany, in a series of familiar letters, with illustrative engravings*. Boston: Published by J. Belcher and J.W. Burditt & co., 1811.
xii, 216 p. 11 (i.e. 12) plates, fold. tables. 18 cm. 1st American from the 5th London edition. Bookplate of Joseph Jones, M.D. Full contemporary calf, spine taped.
BM V p.2251. Jackson p.36. Plesch p.454. Pritzel 9925.

590

WALCOTT, MARY MORRIS (VAUX) (1860-1940). *North American wild flowers*. Washington, D.C.: The Smithsonian Institution, 1925.

5 v. col. plates. In portfolio. C.1, 32 cm. Paperbound in green cloth portfolios. C.2, 37 cm. Deluxe edition in quarter black morocco/beige cloth portfolios. C.2 is no.457.
Nissen BBI 2094.

* 591

WALCOTT, MARY MORRIS (VAUX) (1860-1940). *Wild flowers of America.* 400 flowers in full color based on paintings by Mary Vaux Walcott, with additional paintings by Dorothy Falcon Platt; edited with an introduction and detailed descriptions by H.W. Rickett. New York: Crown Publishers, 1953.
71 p. 400 col. plates. 31 cm. Red cloth binding.

* 592

WALTON, ELIJAH (1832-1880). *Flowers from the upper Alps, with glimpses of their homes.* The descriptive text by T.G. Bonney. London: W.M. Thompson, 1869.
viii, [12]p. 12 col. plates. 33 cm. Plates accompanied by guard sheets. Tan cloth binding, gilt decoration, gilt edges.
Nissen BBI 2104.

593

WASSON, VALENTINA PAVLOVNA (1901-). *Mushrooms, Russia, and history.* By Valentina Pavlovna Wasson and R. Gordon Wasson. New York: Pantheon Books, [c1957].
2 v. (XX, 432 p.) illus., 82 plates (incl. ports.; part col.) maps. 33 cm. In case. "Of this book there have been made 512 copies..." LSU owns copy no.336. "Mushroom stones of middle America, arranged by Stephen F. de Borhegyi" (fold. leaf) in pocket, v.2. Original green cloth bindings.

594

WAWRA, HEINRICH, RITTER VON FERNSEE (1831-1887). *Botanische ergebnisse der reise Seiner Majestät des kaisers von Mexico Maximilian I. nach Brasilien* (1859-60). Auf allerhöchst dessen anordnung beschrieben und hrsg. von d. Heinrich Wawra. Wien: C. Gerold's sohn, 1866.
2 p.l., xvi, 234 p., 2 l. 104 plates (part col.) 49 x 36.5 cm. Tan cloth binding with original paper cover inset.
BM V p.2275. Nissen BBI 2112. Pritzel 10020.

595

WEST, WILLIAM (1848-1914). *A monograph of the British Desmidiaceæ.* By W. West and G.S. West. London: The Ray Society, 1904-23.
5 v. plates (part col.) 22 cm. Vol. V by Nellie Carter. (The Ray Society [publications vols. 82, 84, 88, 92, 108]). Original black stamped cloth binding.
BM V p.2296 (Ray Society Pub.)

* 596

WIED-NEUWIED, MAXIMILIAN ALEXANDER PHILIPP, PRINZ VON (1782-1867). *Beitrag zur Flora Brasiliens.* Mit Beschreibungen von Nees v. Esenbeck u. von Martius. Mit 8 Kupfertafeln. Bonn, 1824.
54 p. 8 plates (5 fold.) 26 cm. (Nova acta physico-medica, Bd.12, Pt.1). Full brown morocco.
BM V p.2315. Pritzel 10223.

* 597

Wild flowers and their teachings. Bath, [Eng.]: Binns and Goodwin, 1845.
ix, 92 p., [36] leaves of plates. illus. 22 cm. The 36 plates are of pressed, dried, mounted specimens, each on a plate printed especially for that flower. LSU copy imperfect: flowers on plate [1] are partially wanting. Full red morocco, gilt decoration.

598

WILLMOTT, ELLEN ANN (1860?-1934). *The genus Rosa.* By Ellen Willmott. Drawings by Alfred Parsons. London: John Murray, 1914.
2 v. illus., plates (part col.) 39 cm. Issued in 25 parts, 1910-14. Paged continuously. Dark green cloth binding.
BM VIII p.1439. Nissen BBI 2166.

* 599

WITHERING, WILLIAM (1741-1799). *An arrangement of British plants; according to the latest improvements of the Linnaean system. To which is prefixed, an easy introduction to the study of botany.* Illustrated by copper-plates. 3d edition. Birmingham: Printed for the author, by M. Swinney; London: Sold by G.G. & J. Robinson and B.& J. White, 1796.
4 v. 31 plates (1 col.) 23 cm. Manuscript note on title page of v.1: "Ornatissimo viro et solertissimo Jacobo Beattie P.P. hocce qualecunque amoris et observantia signum dedit Joannes Brine, M.D. 1809." Vol.4, p.418 incorrectly numbered 420. Half red cloth, marbled boards.
BM V p.2343. Jackson p.234. Plesch p.462. Pritzel 10360.

600

WOLLE, FRANCIS (1817-1893). *Desmids of the United States and list of American pediastrums, with nearly fourteen hundred illustrations on sixty-four colored plates*. New and enlarged edition. Bethlehem, Pa.: Moravian publication office, 1892.

xiv, 182 p., 1 l. front. (port.) LXIV col. plates. 24.5 cm. Each plate accompanied by leaf with descriptive letterpress. Original red cloth binding.
BM V p.2350 (1884 ed.) Nissen BBI 2179 (1887 ed.)

601

WOLLE, FRANCIS (1817-1893). *Fresh-water algae of the United States (exclusive of the diatomaceae) complemental to Desmids of the United States; with 2300 illustrations covering one hundred and fifty-one plates, a few colored, including nine additional of desmids*. Bethlehem, Pa.: Comenius Press, 1887.

2 v. illus. plates. 24.5 cm. Tan library binding.
BM V p.2350

602

WOODVILLE, WILLIAM (1752-1805). *Medical botany: containing systematic and general descriptions, with plates of all the medicinal plants, comprehended in the catalogues of the materia medica, as published by the Royal Colleges of physicians of London, Edinburgh, and Dublin; together with the principal medicinal plants not included in those pharmacopoeias. Accompanied with a circumstantial detail of the medicinal effects, and of the diseases in which they have been most successfully employed. 3d edition in which thirty-nine new plants have been introduced. The botanical descriptions arranged and corrected by Dr. William Jackson Hooker, who has added an index following the arrangement of Jussieu. The new medico-botanical portion supplied by G. Spratt, under whose immediate inspection the whole of the plates have been coloured*. London: John Bohn, 1832.

5 v. hand col. plates. 26 cm. Stamped on title pages: State Library, Louisiana. Tan library binding.
BM V p.2355. Hunt 716 (1st ed.) Jackson p.201. Nissen BBI 2183. Plesch p.463 (1st. ed.) Pritzel 10398.

* 603

ZEITSCHRIFT FÜR PILZFREUNDE. *Populäre Mitteilungen über essbare und schädliche Pilze*. Herausgegeben unter Mitwirkung von Botanikern, Forstmännern, Pilzzüchtern etc. Redigiert von Osmar Thüme. [v.] 1-2, 1883-85. Dresden: A. Köhler, 1883-85.

2 v. in 1. col. plates. 24 cm. Label mounted on lining paper: Dr. Martin Sändig, OHG; Wissenschaftl, Antiquariat; Wiesbaden, Gustav-Freytag-Str. 5. Quarter dark red cloth, marbled boards.
BM V p.2385.

* 604

ZWINGER, THEODOR (1658-1724). *Theatrum botanicum: das ist, Neu voll-kommenes Kräuter-buch; worinnen allerhand Erdgewächse der Bäumen, Stauden und Kräutern; welche in allen vier Theilen der Welt; sonderlich aber in Europa herfür kommen; neben ihren sonderbahren Eigenschafften, Tugenden und fürtrefflichen Würckungen; auch vielen herzlichen Artzney-mittlen und deren Gebrauch; wider allerley Kranckheiten an Menschen und vieh; mit sonderbahrem Fleiss auff eine gantz neue Art und Weise; dergleichen bissher in keinem Kräuter-buch gesehen noch gefunden worden; beschrieben auch mit schönen Theils neuen figuren gezieret und neben denen ordenlichen so wohl Kräuter-als Kranckheit-registern; mit nutzlichen Marginalien vorgestellet sind; Allen Aertzten Wund-ärtzten Apotheckern Gärtnern Hauss-vättern und Hauss-müttern sonderlich auch denen auff dem Land wohnenden Krancken und presthafften Persohnen höchst nutzlich und ergetzlich. Erstens zwar an das Tagliecht gegeben von Herren Bernhard Verzascha, anjetzo aber in eine gantz neue Ordnung gebracht, auch mehr als umb die Helffte vermehret und verbessert durch Theodorum Zvingerum*. Basel: Jacob Bertsche, 1696.

[8], 995, [19]p. 34 cm. Title page in red and black. LSU copy imperfect: various pages mutilated. Manuscript notes on pp.996, 998-1002, 1113-1114. Full black morocco.
BM V p.2403. Nissen BBI 1311. Plesch p.467 (1744 ed.) Pritzel 10532.

PART VI

Zoology (excluding Ornithology)

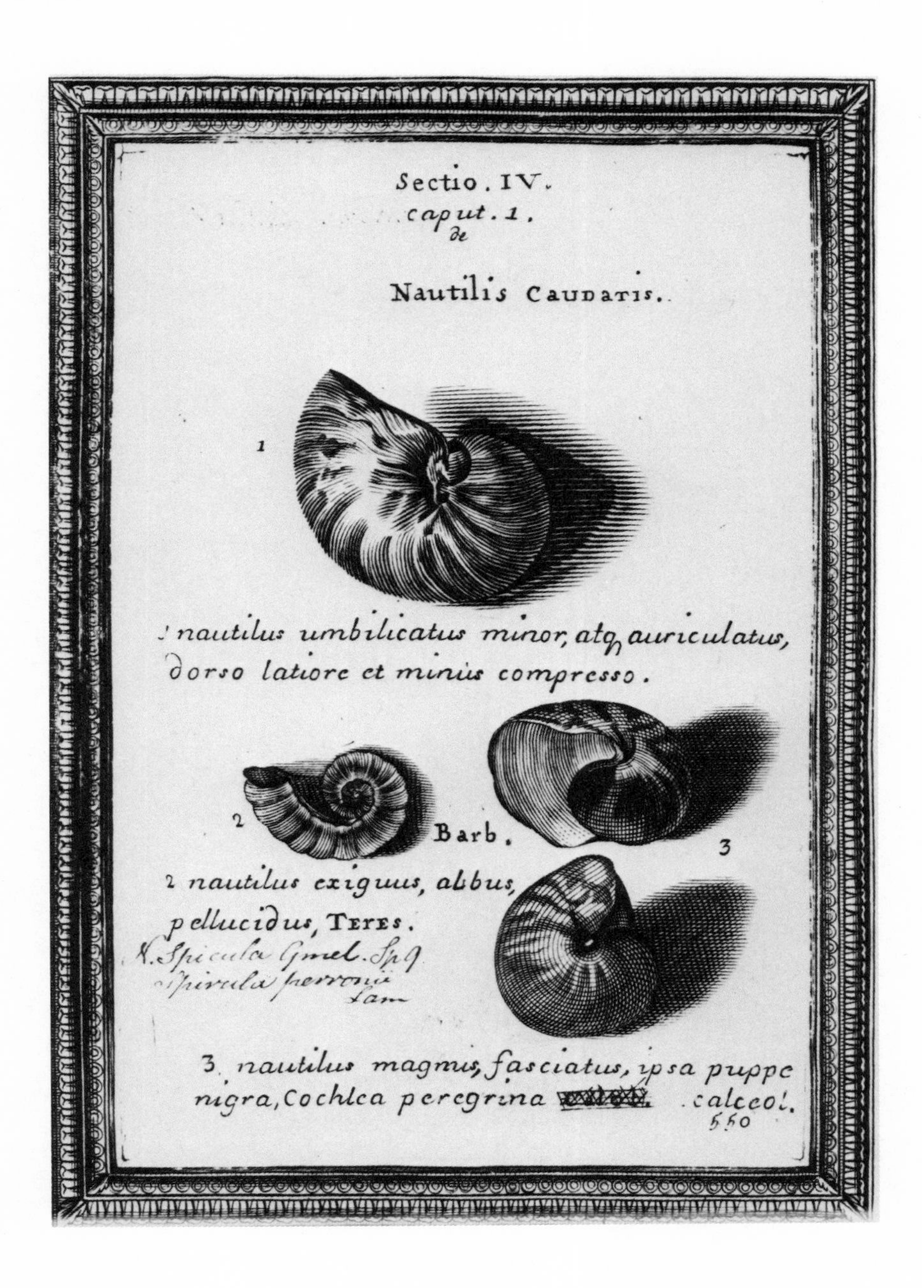

The Nautilus shell engraving from Martin Lister's *Historiae sive Synopsis Methodicae Conchyliorum et Tabularum Anatomicarum*. Oxford, 1770.

605

ADAMS, CHARLES BAKER (1814-1853). *Catalogue of shells collected at Panama, with notes on their synonymy, station, and geographical distribution. Read before the Lyceum of natural history, May 10th, 1852.* New York: R. Craighead, printer, 1852.

viii, 334 p., 1 l. 28 x 22 cm. (From Annals of Lyceum of natural history of N.Y., vol. V). Original brown cloth binding.
BM I p.10.

606

AGASSIZ, LOUIS (1807-1873). *Contributions to the natural history of the United States of America.* Boston: Little, Brown and company; London: Trübner & co., 1857-62.

4 v. illus., 77 plates (4 col., 2 fold.) 33 cm. The complete work was to include 10 vols., but only 4 were published. First monograph (v.1-2) in 3 parts; second monograph (v.3-4) in 5 parts. Black library binding.
BM I p.18. Nissen ZBI 37.

607

AGASSIZ, LOUIS (1807-1873). *Principles of Zoölogy; touching the structure, development, distribution, and natural arrangement of the races of animals, living and extinct, with numerous illustrations...* By Louis Agassiz, and A.A. Gould. Revised edition. Boston: Gould and Lincoln, 1851.

250 p. illus., map. 20 cm. In ms. on verso of title page: T.L. Tanez. Original green cloth binding.
BM I p.19 (1848 ed.)
Garland-Bullard Memorial Collection. Gift of Rietta Garland Albritton and Florence Garland Brookes.

608

ALDER, JOSHUA (1792-1867). *The British Tunicata; an unfinished monograph, by the late Joshua Alder and the late Albany Hancock.* Edited by John Hopkinson with a history of the work by the Rev. A.M. Norman. London: Printed for the Ray Society, 1905-12.

3 v. fronts. (ports.), illus., plates (part col.) 22.5 cm. (The Ray Society [Publications: 83, 86, 93]). Original blue cloth with "R.S." in gilt on front cover.
BM IV p.1655. Nissen ZBI 62.

609

ALDER, JOSHUA (1792-1867). *A monograph of the British nudibranchiate Mollusca.* By Joshua Alder and Albany Hancock. With a supplement by Sir Charles Eliot. London: Printed for the Ray Society, 1845-1910.

8 v. plates (part col.) 35 cm. (The Ray Society... [Publications, v.2, 5, 9, 15, 19, 23, 27, 90]). Blue library binding.
BM I p.26. Nissen ZBI 63.

* 610

ALDROVANDI, ULISSE (1522-1605?). Ulyssis Aldrovandi... *De piscibus libri V. et de cetis lib. unus; Ioannes Cornelius Uterverius, Marc Antonius Bernia in lucem restituit; ad illustrissimum et reverendissimum D.D. Francisc.* Vitellium archiepisc: thessalonic: apud sereniss. venetor rempubl. sedis apostolicae nuncium; cum indice copiosissimo superiorum permissu. Bononiae: Apud N. Thebaldinum, 1638.

732 p. illus. 38 cm. LSU copy imperfect: p.15 incorrectly numbered 13, p.20 incorrectly numbered 16, p.247 incorrectly numbered 243, p.386 incorrectly numbered 368, pp.558-559 incorrectly numbered 552-553. Vellum binding.
BM I p.26. Nissen ZBI 70.

* 611

ALDROVANDI, ULISSE (1522-1605?). *Vlyssis Aldrovandi De qvadrvpedibvs solidipedibvs volvmen integrvm.* Ioannes Cornelivs Vterverivs collegit, & recensuit. Bononiae: Apud N. Thebaldinum, 1639 [i.e. Typis I.B. Ferronij, sumptibus M.A. Berniae, 1648].

495 p. illus. 37 cm. Stamped on title page: Royal Medical Society, Edinburgh. LSU copy imperfect: pp.383-384 incorrectly numbered 347-348. Full calf binding, spine restored.
BM I p.26 (1616 ed.) Nissen ZBI 72 (1649 ed.)

612

ALLMAN, GEORGE JAMES (1812-1898). *A monograph of the fresh-water Polyzoa, including all the known species, both British and foreign.* London: The Ray Society, 1856.

viii, 119 p., 1 l. illus., 11 plates (10 col.) 36 cm. (The Ray Society...[Publications, v.28, 1856]). Each plate preceded by leaf with descriptive letterpress not included in paging. Blue cloth binding.
BM I p.32. Nissen ZBI 92.

* 613

AMARAL, AFRÂNIO DO (1894-   ). *Serpentes do Brasil: iconografia colorida; Brazilian snakes: a color iconography.* Ilustrado com 582 gravuras, sendo 164 tábuas em várias cores e 418 desenhos em duplo

tom. [São Paulo]: Edições Melhoramentos, 1977.
246 p. illus. (some col.) 21 x 28 cm. Portuguese and English. Grey paper boards, dustjacket.

614

American Journal of Conchology. Edited by George W. Tryon, jr.... v.1-7; Feb. 1865-May 1872. Philadelphia: G.W. Tryon, jr., [etc., etc.], 1865-72.
7 v. illus., plates (part col.) ports., tab. 22.5 cm. Vols.3-7 published by the Conchological section of the Academy of Natural Sciences of Philadelphia. From the library of L.S. Frierson, conchologist. Various bindings.
BM I p.38.

615

ANDERSON, JOHN (1833-1900). *Zoology of Egypt*. London: B. Quaritch, 1898- . [Codicote, England: Wheldon & Wesley, 1965- .]
v. illus., plates, map. 35 cm. Vol.1 has stamped on title page: Leiden's Rijks-Museum van Naturrlijke Historie. Vol.3 reprint of 1907 edition, published for the Egyptian government by H. Rees. Size of volumes vary. LSU owns v.1, v.3, text and atlas. Blue cloth binding.
Nissen ZBI 110.

*616

ANDERSON, RUDOLPH MARTIN (1876-1961). *Methods of collecting and preserving vertebrate animals*. 4th edition, revised. [Ottawa: R. Duhamel, Queen's Printer], 1965[reprinted 1975].
vii, 199 p. illus. 25 cm. (National Museum of Canada. Bulletin, no.69. Biological series, no.18). Paperbound.

*617

ANNANDALE, NELSON (1876-1924). *Fasciculi Malayenses; anthropological and zoological results of an expedition to Perak and the Siamese Malay States, 1901-1902, undertaken by Nelson Annandale and Herbert C. Robinson*. London, New York: Published for the University Press of Liverpool by Longmans, Green, 1903- .
4 v. plates (part col.) 27 cm. LSU lacks v.4. V.1, pt.3, has imprint: London, Published for the University Press of Liverpool by Williams & Norgate. Paperbound.
——— *Supplement*. xliii, [1]: 4 plates, 1 map. LSU lacks supplement.
BM VI p.28. Nissen ZBI 131.

618

ANTELME, ADRIEN. *Histoire naturelle des insectes et des mollusques*. Sous la direction de Geoffroy-St.Hilaire. Paris: Librairie Francaise et Etrangere, 1841.
2 v. illus. 18 cm. Quarter morocco/black cloth, marbled endpapers.
BM I p.52. Nissen ZBI 135.

619

ARTEDI, PETER (1705-1735). *Ichthyologia...with an introduction by A.C. Wheeler*. Reprint of *Ichthyologia sive opera omnia de piscibus scilicet...* Lugduni Batavorum: Apud Conradum Wishoff, 1738. Weinheim: J. Cramer; Codicote, Herts [Eng.]: Wheldon & Wesley; New York: Hafner, 1962.
xxiii p., facsim.: 5 pts. in 1 v. (554 p.) 22 cm. (Historiae naturalis classica, t.15). *The life and work of Peter Artedi* by Alwyne C. Wheeler: p.[vii]-xx. Blue cloth binding.
BM I p.64 (orig. ed.)

*620

AUDEBERT, JEAN BAPTISTE (1759-1800). *Histoire naturelle des singes, peints d'après nature*. Paris: Chez l'auteur, H.J. Jansen, an 6ème de la République Françoise, 1797.
iii, [174]p. 63 plates (61 col.) 53 cm. Originally issued in 10 parts. Added title page bound before section on Makis reads: *Histoire naturelle des singes et des makis*. Paris: Chez Desray, an 8 [1800]. Half brown calf/marbled boards.
BM I p.71, VI p.6. Nissen ZBI 156.

621

AUDUBON, JOHN JAMES (1785-1851). *The imperial collection of Audubon animals; the quadrupeds of North America*. Original text by John James Audubon, and John Bachman. Edited and with new text by Victor H. Cahalane. Foreword by Fairfield Osborn. Illustrated by John James Audubon and John Woodhouse Audubon. Maplewood, N.J.: Hammond, inc., [c1967].
xvi, 307 p. col. illus., port. 33 cm. First published under title: *The viviparous quadrupeds of North America*. Ivory imitation leather.

622

AUDUBON, JOHN JAMES (1785-1851). *The quadrupeds of North America*. By John James Audubon and the Rev. John Bachman. New York: George R. Lockwood, [c1849].
3 v. 155 col. plates. 28 cm. Drawn on stone by R. Trembly. Printed and colored by J.T. Bowen, Philadelphia. First published under title: *The vivaparous quadrupeds of North America*. On cover of v.1-3, c.2: Franklin Fairbanks from the

Teachers & Scholars of No. Church S. School. Clipping relating to Audubon's *Ornithological biography* mounted on p.[2] of cover of v.1, c.1. Clipping with bibliographical information mounted on fly leaf in v.1, c.1. Stamp of the Sportsmen's Association of Western Penna. on fly leaf, half title page, and p.100 of v.1, c.1; title page and p.100 of v.2, c.1; title page and p.100 of v.3, c.1. Full morocco, restored tooled spines, marbled endpapers.
Nissen ZBI 163.

623

AUDUBON, JOHN JAMES (1785-1851). *The quadrupeds of North America.* By John James Audubon and the Rev. John Bachman. New York: V.G. Audubon, 1849-54.

3 v. CLV col. plates. 27 cm. Plates LXI, LXIV, LXVII, LXXI-LXXIII, LXXVI-LXXVIII, LXXXI-LXXXII, LXXXVI, XC-XCVIII, C-CLV drawn from nature by J.W. Audubon. Drawn on stone by R. Trembly. Printed by Nagel and Weingartner. The first edition was published under title *The vivaparous quadrupeds of North America,* and consisted of 3 folio volumes, containing CL colored plates, 1845-48, with text in 3 quarto volumes, 1846-53. Full green morocco, gilt ornamentation, marbled endpapers.
BM I p.71. Nissen ZBI 163.

624

AUDUBON, JOHN JAMES (1785-1851). *The quadrupeds of North America.* By John James Audubon and the Rev. John Bachman. New York: V.G. Audubon, 1851-54.

3 v. 155 col. plates. 28 cm. Plates LXI, LXIV, LXVII, LXXI-LXXIII, LXXVI-LXXVIII, LXXXI-LXXXII, LXXXVI, XC-XCVIII, C-CLV drawn from nature by J.W. Audubon. Drawn on stone by R. Trembly. Printed by Nagel and Weingartner. Colored by N. Lawrence. The first edition was published under title *The vivaparous quadrupeds of North America,* consisted of 3 folio volumes, containing CL colored plates, 1845-48, with text in 3 quarto volumes, 1846-53. Bookplate of Mr. Bartlett. Half green morocco/marbled boards, marbled endpapers.
Nissen ZBI 163.

625

BAIRD, WILLIAM (1803-1872). *The natural history of the British Entomostraca.* London: The Ray Society, 1850.

viii, 364 p. XXXVI plates (part col.) 22 cm. (The Ray Society. [Publications, v.17, 1849]). In ms. on title page: J.F. Ogilvie. Original blue cloth binding with "R.S." in gilt on front cover.
BM I p.88. Nissen ZBI 198.

*626

BANFIELD, ALEXANDER WILLIAM FRANCIS (1918-). *The mammals of Canada.* Illustrated by Allan Brooks, Claude Johnson, John Crosby, Charles Douglas, Robert Thompson, and John Tottenham; cartography by Geoffrey Matthews and Jennifer Wilcox. Toronto; Buffalo: Published for the National Museum of Natural Sciences, National Museums of Canada by University of Toronto Press, [1974].

xxv, 438 p., [13] leaves of plates. illus. (some col.) 28 cm. Grey cloth binding, dustjacket.

627

BARBUT, JAMES. *The genera vermium exemplified by various specimens of the animals contained in the orders of the Intestina et Mollusca Linnæi.* Accurately drawn from nature by James Barbut with explanations in English and French. London: Printed for the author by J. Dixwell, and sold by J. Sewell [etc.], 1783-[88].

2 v. 25 plates (24 col.) 27.5 cm. English and French. LSU owns v.2. Marbled boards, taped spine.
BM I p.97. Nissen ZBI 221.

628

BARBUT, JAMES. *Les genres des insectes de Linné; constatés par divers échantillons d'insectes d'Angleterre, copiés d'après nature.* London: Imprimé par J. Dixwell, 1781.

2 p.l., xvii p., 1 l., [20]p. 22 plates (20 hand col., 3 fold.) 27 cm. In ms. on p.[2] of cover: Elizabeth Browning, the gift of M. Meen, Jan. 1830. In ms. on frontispiece: Elizabeth Browning, Jan. 1830. Library of Joseph Jones, M.D. Half red morocco/marbled boards.
BM I p.97. Nissen ZBI 220.

629

BATE, CHARLES SPENCE (1818-1889). *A history of the British sessile-eyed Crustacea.* By C. Spence Bate and J.O. Westwood. London: J. Van Voorst, 1863-68.

2 v. illus. 23 cm. In ms. on fly leaf: M. Ian Crichton, May 4th 1933. Label on p.[2] of cover of both vols.: W.C. Crawford. Original red cloth binding.
BM I p.109. Nissen ZBI 252.

630

BENNETT, EDWARD TURNER (1797-1836). *The gardens and menagerie of the Zoological Society delineated.*

Published, with the sanction of the council, under the superintendence of the secretary and vice-secretary of the Society... Chiswick: Printed by C. Whittingham; London: T. Tegg [etc.], 1830-31.

2 v. illus. 22 cm. Preface signed: E.T. Bennett... Imprint of v.2: Chiswick, Printed by C. Whittingham; London, J. Sharpe. On cover: Illustrated by Harvey. Half calf/cloth, gilt edges.
BM I p.135. Nissen ZBI 314.

631

BENNETT, EDWARD TURNER (1797-1836). *The Tower menagerie: comprising the natural history of the animals contained in that establishment; with anecdotes of their characters and history.* Illustrated by portraits of each, taken from life, by William Harvey; and engraved on wood by Branston and Wright. London: Printed for R. Jennings [etc., etc.,], 1829.

xviii, 241, [1]p. illus. 22 cm. Half calf/cloth, gilt edges.
Nissen ZBI 313. Yale p.26.

*632

BERGE, FRIEDRICH. *Berge's kleines Schmetterlingsbuch für Knaben und Anfänger.* In der Bearbeitung von H. Rebel. Mit 344 Abbildungen auf 24 Farbentafeln und 97 Abbildungen im Text. Stuttgart: E. Schweizerbart, 1911.

viii, 208 p. illus., 24 col. plates. 23 cm. Grey cloth binding.
BM VI p.80. Nissen ZBI 322.

*633

BERLESE, ANTONIO (1836-1927). *Gli insetti: loro organizzazione, sviluppo, abitudini e rapporti coll'uomo...* Milano: Societa Editrice Libraria, 1909-1925.

2 v. illus. (some col.) 29 cm. Label mounted on p.[2] of cover of each vol.: Societa Editrice Libraria, prezzo netto L. 100. In ms. on fly leaves of each vol.: Warren Whitcomb, Jr. Red cloth binding.
BM VI p.83. Nissen ZBI 333.

634

BEWICK, THOMAS (1753-1828). *A general history of quadrupeds. The figures engraved on wood by T. Bewick.* 4th edition. Newcastle upon Tyne: Printed by and for S. Hodgson, R. Beilby, and T. Bewick, 1800.

x, 525 p. illus. 22 cm. Letterpress by R. Beilby. Green library binding.
Nissen ZBI 351.

*635

BIGLAND, JOHN (1750-1832). *A natural history of animals.* Illustrated by twelve coloured plates, engraved from original drawings. Philadelphia: Published by John Grigg, 1828.

189 p. col. illus. 19 cm. Quarter green morocco/marbled boards.

636

BIGLAND, JOHN (1750-1832). *A natural history of birds, fishes, reptiles, and insects.* Illustrated by twelve coloured plates, engraved mostly from original drawings. Philadelphia: John Grigg, 1828.

1 p.l. x p., 1 l., 179 p. 11 col. plates (incl. front.) 19 cm. Added title page illustrated in colors. Paper over boards, mended spine.
Yale p.30 (1834 ed.)

*637

BINGLEY, WILLIAM (1774-1823). *Animal biography; or, Popular Zoology. Illustrated by authentic anecdotes of the economy, habits of life, instincts, and sagacity of the animal creation.* 5th edition. London: Printed for F.C. and J. Rivington; Jeffery and Son; Longman, Hurst, Rees, Orme and Co., etc., 1820.

4 v. 19 plates. illus. 19 cm. Full calf/marbled endpapers.
BM I p.164 (other eds.)

638

BLACKWALL, JOHN (1790-1881). *A history of the spiders of Great Britain and Ireland.* London: R. Hardwicke, 1861-64.

2 v. col. plates. 36 cm. (Ray Society [Publications: v.31, 1859; v.34, 1862]). Blue cloth binding.
BM I p.169. Nissen ZBI 380.

639

BLAINVILLE, HENRI MARIE DUCROTAY DE (1777-1850). *Manuel de malacologie et de conchyliologie.* Paris, Strasbourg: F.G. Levrault, 1825-27.

viii, 664 p. fold. tables and atlas (2 p.l., [649]-664 p., 109 plates). 22 cm. In ms. on fly leaf: E.C. See. Half green morocco/marbled boards.
BM I p.169. Nissen ZBI 391.

640

BOISDUVAL, JEAN ALPHONSE (1801-1879). *Histoire naturelle des insectes.* Spécies général des lépidoptères par MM. Boisduval et Guénée. Paris: Roret, 1836-58.

7 v. and atlas (24, 58 plates (part col.)) 21 cm. The work was begun by Boisduval, who after the publication of one volume associated with himself Achilles Guénée, by whom v.5-10 were written. Vols.2-4 never published. LSU owns v.5-10. Half black morocco/cloth, spines taped.
BM I p.188. Nissen ZBI 450.

641

The Book of animals, class mammalia (animals that suckle their young). Published under the direction of the Committee of general literature and education appointed by the Society for Promoting Christian Knowledge. 3d edition. London: John W. Parker, 1835.

160 p. front. illus. 14 cm. Advertisements in back. Original blue cloth binding.

642

The Book of reptiles: class Reptilia; with some account of the fossil remains of animals whose species have become extinct. Published under the direction of the Committee General Literature and Education, appointed by the Society for Promoting Christian Knowledge. London: J. W. Parker, 1835.

vii, [1], 132 p. illus. 14 cm. Red library binding.

*643

BORELLI, GIOVANNI ALFONSO (1608-1679). *De motu animalium.* Editio altera, correctior & emendatior. Lugduni in Batavis: Apud J. de Vivie, C. Boutesteyn, D. Gaesbeeck, & P. vander Aa, 1685.

2 pts. in 1 v. 21 cm. Added title page, engraved. Vol.1, p.90 incorrectly numbered 82, p.140 incorrectly numbered 104, p.177 incorrectly numbered 175, p.198 incorrectly numbered 195, p.273 incorrectly numbered 279, p.274 incorrectly numbered 280; v.2, p.134 incorrectly numbered 234, pp.302-303 incorrectly numbered 320-321, p.314 incorrectly numbered 300. Bookplate of James Murphy, M.R.C.V.S., Glasgow. "Carolus Jo. a Jesu, clericorum regularium pauperum matris dei scholarum piarum praepositus generalis, benevolo lectori salutem" (v.2): dated Romae 1681. Full calf, restored.
BM I p.201. Nissen ZBI 465.

644

BOULENGER, GEORGE ALBERT (1858-1937). *The tailless batrachians of Europe.* London: Printed for the Ray Society, 1897-98.

2 v. illus., XXIV plates (part col.) fold. maps. 22.5 cm. (The Ray Society [Publications: 74-75]). Paged continuously. Original blue cloth binding with "R.S." on front cover.
BM I p.213. Nissen ZBI 496.

645

BOWERBANK, JAMES SCOTT (1797-1877). *A monograph of the British Spongiadae.* London: Pub. for the Ray Society by R. Hardwicke, 1864-82.

4 v. 146 plates. 23 cm. (Ray Society [Publications: v.36, 1864; v.38, 1865; v.51, 1874; v.57, 1879]). Vol.4 (supplementary), edited, with additions, by the Rev. A.M. Norman... Original blue cloth with "R.S." in gilt on front cover, gilt edges.
BM I p.218. Nissen ZBI 522.

*646

BRISSON, MATHURIN JACQUES (1723-1806). *Regnum animale in classes IX distributum, sive Synopsis methodica sistens generalem animalium distributionem in classes IX, & duarum primarum classium, quadrupedum scilicet & cetaceorum, particularem divisionem in ordines, sectiones, genera & species. Cum brevi cujusque speciei descriptione, citationibus auctorum de iis tractantium, nominibus eis ab ipsis & nationibus impositis, nominibusque vulgaribus.* Parisiis: apud C.J.B. Bauche, 1756.

vi, 382 p. fold. plate. 27 cm. Added title page in French. Text in Latin and French in parallel columns. Full calf, marbled endpapers.
BM I p.237.

*647

BRITISH MUSEUM (NATURAL HISTORY). DEPT OF ENTOMOLOGY. *An illustrated catalogue of the Rothschild collection of fleas (Siphonaptera) in the British Museum (Natural History).* With keys and short descriptions for the identification of families, genera, species, and subspecies. By G.H.E. Hopkins & Miriam Rothschild. London: British Museum, 1953-

v. illus., group port., fold. map. 29 cm. Vols.4-5 are nos.652 and 706 of the British Museum (Natural History) publications. LSU owns v.1-5. Green cloth binding.

648

BRITISH MUSEUM (NATURAL HISTORY). DEPT OF ZOOLOGY. *Catalogue of the collection of Mazatlan shells in the British Museum.* Collected by Frederick Reigen, described by Philip P. Carpenter. London: Printed by order of the Trustees, [by P.P. Carpenter], 1857.

xii, 552 p. 18.5 cm. Paper boards.
BM I p.249.

*649

BRITISH MUSEUM (NATURAL HISTORY). DEPT OF ZOOLOGY. *Catalogue of the specimens of Heteropterous-Hemiptera in the collection of the British Museum.* By Francis Walker. London: Printed for the Trustees of the British Museum, 1867-73.

8 v. in 3. 23 cm. Parts 1-3 paged continuously. Edited by J. E. Gray. Embossed on title page and last page of each vol.: Free Public Library, Wigan. Brown cloth binding.
BM I p.248.

650

BROWN, GEORGE THOMAS (1827-1906). *Life on the farm. Animal life.* London: Bradbury, 1886.

141 p. 19 cm. (Handbook of the farm series, edited by J. Chalmers Morton). Original brown cloth binding.

651

BROWN, THOMAS (1785-1862). *Conchologist's text-book. Embracing the arrangements of Lamarck and Linnaeus, with a glossary of technical terms. To which is added a brief account of the Mollusca.* 9th edition corrected and enlarged by William Macgillivray... London [etc.]: A. Fullarton and co., [185-?].

xii, 232 p. XXI col. plates. 18 cm. Original red cloth binding.
BM I p.261 (various eds.) Nissen ZBI 612.

652

BROWN, THOMAS (1785-1862). *The elements of conchology: or, Natural history of shells: according to the Linnean system with observations on modern arrangements.* London: Lackington, Allen & co., [etc., etc.], 1816.

4 p.l., 168 p., 1 l. 9 hand col. plates (incl. front.) 22 cm. Tan library binding.
BM I p.261. Nissen ZBI 606.

653

BROWN, THOMAS (1785-1862). *The taxidermist's manual; or, The art of collecting, preparing, and preserving objects of natural history.* 20th edition. London, Edinburgh, and Dublin: A. Fullarton and co., [1851].

xii, 150 p. illus., VI plates. 18 cm. Original red cloth binding.
Yale p.44.

*654

BRUES, CHARLES THOMAS (1879-). *Classification of insects; keys to the living and extinct families of insects, and to the living families of other terrestrial arthropods.* By Charles T. Brues, A.L. Melander, and Frank M. Carpenter. [2d, revised edition.] Cambridge: The Museum, c1954.

v., 917 p. illus. 23 cm. (Bulletin of the Museum of Comparative Zoology at Harvard College, v.108). Red cloth binding.

655

BRUGUIÈRE, JEAN GUILLAUME (1750-1799). *Tableau encyclopédique et méthodique des trois règnes de la nature...* Paris: Chez Panckoucke, 1791-1816.

v. plates (part fold.) 32 x 24.5 cm. LSU owns only v.7. Black library binding.
BM II p.527. Nissen ZBI 4621.

656

BUCKLER, WILLIAM (1814-1884). *The larvae of the British butterflies and moths.* London: Printed for the Ray Society, 1886-1901.

9 v. col. plates. 22.5 cm. (The Ray Society [Publications: v.63, 64, 66, 68, 69, 71, 72, 73, 77]). Vols.1-5 edited by H. Stainton; vol.6 edited by G.T. Porritt. Original black cloth binding with "R.S." in gilt on front cover.
BM IV p.1655. Nissen ZBI 657.

657

BUCKTON, GEORGE BOWDLER (1817-1905). *Monograph of the British aphides...* London: Printed for the Ray Society, 1876-83.

4 v. CXXXIV (i.e.150) plates (141 col.) 22.5 cm. (The Ray Society [Publications: v.52, 1875; v.54, 1877; v.58, 1880; v.60, 1883]). Original blue cloth binding with "R.S." on front cover.
BM I p.277. Nissen ZBI 659.

658

BUCKTON, GEORGE BOWDLER (1817-1905). *A monograph of the Membracidae.* To which is added a paper entitled *Suggestions as to the meaning of the shapes and colours of the Membracidae in the struggle for existence.* London: Lovell Reeve, 1903.

296 p. LX plates (col.) 30 cm. Each plate accompanied by descriptive letterpress. Dark red cloth binding.
BM VI p.154. Nissen ZBI 661.

659

BURROW, EDWARD JOHN (1785-1861). *Elements of conchology, according to the Linnaean system.* New edition. London: J. Duncan, 1844.

xix, [1], 245 p. XXVIII plates. 22 cm. In ms. on fly leaf: Jos. Jones Princeton, ? Hall, Feb. 12th, 1851. From the library of Joseph Jones, M.D. Original olive green cloth binding.
BM I p.291 (earlier eds.) Nissen ZBI 767.

660

BUTLER, EDWARD ALBERT (1845-). *A biology of the British Hemiptera-Heteroptera.* London: H.F. & G. Witherby, 1923.

viii, 682 p., 1 l. illus., VII plates (3 col.) 26 cm. Brown cloth binding.

661

Butterflies and moths. Text by Alfred Werner and Josef Bijok. Norman Riley, editor. Rev. [and enlarged] edition. New York: Viking Press, [c1965].

126 p. illus., map (on lining paper), col. plates. 31 cm. Yellow cloth binding.

662

CABRERA, ANGEL (1879-). *Catálogo de los mamíferos de América del Sur.* Buenos Aires, Impr. y Casa Editora "Coni," 1957-1961.

2 v. (xxii, 732 p.) port. 27 cm. (Revista del Museo Argentino de Ciencias Naturales "Bernardino Rivadavia" e Instituto Nacional de Investigación de las Ciencias Naturales. Ciencias zoológicas, t.4, no.1-2). Printed paper wrappers, uncut.

*663

CAIUS, JOHN (1510-1573). *Joannis Caii Britanni De canibus Britannicis, liber unus; De rariorum animalium & stirpium historia, liber unus; De libris propriis, liber unus; De pronunciatione Graecae & Latinae linguae, cum scriptione nova, libellus; ad optimorum exemplarium fidem recogniti à S. Jebb.* Londini: Impeniis C. Davis, 1729.

xv, 249 p. 18 cm. Stamped on title page and p.[1]: Ex bibliotheca J. Richard D.M. Armorial bookplate of W.G. Peene. J.H. Lupton in ms. on fly leaf. Manuscript notes throughout the text. Page 53 incorrectly numbered 43. Half red morocco/cloth, marbled endpapers.
BM I p.297.

664

CAMERON, PETER (d. 1912). *A monograph of the British phytophagous Hymenoptera (Tenthredo, Sirex, and Cynips, Linné).* London: Printed for the Ray Society, 1882-93.

4 v. 84 plates (part col.) 22.5 cm. (Ray Society [Publications: v.59, 1881; v.62, 1885; v.67, 1890; v.70, 1893]). Original blue cloth binding with "R.S." in gilt on front cover.
BM IV p.1655. Nissen ZBI 794.

*665

Cassell's Natural History. Edited by P. Martin Duncan. London, New York: Cassell [pref. 1883]-89.

6 v. illus. (part col.) 28 cm. Half green calf/marbled boards, marbled endpapers.
BM I p.491.

666

Cassell's Popular Natural History. London: Cassell, Petter & Galpin, [1865-66].

4 v. in 2. col. fronts., illus., col. plates. 27 cm. LSU owns only v.1-2. Half red morocco/marbled boards.

667

CATLOW, AGNES (1807-1889). *The conchologist's nomenclator. A catalogue of all the recent species of shells under the subkingdom Mollusca, with their authorities, synonymes, and references to works where figured or described.* By Agnes Catlow, assisted by Lovell Reeve. London: Reeve brothers, 1845.

viii, 326 p. front. (tables). 23 cm. Brown library binding.
BM I p.327.

668

CHENU, JEAN CHARLES (1808-1879). *Illustrations conchyliologiques; ou, Description et figures de toutes les coquilles connues vivantes et fossiles, classées suivant le système de Lamarck, et comprenant les genres nouveaux et les espèces récemment découvertes.* Par M. Chenu, avec la collaboration des principaux conchyliologistes de la France et de l'étranger. Paris: Fortin, Masson et Cie, [1843-53].

4 v. in 85. illus., plates (part col.) 51 cm. Some parts have imprint: Paris, A. Franck. LSU owns pts.1-55. Unbound in folders.
BM I p.341. Nissen ZBI 877.

669

CHENU, JEAN CHARLES (1808-1879). *Leçons élémentaires sur l'histoire naturelle des animaux conchyliolo-*

gie; précédées d'un aperçu général sur la zoologie. Paris: J.J. Dubochet, Le Chevalier, 1847.

viii, 364 p. illus., 12 col. plates. 26 cm. Quarter red morocco/marbled boards, marbled endpapers.
BM I p.341.

670

CONRAD, TIMOTHY ABBOTT (1803-1877). *Monography of the family Unionidae, or Naiades of Lamarck, (fresh water bivalve shells) of America, illustrated by figures drawn on stone from nature.* Philadelphia: E.G. Dorsey, 1835-1838.

94 p. 50 plates. 24.5 cm. Issued in 10 parts. Continuously paged. Blue library binding, original printed wrappers retained. From the library of L.S. Frierson, Conchologist.
BM I p.375. Nissen ZBI 944.

671

CONRAD, TIMOTHY ABBOTT (1803-1877). *A synopsis of the family of Naiades of North America, with notes, and a table of some of the genera and sub-genera of the family, according to their geographical distribution, and descriptions of genera and sub-genera.* [n.p., 18-?].

27 p. 24 cm. Black library binding.
BM I p.375.

* 672

COPLEY, HUGH. *The game fishes of Africa.* London: H.F. & G. Witherby, [1952].

276 p. illus. 23 cm. "Designed to be a companion to Lydekker's *The game animals of Africa.*" "The fish and fisheries of the lower Ogun River," by P.I.R Maclaren: pp.177-182 from *The Nigerian field* laid in. Blue cloth binding, dustjacket.

673

COUCH, JONATHAN (1789-1870). *A history of the fishes of the British Islands.* London: George Bell and sons, 1877.

4 v. illus. 252 col. plates (incl. fronts.) 26 cm. Bookplate of T.N. Brushfield, M.D., F.S.A. Bright blue cloth binding, gilt decoration.
BM I p.393. Nissen ZBI 979.

* 674

COX, HERBERT EDWARD. *A handbook of the Coleoptera; or, Beetles of Great Britain and Ireland.* London: E.W. Janson, 1874.

2 v. illus. 23 cm. Signed in ms. on fly leaf of each vol.: T.N. Hart-Smith. "An exchange list of British Coleoptera": 8 p. laid in v.1. Holograph letter from Oliver E. Janson to the Rev. T.A. Preston laid in v.1. Brown cloth binding.
BM I p.396.

675

CRUZ LIMA, ELADIO DA (1900-1943). *Mammals of Amazonia.* With forty-two colored plates by the author. Rio de Janerio: Belém do Pará, 1945.

v. col. plates. 38 cm. Regular edition. LSU owns only vol.1. (Rio de Janerio. Museu paraense Emilio Goeldi de historia natural e etnografia. Contribution.) Tan cloth binding.

676

CUVIER, GEORGES, BARON (1769-1832). *The animal kingdom arranged in conformity with its organization...with additional descriptions of all the species hitherto named, and of many not before noticed.* By Edward Griffith...and others... London: Printed for G.B. Whittaker, 1827-32.

15 v. plus index vol. 13 front. (12 col., 1 fold.), plates (part col., part fold.), fold. tables. 25 cm. Issued in parts. Plates mostly colored and many are in two states, both colored and uncolored. The first edition in English. Bookplate of Charles Atwood Kofoid. Half black morocco/marbled boards, marbled endpapers.
BM I p.410. Nissen ZBI 1015.

* 677

CUVIER, GEORGES, BARON (1769-1832). *The animal kingdom, arranged according to its organization, serving as a foundation for the natural history of animals, and an introduction to comparative anatomy. With figures designed after nature: the Crustacea, Arachnides, & Insecta, by M. Latreille.* Translated from the latest French edition. With additional notes, and illustrated by nearly 500 additional plates. London: G. Henderson, 1834-37.

4 v. ports. and atlas (4 v. plates (part col.)) 23 cm. Added title pages engraved with imprint date: 1837. Atlas has engraved title pages only. Illustration statement on engraved title pages of atlas reads: Illustrated by nearly 800 coloured plates. Full calf binding.
BM I p.410. Nissen ZBI 1016.

678

CUVIER, GEORGES, BARON (1769-1832). *Leçons d'anatomie comparée de G. Cuvier.* Recueillies et publiées sous ses yeux par C. Duméril. Paris: Baudouin, an VIII [1800]-an XIV.—1805.

5 v. LII plates, 7 fold. tables. 21 cm. Imprint of v.3-5: Paris, Crochard [etc.] an XIV, 1805. Vols.3-5 ed. by G. L. Duvernoy. In ms. on fly leaf of all vols.: John Cassin, Jany 15th 1858, Philadelphia. In ms. on fly leaf of all vols.: Penkinf? Philadelphia 1872. In ms. on title page of v.1: N.C. Nancrede 1800. Additional ms. notes in v.1. Contemporary calf, spines mended.
BM I p.409. Nissen ZBI 1008.

679

CUVIER, GEORGES, BARON (1769-1832). *Le règne animal distribué d'après son organisation, pour servir de base a l'histoire naturelle des animaux, et d'introduction a l'anatomie comparée.* Avec figures dessinées d'après nature. Nouvelle édition, revue et augmentée. Paris: Déterville, 1829-30.
5 v. illus. 21 cm. V.1, 3, 5, marbled calf, spines taped. V.3, 4, brown library binding.
BM I p.410. Nissen ZBI 107. Zimmer p.155.

680

CUVIER, GEORGES, BARON (1769-1832). *Le règne animal distribué d'après son organisation, pour servir de base à l'histoire naturelle des animaux, et d'introduction à l'anatomie comparée.* Par Georges Cuvier. Édition accompagnée de planches gravées, représentant les types de tous les genres, les caractères distinctifs des divers groupes et les modifications de structure sur lesquelles repose cette classification; par une réunion de disciples de Cuvier, mm. Audouin, Blanchard, Deshayes, Alcide d'Orbigny, Doyère, Dugès, Duvernoy, Laurillard, Milne Edwards, Roulin et Valenciennes. Paris: Fortin, Masson, et cie, [1836-49].
11 v. and atlases. plates (part fold., part col.) 28 cm. Plates 16 of "Annélides," 8 and 19 of "Reptiles" and 19 of "Zoophytes" not issued. Each plate accompanied by leaf with descriptive letterpress. LSU lacks v.10 and atlas. Half red morocco/marbled boards.
BM I p.410. Nissen ZBI 1014. Zimmer p.156.

681

CUVIER, GEORGES, BARON (1769-1832). *Tableau élémentaire de l'histoire naturelle des animaux...* Paris: Baudouin, imprimeur, an 6 [1798].
xvj, 710 p. XIV plates. 19.5 cm. Full brown morocco.
BM I p.409. Nissen ZBI 1012.

*682

DANCE, S. PETER, ed. *The collector's encyclopedia of shells.* Photographs by Ian Cameron. New York: McGraw-Hill, [1974].
288 p. illus. 26 cm. Black cloth binding, dustjacket.

683

DARWIN, CHARLES ROBERT (1809-1882). *The formation of vegetable mould, through the action of worms, with observations on their habits.* New York: Appleton and company, 1898.
1 p.l., vii, 326 p. illus. 20.5 cm. Half red morocco/marbled boards.
BM I p.423 (other eds.)
Garland-Bullard Memorial Collection. Gift of Rietta Garland Albritton and Florence Garland Brookes.

684

DARWIN, CHARLES ROBERT (1809-1882). *A monograph on the sub-class Cirripedia, with figures of all the species.* London: Ray Society, 1851-54.
2 v. illus., 42 plates (3 col.) 22 cm. (The Ray Society. [Publications, v.21, 1851; v.25, 1853]). V.1, original blue cloth binding with "R.S." in gilt on front cover. V.2, blue library binding.
BM I p.422. Nissen ZBI 1041.

685

DARWIN, CHARLES ROBERT (1809-1882). *The various contrivances by which orchids are fertilised by insects.* Second edition, revised. With illustrations. London: John Murray, 1877.
xvi, 300 p. illus. 19.5 cm. Original green cloth binding, gilt lettering.
BM I p.422. Pritzel 2098 (1st ed.)

686

DARWIN, CHARLES ROBERT (1809-1882), ed. *The zoology of the voyage of H.M.S. Beagle, under the command of Captain Fitzroy, R.N., during the years* 1832 *to* 1836. Published with the approval of the lords commissioners of Her Majesty's Treasury. Edited and superintended by Charles Darwin. London: Smith, Elder, and co., 1839-43.
5 v. 166 plates (82 col., 3 fold.) 32 x 25 cm. Issued in parts, 1838-43. LSU owns only pts.1-2. In manuscript on fly leaf: Joseph Jones from C.E. Maxwell, Philadelphia, April 5th 1851. Black library binding.
BM II p.715. Jackson p.224. Nissen ZBI 1391. Zimmer p.157.

*687

DARWIN, CHARLES ROBERT (1809-1882), ed. *The zoology of the voyage of H.M.S. Beagle, during the years*

1832-1836. Wellington, N.Z.: Nova Pacifica, 1980.

3 v. illus. (some col.) 33 cm. Facsimile reprint of first edition published London: Smith, Elder, 1839-43. Limited edition of 750 copies. LSU owns copy no.122. Quarter blue morocco/cloth, marbled endpapers.

*688

DENNY, HENRY (1803-1871). *Monographia Pselaphidarum et Scydmaenidarum Britanniae: or, An essay on the British species of the genera Pselaphus, of Herbst, and Scydmaenus, of Latreille; in which those genera are subdivided, and all the species hitherto discovered in Great Britain are accurately described and arranged, with an indication of the situations in which they are usually found.* Norwich: S. Wilkin, 1825.

1 p.l., vi, 74 p. 14 leaves of plates, col. illus. 24 cm. Green cloth binding.
BM I p.441. Nissen ZBI 1077.

*689

DENTON, SHERMAN FOOTE. *As nature shows them; moths and butterflies of the United States, east of the Rocky Mountains.* With over 400 photographic illustrations in the text and many transfers of species from life. Boston: J.B. Millet Co., [c1900].

3 v. illus., 57 col. plates. 26 cm. "A limited edition of 500 copies, of which this is copy no.B 5." Vol.3 without edition statement. "The colored plates, or nature prints, used in the work, are direct transfers from the insects themselves...the scales of the wings of the insects are transferred to the paper while the bodies are printed from engravings and afterward colored by hand." Label mounted on lining paper of each vol.: Harry F. Marks, Bookseller, 31 West 47th St., N.Y.C. Half brown morocco/cloth binding.
Nissen ZBI 1079.

690

DESMAREST, ANSELME GAETAN (1784-1838). *Mammalogie, ou Description des espèces de mammifères.* Paris: Mveuve Agasse, 1820-22.

2 v. and atlas of 126 plates (5 fold.) 29 x 23.5 cm. Atlas wanting. Black library binding.
BM I p.445. Nissen ZBI 4621.

*691

DE SYLVA, DONALD P. *Systematics and life history of the great barracuda, Sphyraena barracuda (Walbaum)...* Coral Gables, Fla.: University of Miami Press, [1970].

vii, 179 p. illus., maps (1 fold.) 23 cm. (Studies in tropical oceanography, no.1). Dissertation, Cornell. Blue cloth binding.

*692

DONOVAN, EDWARD (1768-1837). *The natural history of British fishes: including scientific and general descriptions of the most interesting species, and an extensive selection of accurately finished coloured plates, taken entirely from original drawings, purposely made from the specimens in a recent state, and for the most part whilst living.* London: Printed for the author, and F. and C. Rivington, 1802-08.

5 v. in 1. 120 leaves of hand-colored plates. 24 cm. Each plate accompanied by descriptive letterpress. Half brown calf/marbled boards, gilt edges.
BM I p.473. Nissen ZBI 1141.

693

DONOVAN, EDWARD (1768-1837). *The natural history of British insects; explaining them in their several states, with the periods of their transformations, their food, oeconomy, &c. Together with the history of such minute insects as require investigation by the microscope. The whole illustrated by coloured figures, designed and executed from living specimens.* London: Printed for the author, and for F. and C. Rivington, 1792-1813.

16 v. in 8. 576 plates (part col.) 25 cm. Vols.8-9 have imprint: London, Printed by D. Bye and H. Law for the author, and for F. and C. Rivington. Armorial bookplate of George A. C. May mounted on lining paper of each volume. Signed in ms. in v.2, 5-8: Elizabeth Sinclaire, London, 1802. LSU owns 9 vols. Vol.1 imperfect: plates 17-18 wanting, plates 1-3 duplicated; v.5 imperfect: plate 173 wanting; v.6 imperfect: plate 191 wanting. Plates 333-336, 343-351, 454-456, 466 bound without text at end of v.9. Half blue morocco/marbled boards, marbled endpapers; restored by St. Crispin's bindery, Dallas.
BM I p.473. Nissen ZBI 1142.
Gift of Leo D. Newsom.

*694

DONOVAN, EDWARD (1768-1837). *The natural history of British insects; explaining them in their several states, with the periods of their transformations, their food, oeconomy, &c. Together with the history of such minute insects as require investigation by the microscope. The whole illustrated by coloured figures, designed and executed from living specimens.* London: Printed by Bye and Law for the author, and for F. and C. Rivington, 1793-1802 [v.1, 1802].

10 v. 360 hand col. plates. 26 cm. Vol.8 has imprint:

Printed by Law and Gilbert for the author and for F. and C. Rivington. Quarter cloth binding/paper boards.
BM I p.473. Nissen ZBI 1142.

695

DONOVAN, EDWARD (1768-1837). *Natural history of the insects of India, containing upwards of two hundred and twenty figures and descriptions.* A new edition brought down to the present state of the science, with systematic characters of each species, synonyms, indexes, and other additional matter, by J.O. Westwood. London: Henry G. Bohn, 1842.

vi, 102 p. lviii col. plates. 31 cm. Bookplate of Cecil Reid. Half red morocco/marbled boards.
BM I p.473. Nissen ZBI 1144.

696

DOUGLAS, JOHN WILLIAM (1814-1905). *The British Hemiptera. Vol.I Hemiptera-Heteroptera.* By John William Douglas and John Scott. London: Pub. for the Ray Society by R. Hardwicke, 1865.

xii, 627, [1]p. 21 leaves of plates. 22.5 cm. Brown library binding
BM IV p.1655

697

DRAPARNAUD, JACQUES-PHILIPPE RAYMOND (1772-1805). *Histoire naturelle des mollusques terrestres et fluviatiles de la France, ouvrage posthume de Jacques-Philippe-Raymond Draparnaud...avec XIII planches.* Paris: Plassan [An XIII, 1805].

viii, 164 p. plates. 29 cm. In ms. on title page: John Ritchie, Jr. '85. Ms. notes bound between plates and annotations on pages. Brown library binding.
BM I p.477. Nissen ZBI 1152.

698

DRESDEN. K. ZOOLOGISCHES UND ANTHROPOLOGISCH-ETHNOGRAPHISCHES MUSEUM ZU DRESDEN. *Mitttheilungen aus dem K. Zoologischen Museum.* Hrsg. mit unterstützung der general direction der königlichen sammlungen für kunst und wissenschaft, von dr. A.B. Meyer. Dresden: R. v. Zahn [etc.], 1875-78.

3 v. XXXV plates (part fold., part col.) 34.5 cm. LSU lacks v.2. In ms. on cover of v.1: From the Royal collections of art & leisure, Dresden, May 24th, 1876. Original printed paper over boards.

*699

DRURY, DRU (1725-1803). *Illustrations of natural history. Wherein are exhibited upwards of two hundred and forty figures of exotic insects, according to their different genera; very few of which have hitherto been figured by any author, being engraved and coloured from nature, with the greatest accuracy, and under the author's own inspection, on fifty copperplates. With a particular description of each insect: interspersed with remarks and reflections on the nature and properties of many of them. To which is added a translation into French.* London: Printed for the author and sold by B. White, 1770-82.

3 v. 152 plates (150 hand col.) 30 cm. English and French in parallel columns. Manuscript note on fly leaf: Knowsley Garden Library, C.a. Shelf 1, no.11-13. Manuscript note on slip mounted on plate 48 in v.3. Plate 2 in v.2 duplicated, 1 in color and 1 in black and white. Armorial bookplate in each volume with motto: Sans changer. Full calf, gilt decoration, marbled endpapers, marbled edges.
BM I p.481. Nissen ZBI 1160.

700

DUJARDIN, FÉLIX (1801-1860). *Histoire naturelle des zoophytes. Infusoires, comprenant la physiologie et la classification de ces animaux, et la manière de les étudier à l'aide du microscope.* Paris: Roret, 1841.

xii, 684 p., and atlas (2 p.l., 14, [2]p., 22 (i.e.23) plates). 22 cm. Stamped on fly leaf: Dr. Louis Planchon, Montpellier, No. DL. Quarter green morocco/marbled boards, marbled endpapers.
BM I p.487. Nissen ZBI 1189.

701

DURIN, BERNARD. *Insects etc.: an anthology of arthropods featuring a bounty of beetles.* Paintings by Bernard Durin. With a literary anthology introduced and selected by Paul Armand Gette. Entomological commentaries by Gerhard Scherer. Translated by Georg Zappler. New York: Hudson Hill Press, 1981.

108 p. col. illus. 34 cm. Translation of: *Käfer und andere Kerbtiere.* Orange cloth binding, dustjacket.

702

EDWARDS, WILLIAM HENRY (1822-1909). *The butterflies of North America.* 1st-3d series. First series published in Philadelphia by the American Entomological Society in 1868-72; with first series is bound: His *Synopsis of North American Butterflies.* Text reprinted in Boston by Houghton Mifflin in 1888. Boston: 1872-97.

3 v. 30 cm. Issued in parts. LSU owns series I & II in 2 v.

V.1, half green morocco/cloth binding; v.2, rebound in library binding.
BM II p.512. Nissen ZBI 1234.

703

ELLIOT, DANIEL GIRAUD (1835-1915). *A review of the Primates.* New York: American Museum of Natural History, 1912.

3 v. fronts. (v.1, 3) plates (part col.) 28 cm. American Museum of Natural History. Monograph series [no.1], Vol.I-III). "The date 1912 on the title-pages...should be corrected to June 1913. Although all the text, except the Appendix in volume III, was printed in 1912, unexpected delay in the preparation of the colored plates prevented the issue of the work till June 15, 1913." "Corrections" inserted in each vol. Copy 1, black library binding; c.2, unbound in original printed wrappers.
Nissen ZBI 1280.
Copy 2, gift of Dr. and Mrs. A. Brooks Cronan.

704

ELLIS, JOHN (1710-1776). *Essai sur l'histoire naturelle des corallines, et d'autres productions marines du même genre, qu'on trouve communement sur les cotes de la Grande-Bretagne et d'Irlande; auquel on a joint une description d'un grand polype de mer, pris auprès du pole arctique, par des pêcheurs de Baleine, pendant l'été de* 1753. Traduit de l'anglois [par J.N.S. Allamand]. A la Haye: Chez P. de Hondt, 1756.

xvi, 125 p. front., XXXIX plates (5 fold.) 28 cm. Title page in red and black. Brown library binding.
BM II p.523. Nissen ZBI 1281.

*705

ENGRAMELLE, MARIE DOMINIQUE JOSEPH (1727-1781). *Papillons d'Europe, peints d'après nature.* Paris: P.M. Delaguette, Basan & Poignant, 1779-92.

v. inplates (part col.) 32 cm. LSU owns 10 vols. LSU v.4 imperfect: p.171 mutilated. Armorial bookplate mounted on lining paper of v.7. Full calf/marbled boards. V.7, half brown calf.
BM II p.539. Nissen ZBI 1300.

*706

ENTOMOLOGICAL SOCIETY OF OXFORD. *An accentuated list of the British lepidoptera, with hints on the derivation of the names.* London: John Van Voorst, 1858.

xliv, 118 p. 23 cm. "Published by the Entomological Societies of Oxford and Cambridge. In ms. on fly leaf: T.P.A. Birtwhistle. Green cloth binding.

707

EVANS, ARTHUR HUMBLE. *A vertebrate fauna of the Shetland Islands.* By Arthur H. Evans and T.E. Buckley. Edinburgh: David Douglas, 1899.

xxix, 248 p. illus., plates, map. 23 cm. (A vertebrate fauna of Scotland, [v.8]). Illustrated title page by O.A.J.L. Original green cloth binding.
BM II p.548. Yale p.90.

708

FAYRER, JOSEPH, SIR, BART. (1824-1907). *The Thanatophidia of India; being a description of the venomous snakes of the Indian peninsula, with an account of the influence of their poison on life and a series of experiments.* 2d edition, revised and enlarged by J. Fayrer... London: J. and A. Churchill, 1874.

xi, [1], 178 p. illus. 31 plates (28 col., incl. 2 double) 45.5 x 33.5 cm. Original red cloth binding.
BM II p.560. Nissen ZBI 1339.

709

FORBES, EDWARD (1815-1854). *A monograph of the British naked-eyed Medusæ: with figures of all the species.* London: Printed for the Ray Society, 1848.

4 p.l., 104 p. XIII col. plates. 35 cm. (The Ray Society [Publications, v.12]). Half green morocco/marbled boards.
BM II p.591. Nissen ZBI 1405.

*710

FORD, ALICE ELIZABETH (1906-   ), ed. *Audubon's butterflies, moths, and other studies.* New York: Studio Publications in association with Crowell, [1952].

120 p. plates (part col.) 26 cm. "Audubon's sketchbook...fifteen pages of insects and reptiles...is reproduced for the first time in this book." Quarter yellow/green cloth binding.

711

Forty drawings of fishes made by the artists who accompanied Captain James Cook on his three voyages to the Pacific, 1768-71, 1772-75, 1776-80; some being used by authors in the description of new species. Text by P.J.P. Whitehead. London: Trustees of the British Museum (Natural History), 1968.

xxxi, 144 p. col. illus., facsims, 7 ports. 45 cm. (British Museum (Natural History) Publication, no.670). Blue cloth binding.

*712

FRIEDENTHAL, HANS WILHELM KARL (1870-). *Tierhaaratlas.* Mit 989 Abbildungen auf 16 mehrfarbigen und 19 einfarbigen Tafeln. Jena: G. Fischer, 1911.

19, [65]p. 35 plates (16 col.) 40 cm. Marbled boards.
Nissen ZBI 1434.

713

FRITSCH, GUSTAV THEODOR (1838-). *Die elektrischen Fische. Nach neuen Untersuchungen anatomisch-zoologisch dargestellt.* Leipzig: Viet, 1887-90.

2 v. illus. 41 cm. Printed boards.
BM II p.626. Nissen ZBI 1445.

*714

FROOM, BARBARA. *The turtles of Canada.* Toronto: McClelland and Stewart, c1976.

120 p. illus. (some col.) 23 cm. Green paper over boards, dustjacket.

715

FUERTES, LOUIS AGASSIZ (1874-1927). *Album of Abyssinian birds and mammals from paintings.* By Louis Agassiz Fuertes. [Chicago: Field Museum of Natural History, 1930].

2 p.l. 32 col. plates. 30.5 x 26 cm. Issued in portfolio. Descriptive letterpress on verso of each plate.
Nissen VBI 344. Yale p.103.

716

GARRETT, ANDREW. *Andrew Garrett's Fische der Südsee, von Albert Günther.* Lehre: J. Cramer; New York: Stechert-Hafner Service Agency, 1966.

3 v. in 1. (iv, 515 p.) illus., 180 plates. 28 cm. Reprint of the edition issued in 9 pts. in 1873-1910 in the series *Journal des Museum Godeffroy* published in Hamburg by L. Friederichsen, as its v.2, 4, and 6 (consisting of its Heft 3, 5, 7, 9, 11, 13, 15-17). Original blue cloth binding.
Nissen ZBI 181.

*717

GEOFFROY SAINT-HILAIRE, ÉTIENNE (1772-1844). *Histoire naturelle des mammifères, avec des figures originales, coloriées, dessinées d'après des animaux vivans; publiée sous l'autorité de l'administration du Muséum d'histoire naturelle.* Par M. Geoffroy Saint-Hilaire et par M. Frédéric Cuvier. Paris: A. Belin, 1824-47.

4 v. 431 col. plates. 53.5 cm. Dates of publication: v.1, 1824; v.2, 1824; v.3, 1847; v.4, n.d. Lithographed by C. de Last. LSU owns v.1-3. In ms. on title page of v.1: Presented to the Louisiana Seminary of Learning and Military Academy near Alexandria by Cadet Andre Guyol of New Orleans, 1867. Modern half dark red calf/marbled boards. Binding by William A. Payne.
BM II p.656.

718

GERVAIS, PAUL (1816-1879). *Histoire naturelle des mammifères, avec l'indication de leurs moeurs, et de leurs rapports avec les arts, le commerce et l'agriculture.* Paris: L. Curmer, 1854-55.

2 v. col. front., illus., plates (part col.) 27.5 cm. LSU owns only v.1. Original blue cloth binding, decorated in gilt and colors.
BM II p.665.

*719

GESNER, KONRAD (1516-1565). Conr. Gesneri... *Historiae animalium liber II. qui est de quadrupedibus oviparis; nunc denuo recognitus ac pluribus in locis ab ipso authore ante obitum emendatus & auctus, atqe aliquot novis iconibus & descriptionibus locupletatus, ac denique brevibus in margine annotationibus illustratus...* [2nd ed.] Francofurdi: Ex officina typographica I. Wecheli, impensis R. Cambieri, 1586.

119 p. illus. 38 cm. LSU copy imperfect: p.22 incorrectly numbered 25. Half vellum/marbled boards.
BM II p.668. Nissen ZBI 1550.

720

GESNER, KONRAD (1516-1565). *Thierbuch, das ist ein kurtze bschreybung [!] aller vierfüssigen thieren so auff der erdē vnd in wassern wonend sampt jrer waren conterfactur...* Zürich: C. Froschower, 1563. [Zürich: Josef Stocker-Schmid Dietikon, c1965].

4 p.l., clxxii numb.1., 13 p. illus. 40 cm. Translation of books I and II of the author's *Historia animalium.* A facsimile of the first German Froschauer edition. 500 numbered examples. Nos.1-35 hand-printed on hand-made "Richard de Bas" paper, with watermark "JS". LSU owns copy no.26. *"Zum Gebrauch und Verständnis des Tierverzeichnisses"* and *"Synoptisches Verzeichnis der Tiernamen"* by Vinzenz Ziswiler: 13 p. at end. In portfolio.
BM II p.668 (1606 ed.) Nissen ZBI 1552.

*721

GOLDSMITH, OLIVER (1728-1774). *A history of the earth, and animated nature.* Dublin: Printed by

James Williams, 1782.

8 v. plates. 19 cm. In ms. on fly leaf: John McDonald. LSU owns only v.6. Full calf binding.
BM II p.693 (1816 ed.) Nissen ZBI 1621 (various eds.)

*722

GOLDSMITH, OLIVER (1728-1774). *A history of the earth, and animated nature...illustrated with eighty-five copperplates.* A new edition, with corrections and alterations. Philadelphia: Grigg & Elliot, 1839.

4 v. illus. 22 cm. Added title page, engraved with imprint date, 1830. Stamped on engraved title page in each vol.: M.J. Bray, M.D. Full calf binding.
BM II p.693 (1816 ed.) Nissen ZBI 1624 (various eds.)

723

GOSSE, PHILIP HENRY (1810-1888). *A year at the shore; with thirty-six coloured illustrations...* London: Strahan & co., 1865.

xii, 330 p. XXXVI col. plates (incl. front.) 18 cm. Original green cloth binding.
BM II p.697. Nissen ZBI 1655.

*724

GOULD, JOHN (1804-1881). *The mammals of Australia.* London: Printed by Taylor and Francis, published by the author, 1863.

3 v. 182 col. plates. 56 cm. Each plate accompanied by leaf with descriptive letterpress. Issued in parts, 1845-1863. Copy 2 in 13 parts, as originally issued. Full green morocco, gilt ornamentation.
BM II p.701. Nissen ZBI 1661.

725

GRAFF, LUDWIG VON (1851-1924). *Das Genus Myzostoma (F.S. Leuckart).* Leipzig: Wilhelm Engelmann, 1877.

viii, 82 p. 11 leaves, 11 plates. illus. 35 cm. In ms. on fly leaf: Carl Chun. Marbled boards, mended spine.
BM II p.704. Nissen ZBI 1670.

*726

GRAY, JOHN EDWARD (1800-1875). *Gleanings from the menagerie and aviary at Knowsley Hall.* Knowsley, [Eng.], 1846.

[14]p. 17 col. plates. 57 cm. "Printed for private distribution." "Plates are selected from the series of drawings made by Mr. Edward Lear from the living animals in the Right Honourable the Earl of Derby's menagerie at Knowsley Hall... Lithographed...by Mr. J.W. Moore, and coloured by Mr. Bayfield." Presentation copy to E. Paget. Letter from Mr. John S. McIlhenny to Mr. Graves laid in. Quarter black morocco/cloth binding.
Anker 189. BM II p.713. Nissen ZBI 1691. Zimmer p.273.

*727

GRAY, JOHN EDWARD (1800-1875). *Illustrations of Indian zoology; chiefly selected from the collection of Major-General Hardwicke.* London: Published by Treuttel, Wurtz, Treuttel, Jun. and Richter, 1830-1834.

2 v. 202 col. plates, port. 50 cm. Vol.2 published by A. Richter and Parbury, Allen. Half calf/marbled boards, marbled endpapers.
BM II p.712. Nissen ZBI 1694. Yale p.116.

728

GREY OWL (1888-). *Pilgrims of the wild.* By Wa-sha-quon as in (Grey Owl). London: L. Dickson & Thompson, [1935].

xxii, 281, [1]p. front., plates, ports. 22 cm. Original black cloth binding.

729

GÜNTHER, ALBERT CARL LUDWIG GOTTHILF (1830-1914). *Report on the shore fishes, deep-sea fishes, pelagic fishes, collected by H.M.S. Challenger.* Reprint, Weinheim: Cramer; New York: Hafner, 1963.

3 pts. in 1. illus. 25 cm. atlas (plates). 33 cm. Originally issued as *Challenger zoological reports*, VI, LVII and LXXVII. (Historiae naturalis classica, t.28). Original blue cloth binding.
BM II p.746 (orig. ed.) Nissen ZBI 1743.

730

GÜNTHER, ALBERT CARL LUDWIG GOTTHILF (1830-1914). *The reptiles of British India.* London: Published for the Ray Society by R. Hardwicke, 1864.

1 p.l., xxvii, 452 p. illus., XXVI plates. 37 cm. (The Ray Society [Publications: v.35, 1863]). Blue cloth binding.
BM IV p.1655. Nissen ZBI 1744.

731

HALDEMAN, SAMUEL STEHMAN (1812-1880). *A monograph of the freshwater univalve mollusca of the United States, including notices of species in other parts of North America.* Philadelphia: E.G. Dorsey, 1842.

8 v. in 1. Various pagings. front. (port.), plates (part col.)

24.5 cm. In ms. on title page: John Ritchie, Jr. Half black morocco/marbled boards.
BM II p.769. Nissen ZBI 1802.

732

HANSEN, HANS JACOB (1855-1936). *The Choniostomatidae, a family of Copepoda, parasites on Crustacea Malacostraca.* [Translated from the Danish manuscript by Louise von Cossel]. Copenhagen: A.F. Høst, 1897.

205 p. 13 plates. 28 cm. Author's presentation copy to Mr. L.A. Borradaile. Green cloth binding.
BM II p.783. Nissen ZBI 1829.

733

HARLAN, RICHARD (1796-1843). *Fauna americana: being a description of the mammiferous animals inhabiting North America.* Philadelphia: A. Finley, 1825.

x, 318 p. 21.5 cm. Brown library binding.
BM II p.786.

734

HARVARD UNIVERSITY. MUSEUM OF COMPARATIVE ZOOLOGY. *Memoirs.* v.1-55, 1864-1938.

55 v. illus., plates (part col.), photos, maps, tables, diagrs. 35.5 cm. LSU lacks v.3, 15, 22, and a few scattered nos. Bindings vary.
BM I p.302. Nissen ZBI 4681.

*735

HENNEGUY, LOUIS FÉLIX (1850-1928). *Les insectes, morphologie, reproduction, embryogénie...* Leçons recueillies par A. Lécaillon & G. Poirault... Paris: Masson et cie, 1904.

xviii p., 1 l., 804 p. illus., IV col. plates. 28 cm. In ms. on fly leaf: Warren Whitcomb Jr. Red cloth binding.
BM VI p.453.

*736

HERSHKOVITZ, PHILIP. *Living New World monkeys (Platyrrhini): with an introduction to Primates.* Chicago: University of Chicago Press, 1977- .

v. 31 cm. Grey cloth binding, dustjacket.

737

HEUDE, PIERRE (1836-1902). *Conchyliologie fluviatile de la province de Nanking.* Paris: F. Savy, [1885].

10 v. in 1. 80 plates. 22.5 x 32.5 cm. Label mounted over imprint: Paris, E. Guilmoto, éditeur. From the library of L.S. Frierson, Conchologist. Label on p.[2] of cover: T.J. Leaton, General Book Bindery, 1138 Jewell St., Shreveport, La. Black cloth binding, decorated endpapers.
BM II p.838. Nissen ZBI 1928.

*738

HOLBROOK, JOHN EDWARDS (1794-1871). *North American herpetology; or, A description of the reptiles inhabiting the United States.* Philadelphia: J. Dobson; London: R. Baldwin; [etc., etc.], 1842.

5 v. col. plates. 31 cm. First edition published 1836-38 in 3 vols. Ex libris Henry H. Ficken. "Description of a new species of salamander," by Lewis R. Gibbes: 2 leaves and plate from the *Boston journal of natural history*, v.5, no.1, laid in v.1. "The coral snake and its imitators": pp.65-68 detached from the *Bulletin of the Charleston Museum*, v.15, no.7, Nov. 1919, laid in v.1. "How to know deadly snakes and how to avoid them": clipping laid in v.5. Portrait of the author detached from the *New York Zoological Society Bulletin*, v.35, no.4, July-August, 1932, p.136 laid in v.5. Half red morocco/marbled boards.
BM II p.861. Nissen ZBI 1980.

739

HOPE, FREDERICK WILLIAM (1797-1862). *The coleopterist's manual...* London: H.G. Bohn [etc.], 1837-40.

3 v. col. fronts., 8 plates (part col.) 23.5 cm. LSU owns only pt.2. Black library binding.
BM II p.873.

740

Hortus sanitatis (maior). An early English version of *Hortus sanitatis*; a recent bibliographical discovery by Noel Hudson. A translation by Laurence Andrew of *Der Dieren palleys*, a version of the *Hortus sanitatis*. "A facsimile of *The noble lyfe & natures of man, of bestes, serpentys, fowles & fisshes.* Antwerp, [Jan van Doesborgh], c1521." London: B. Quaritch, [1954].

xiii, 164, xvii-xxx p. 29 cm. Red cloth binding.
Nissen ZBI 4732.

*741

HOUGHTON, WILLIAM (1829-1897). *British freshwater fishes.* Illustrated with a coloured figure of each species drawn from nature by A.F. Lydon, and numerous engravings. London: W. Mackenzie [pref. 1879].

2 v. (xxvi, 204 p.) illus., col. plates. 37 cm. Brown pictorial cloth binding.
BM II p.880. Nissen ZBI 2009.

742

HUDSON, CHARLES THOMAS (1828-1903). *The Rotifera; or wheel-animalcules.* By C.T. Hudson assisted by P.H. Gosse. In two volumes, with supplement. London: Longmans, Green, 1886-89.

3 v. in 2. illus., 38 fold. plates. 29 cm. Plates (numbered A-D, I-XXXIV) partly colored. Supplement in v.2 (62 p.) Original green cloth binding.
BM II p.885. Nissen ZBI 2024.

743

HUDSON, GEORGE VERNON. *The butterflies and moths of New Zealand.* Wellington, N.Z.: Ferguson & Osborn, limited, 1928.

xi, 386, [2]p. col. front., plates (part col.) 30.5 cm. Each plate accompanied by descriptive letterpress. Green cloth binding.

——— *A supplement*...with 10 coloured plates. Wellington, N.Z.: Ferguson & Osborn, 1939. (3 p.l., [387]-481, [2]p. plates (part col.) 30 cm.) Half red morocco/cloth.
Nissen ZBI 2025.

744

HUDSON, WILLIAM HENRY (1841-1922). *The naturalist in La Plata.* London: Chapman and Hall, ld., 1892.

vii, [1], 388 p. illus., 3 plates (incl. front.) 21.5 cm. Original green pictorial cloth binding.
BM II p.885. Nissen ZBI 2033. Yale p.139.

745

HUMPHREYS, HENRY NOEL (1810-1879). *British butterflies and their transformations.* Arranged and illustrated in a series of plates by H.N. Humphreys. With characters and descriptions by J.O. Westwood. New edition, revised and corrected by the author. London: T. Sanderson, 1857.

xii, 137 p. 42 col. plates. 30 cm. Added title page in color. Red cloth binding.
BM II p.892. Nissen ZBI 2049.

746

HUMPHREYS, HENRY NOEL (1810-1879). *British moths and their transformations.* Arranged and illustrated in a series of plates by H.N. Humphreys. With characters and descriptions by J.O. Westwood. New edition, revised and corrected by the author. London: T. Sanderson, 1857.

2 v. 125 col. plates. 30 cm. LSU owns only v.1. Red cloth binding.

*747

HUMPHREYS, HENRY NOEL (1810-1879). *The genera and species of British butterflies.* Described and arranged according to the system now adopted in the British Museum. Illustrated by plates, in which all the species and varieties are represented, accompanied by their respective caterpillars, and the plants on which they feed. London: P. Jerrard, [1859].

xii, 66, 4 p. 33 col. plates. 28 cm. Half title page in color. LSU copy imperfect: p.47 wanting; p.27 inserted in place of p.47. Red cloth binding with gilt decoration, gilt edges.
Nissen ZBI 2053.

748

HUXLEY, THOMAS HENRY (1825-1895). *An introduction to the classification of animals.* London: J. Churchill & sons, 1869.

4 p.l., 147, [1]p. illus. 23 cm. Original brown cloth binding, mended.
BM II p.898. Yale p.142.

749

HUXLEY, THOMAS HENRY (1825-1895). *The oceanic Hydrozoa; a description of the Calycophoridae and Physophoridae observed during the voyage of H.M.S. "Rattlesnake," in the years* 1846-1850. London: Printed for the Ray Society, 1859.

x, 143 p., 12 l. XII plates. 36 cm. (The Ray Society [Publications, 30, 1858]). Blue library binding.
BM II p.897. Nissen ZBI 2065.

*750

Indian zoology. 2d edition. London: Printed by H. Hughes, for R. Faulder, 1790.

viii, 161 p. 16 plates. 27 cm. Engraved title page with vignette. Signed in ms. on title page: Robert Heron, 1795. LSU copy imperfect: plates 1, 9, and 16 wanting. "Advertisement" dated March 1, 1791. *"An essay on India, its boundaries, climate, soil, and sea.* Translated from the Latin of John Reinhold Forster, by John Aikin, M.D.": pp.[1]-27. *"The Indian faunula* [arranged by John Latham and Hugh Davies]": pp.[57]-161. Disbound.
Anker 395. BM IV p.1543. Nissen VBI 714. Zimmer p.488 (1st ed.)

*751

JAEGER, BENEDICT. *The life of North American insects.* By B. Jaeger. Assisted by H.C. Preston. With numerous illustrations, from specimens in the cabinet of the author. New York: Harper & Brothers, 1859.

xiv, 319 p. illus. 19 cm. Brown cloth binding with gilt decoration.
BM III p.919 (1854 ed.)

*752

JARDINE, WILLIAM, SIR, BART. (1800-1874). *British Salmonidae.* London: Decimus, 1979.

[14] leaves, [12] leaves of plates, illus. (some col.) 65 cm. "First published privately in Edinburgh and issued in parts between 1839 and 1841." "This facsimile edition is limited to 500 copies of which this is number 41." Issued in slipcase. *Sir William Jardine and the British Salmonidae,* by Alwyne Wheeler: [4]p. laid in. Quarter brown morocco with label on front cover, green cloth, marbled endpapers.
BM II p.927. Nissen ZBI 2092 (orig. ed.)

753

JAY, JOHN CLARKSON (1808-1891). *A catalogue of the shells, arranged according to the Lamarckian system; together with descriptions of new and rare species, contained in the collection of John C. Jay...* 3d edition. New York and London: Wiley & Putnam, 1839.

125, [1]p. X plates. 28.5 x 23.5 cm. Black library binding.
BM II p.929.

754

JAY, JOHN CLARKSON (1808-1891). *A catalogue of the shells arranged according to the Lamarckian system, with their authorities, synonymes, and references to works where figured or described, contained in the collection of John C. Jay.* 4th edition. New York: Printed by R. Craighead, 1850.

2 p.l., 459, [1]p. 29.5 x 23 cm. Tan cloth binding.
BM II p.929.

*755

JERDON, THOMAS CLAVERHILL (1811-1872). *The mammals of India; a natural history of all the animals known to inhabit continental India.* London: John Wheldon, 1874.

xxxi, 335 p. 26 cm. Stamped on title page: Taru, no. 3643/111, Mori Gate, Delhi 110006, India. Half black calf/cloth binding.

756

JOHNSTON, GEORGE (1797-1855). *A history of British sponges and lithophytes.* Edinburgh: W.H. Lizars [etc., etc.], 1842.

xii, 264 p. illus., XXV plates (incl. front.) 22 cm. Bookplate of Basil Woodd Smith. Original brown cloth binding.
BM II p.938. Nissen ZBI 2123.

757

JOHNSTON, GEORGE (1797-1855). *A history of the British zoophytes.* 2d edition. London: J. Van Voorst, 1847.

2 v. illus., LXXXIV (i.e.73) plates. 22 cm. Vol.1, text; v.2, plates. Each plate accompanied by descriptive letterpress. Original brown cloth binding.
BM II p.938. Nissen ZBI 2124.

758

Journal de conchyliologie, comprenant l'étude des mollusques vivants et fossiles... t.1-4, fév. 1850-nov. 1853; t.5-8 (2. sér., t.1-4) juillet 1856-oct. 1860; v.9-46 (3. sér., t.1-38) jan. 1861-oct.1898; v.47- . (4. sér., t.1-) avril 1899- . Paris: Petit de La Saussaye, [etc.] 1850-19.

v. illus. plates (part. col.) ports. tables. 22 cm. No numbers issued 1854-55. LSU owns 68 v. in 46, 1850-1923 and index to vol.1-40. Various bindings.
BM II p.945.

759

Journal of Zoology. 1830/31- . London: 1833- .

v. illus. 23-26 cm. Title varies: 1830/31-Aug. 1945 as *Proceedings of the scientific meetings (general meetings for scientific business) of the Zoological Society of London.* Dark green morocco bindings. From 1901, dark green cloth binding.
BM V p.2397. Nissen ZBI 4744. Yale p.320.

760

JURINE, LOUIS (1751-1819). *Histoire des monocles qui se trouvent aux environs de Genève.* Genève, Paris: J.J. Paschoud, 1820.

xvi, 258, [2]p. 22 col. plates. 28 x 21.5 cm. Paper boards.
BM II p.952. Nissen ZBI 2146.

761

KENT, WILLIAM SAVILLE (d. 1908). *A manual of the Infusoria: including a description of all known flagellate, ciliate, and tentaculiferous Protozoa, British and foreign, and an account of the organization and affini-*

ties of the sponges. London: David Bogue, 1880-82.
3 v. col. front., illus., 51 (i.e.52), plates. 27 cm. Vols.1-2, text; v.3, plates. On p.[2] of cover of v.1 in ms.: W.H. Anstead. Red cloth binding, gilt edges.
BM II p.969.

762

KOCH, CARL LUDWIG (1778-1857). *Die Myriapoden.* Getreu nach der Natur abgebildet und beschrieben von C.L. Koch. Halle: Schmidt, 1863.
2 v. 119 col. plates. 25 cm. Dark green cloth binding.
BM II p.1003. Nissen ZBI 2254.

763

LA BLANCHERE, HENRI MARIE PIERRE RENE MOULLIN DE (1821-1880). *La pêche et les poissons; nouveau dictionnaire général des pêches, publié sous les auspices de LL. EE. MM. le Ministre de la Marine et des Colonies, le Ministre du Commerce et de l'Agriculture et le Ministre de l'Instruction publique...précedé d'une préface par Aug. Duméril...* 1,100 illustrations dessinées et coloriées par A. Mesnel... Paris: C. Delagrave et cie., 1868.
xv, [1], 859 p. illus., col. plates. 27.5 cm. Bookplate of David L. Blondheim. Half blue morocco/marbled boards.
Nissen ZBI 2334.

764

LACÉPÈDE, BERNARD GERMAIN ÉTIENNE DE LA VILLE SUR ILLON, COMTE DE (1756-1825). *Histoire naturelle des quadrupèdes ovipares et des serpens.* Paris: Hôtel de Thou, 1788-1789.
2 v. illus. 28 cm. Note below imprint: Sous le privilege de l'Académie Royale des Sciences. LSU owns only v.2 (2 p.l., 8, [5]-[2]-144, 527 p.) Stamped on fly leaf: Kraig K. Adler Library. In ms. on title page: Kraig Adler 1962. Half calf/paper boards.
BM III p.1041. Nissen ZBI 239.

765

LAMARCK, JEAN BAPTISTE PIERRE ANTOINE DE MONET DE (1744-1829). *Philosophie zoologique, ou, Exposition des considérations relatives à l'histoire naturelle des animaux...* Nouvelle edition. Paris: J.B. Baillière, 1830.
2 v. 21 cm. Marbled boards, original wrappers retained.
BM III p.1048.

766

LEA, ISAAC (1792-1886). *Observations on the genus Unio, together with descriptions of new genera and species in the families Naiades, Conchae, Colimacea, Lymnaeana, Melaniana and Peristomiana: consisting of Four memoirs read before the American Philosophical Society from 1827 to 1834, and originally published in their transactions.* With numerous coloured plates. Philadelphia: Printed by James Kay, Jun., & co., [1834-1874].
13 v. in 8. 293 plates (part col.) 34 x 26 cm. (v.1-5: 28.5 x 22.5 cm.; v.2-5; v.6-13: 34.5 x 26.5 cm.) Manuscript notes throughout text. In v.9-10 engraved portrait of Isaac Lea and holograph letter from Lea to Prof. J. Leidy. Vols.1-2, 3-5, 6-7 autographed presentation copies to Major John LeConte. Half black morocco/marbled boards.
——— *Index.* By Isaac Lea. Philadelphia: Printed for the author by T.K. Collins, 1867-74. (3 v. in 1. 34 x 26 cm.) Half black morocco/marbled boards.
BM III p.1071. Nissen ZBI 2403.

767

LEA, ISAAC (1792-1886). *Observations on the Naîades, and descriptions of new species of that and other families. Read before the American Philosophical Society, May 7, 1830.* [Philadelphia, 1834.]
Caption-title, 63-123 p. [16] plates. 30 cm. Reprinted from *Transactions of the American Philosophical Society*, n.s. v.4, 1834. Black cloth binding.
BM III p.1071.

768

LEA, ISAAC (1792-1886). *Publications of Isaac Lea on recent conchology, extracted from a list of American writers on recent conchology.* By George W. Tryon, January 1, 1861 [and other papers on conchology extracted from various journals. n.p., n.d.].
1 v. (various pagings). 23 cm. Half brown morocco/marbled boards.
BM V p.2146.

*769

LE MOULT, EUGÈNE. *Les morpho d'Amérique du Sud et Centrale; historique, or morphologie, systématique. [Par] E. Le Moult & P. Réal. [Paris]: Éditions du Cabinet entomologique E. Le Moult,* 1962-63.
2 v. illus. (part col.), maps. 33 cm. (Novitates entomologicae. Supplement). Vol.2 consists of plates in color and black and white. Paperbound.

*770

LESKE, NATHANAEL GOTTFRIED (1751-1786). *Ichthyologiae Lipsiensis specimen.* Lipsiae: Apud Siegfried Lebrecht Crusium, 1774.

82 p. 21 cm. Colophon: Lipsiae, Ex Officina Breitkopfiana. Stamped on title page and verso of title page: Museum Zoologie, Univ. Lipsiensi. Manuscript notes on fly leaf. Half calf/marbled boards, marbled endpapers.
BM III p.1094.

*771

LINNÉ, CARL VON (1707-1778). *Caroli Linnaei Entomologia, faunae suecicae descriptionibus aucta: DD. Scopoli, Geoffroy de Geer, Fabricii, Schrank, &c., speciebus vel in systemate non enumeratis, vel nuperrime detectis, vel speciebus Galliae australis locupletata, generum specierumque rariorum iconibus ornata; curante & augente Carolo de Villers.* Lugduni: sumptibus Piestre et Delamolliere, 1789.

4 v. 11 (i.e.12) fold. plates. 23 cm. Paper wrappers.
BM III p.1136

772

LISTER, MARTIN (1638?-1712). *Historiae sive Synopsis methodicae conchyliorum et tabularum anatomicarum.* Editio altera. Recensuit et indicibus auxit Gulielmus Huddesford... Oxonii: E typographeo Clarendoniano, 1770.

iv, p., 1059 plates; 7 p., 22 plates; 12, 77 p. 41 cm. "Susanna et Anna Lister figuras delinearunt." Indices at end have special title page. Half tan morocco, marbled endpapers.
BM III p.1155. Nissen ZBI 2529.

773

LOISEL, GUSTAVE ANTOINE ARMAND (1864-). *Histoire des ménageries de l'antiquité à nos jours...* Paris: O. Doin et fils [etc.], 1912.

3 v. illus. (plans), 60 plates. 25.5 cm. Green cloth binding.

*774

LYDEKKER, RICHARD (1849-1915). *Wild oxen, sheep, & goats of all lands, living and extinct.* London: Rowland Ward, 1898.

xiv, 318 p., 27 leaves of plates. illus. 31 cm. ([His Game mammals of the world; v.2]). "This edition consists of five hundred copies, numbered and signed, of which this is no.111." Ex libris John Arthur Brooke. Original green cloth binding.
BM III p.1198.

*775

MACGILLIVRAY, WILLIAM (1796-1852). *Lives of eminent zoologists, from Aristotle to Linnaeus; with introductory remarks on the study of natural history, and occasional observations on the progress of zoology.* Second edition. Edinburgh: Oliver & Boyd; London: Simpkin & Marshall, 1834.

391 p. front. [1]-391 [1]p. (port.) 18.2 cm. Half tan calf/ marbled boards, red spine labels, speckled edges.
BM VII p.772.

776

MANTELL, GIDEON ALGERNON (1790-1852). *The invisible world revealed by the microscope; or, Thoughts on animalcules.* A new edition. London: J. Murray, 1850.

3 p.l., xvi, 144 p. illus., XII col. plates. 17.5 x 14 cm. Original brown cloth binding, mended.
BM IV p.1235 (1st ed.)

777

MARTINI, FRIEDRICH HEINRICH WILHELM (1729-1778). *Systematisches conchilien-cabinet von Martini und Chemnitz.* In verbindung mit drdr. Philippi, Pfeiffer, Römer, Dunker, Kobelt, H.C. Weinkauf, S. Clessin, Brot. und von Martens, neu herausgeben und vervollständigt von dr. H.C. Küster... Nürnberg: Bauer & Raspe, 183 - .

v. in col. plates. 27.5 x 23 cm. Continued after the editor's death by H.C. Weinkauff and Wilhelm Kobelt. First published, Nuremberg, 1769-1829, under title: *Neues systematisches conchylien-cabinet.*

——— *Index by monographs and genera* [comp. by William H. Dall], [n.p., 1907]. [37] l. 27.5 cm. LSU owns only v.9, nos.1 & 2. Half red morocco, cloth binding.
BM III p.1252. Nissen ZBI 2723.

*778

MARTYN, THOMAS (fl. 1760-1816). *The English Entomologist: exhibiting all the coleopterous insects found in England including upwards of 500 different species, the figures of which have never before been given to the public. The whole accurately drawn and painted after nature, (arranged and named according to the Linnean system)...* London: His Academy for Illustrating and Painting Natural History, 1792.

33 p. 44 leaves of plates, illus. 35 cm. Plates 1-42 are hand-colored. LSU copy imperfect: plates 43-44 wanting. Half green morocco/cloth binding.
BM III p.1259. Nissen ZBI 2725.

779

MAXWELL-LEFROY, HAROLD (1877-1925). *Indian insect life; a manual of the insects of the plains (tropical*

India) by H. Maxwell-Lefroy. Assisted by F.M. Howlett. [Published under the authority of the government of India.] Agricultural research institute, Pusa. Calcutta & Simla: Thacker, Spink & co.; [etc., etc.], 1909.

xii, 786 p. front. (map), illus., LXXXIV plates (part col.) 25.5 cm. Each plate accompanied by a guard sheet with descriptive letterpress. Green cloth binding.
BM VI p.502. Nissen ZBI 2741.
Gift of Leo D. Newsom.

780

MAYO, ELIZABETH (1793-1865). *Lessons on shells, as given in a Pestalozzian school, at Cheam, Surrey, by the author of "Lessons on Objects."* Illustrated by ten plates, drawn from nature. New York: Peter Hill & co., 1834.

vi, 1 l, 218 p. 10 plates. 16 cm. Bookplate: Library of Hillyer Rolston Speed. Explanatory leaf opposite each plate. Black library binding.
Nissen ZBI 2754.

*781

MC ALLISTER, D. E. *Fishes of Canada's national capital region = Poissons de la région de la capitale du Canada.* [By] D.E. McAllister & B.W. Coad; assisted by J. Aniskowicz...[et al.]; jointly sponsored by National Museum of Natural Sciences...[et al.] Ottawa: Dept. of the Environment, Fisheries and Marine Service, 1974.

200 p. [4] leaves of plates, illus. 25 cm. (Miscellaneous special publication — Fisheries and Marine Service, Dept. of the Environment, 24). English and French. Paperbound.

782

MC COOK, HENRY CHRISTOPHER (1837-1911). *American spiders and their spinningwork. A natural history of the orbweaving spiders of the United States with special regard to their industry and habits.* [Philadelphia]: The author, Academy of Natural Sciences of Philadelphia, 1889-93.

3 v. front. (v.3, port.), illus., 35 col. plates. 28 cm. Green library binding.
BM III p.1206. Nissen ZBI 2625.

783

MC ILHENNY, EDWARD AVERY (1872-1949). *The alligator's life history.* Illustrated with photographs taken by the author, E.A. McIlhenny. Boston: The Christopher publishing house, [c1935].

2 p.l., 7-117 p. plates, port. 24.5 cm. Original red cloth binding.

*784

MERIAN, MARIA SIBYLLA, FRAU J.A. GRAFF (1647-1717). *Histoire générale des insectes de Surinam et de toute l'Europe, contenant leurs descriptions, leurs figures, leurs differentes metamorphoses, de même que les descriptions des plantes, fleurs & fruits, dont ils se nourrissent...par Mademoiselle Marie Sybille de Merian, en deux parties in folio.* 3d edition, revue, corrigée, & considérablement augmentée, par M. Buch'oz...a laquelle on a joint une troisieme partie qui traite des plus belles fleurs, telles que des plantes bulbeuses, liliacées, carophillées, &c. avec leur description exacte, leur culture, & leurs propriétés... Paris: L.C. Desnos, 1771.

3 v. in 2. 189 p. 50 cm. Vol.1, a new edition in Latin and French, the French translation by Jean Rousset de Missy, of the author's *Metamorphosis insectorum Surinamensium,* first published in Latin and Dutch, Amsterdam, 1705; vol.2, a translation by Jean Marret of the author's *Der raupen wunderbare verwandelung, und sonderbare blumennahrung,* Nuremberg, etc., 1679-83. In ms. on title page of v.1-2 John Gebhard Jun., Schoharie, N.Y. In ms. on p.[viii] of v.1-2 and verso of title page v.3: Mr. J. Albert Lintner, will please accept these volumes describing the Natural History of Surinam, by Marie Sybilla Merian, as a slight token of the esteem and friendship of his Humble servant, John Gebhard, Jun. Schoharie, Jany. 1, 1872. Full calf binding, gilt decorative spine restored, marbled endpapers.
BM III p.1299 (other eds.) Nissen BBI 1342.

785

MERIAN, MARIA SIBYLLA, FRAU J.A. GRAFF (1647-1717). *Leningrader Aquarelle...* Hrsg. von Ernst Ullmann, unter Mitarbeit von Helga Ullmann, Wolf-Dietrich Beer, und Boris Vladimirovic Lukin. Leipzig: Edition Leipzig, [c1974].

2 v. illus. (some col.) 49 cm. "Einmalige limitierte Weltauflage von 1750 arabisch numerierten Exemplaren 50 unverkaufliche Exemplare ... Dieses Examplar tragt die Nr. 1360." Text in German, English, French, and Russian. "Der Text wurde von George Baurley, Leipzig, ins Englische, von Daniel Poncin, Poitiers, ins Franzosische und von Tatjana Lukina, Leningrad, ins Russische ubersetzt." "Die Originale...Aquarelle der Maria Sibylla Merian befinden sich im Besitz des Archivs und der Bibliothek der Akademie der Wissenschaften der UdSSR in Leningrad." "Das Bezugsgewebe im Dessin 'Bouquet' wurde in Einzelfertigung von Gerhard Hesse, Leipzig, handmarmoriert Damit ist jeder Bezugsbogen ein Unikat." Vellum spine, marbled boards.

786

MICHAEL, ALBERT DAVIDSON (1836-). *British Oribatidae.* London: The Ray Society, 1884-88.

2 v. 62 plates (part col.) 22.5 cm. Paged continuously. (The Ray Society [Publications: v.61, 1883; v.65, 1888]). Original blue cloth binding with "R.S." on front cover.
BM IV p.1655. Nissen ZBI 2808.

787

MICHAEL, ALBERT DAVIDSON (1836-). *British Tyroglyphidae*. London: The Ray Society, 1901-03.

2 v. XXXIX plates (part col.) 22.5 cm. (The Ray Society [Publications: v.79, 1901; v.81, 1903]). Original black cloth binding with "R.S." in gilt on front cover.
BM IV p.1655. Nissen ZBI 2809.

788

MICHELET, JULES (1798-1874). *L'insecte*. Cinquieme édition. Paris: L. Hachette et cie, 1858.

2 p.l., xxxix, 404 p. 18 cm. Label on p.[2] of cover: Hebert & Escousse, 131 Chartres Street, New Orleans. Marbled boards, taped spine, marbled endpapers.

*789

MILLAIS, JOHN GUILLE (1865-1931). *The mammals of Great Britain and Ireland*. London, New York: Longmans, Green, 1904-06.

3 v. plates (part col.) 36 cm. "Only one thousand and twenty-five copies of this volume have been printed. This copy is number 991." Photogravures by Millais, H. Grönvold, G.E. Lodge and E.S. Hodgson; coloured plates by Millais, Archibald Thorburn, G.E. Lodge and H.W.B. Davis, and uncoloured plates by Millais and from photographs. Original blue cloth binding.
BM III p.1312. Nissen ZBI 2819.

790

MILNE-EDWARDS, HENRI (1800-1885). *Histoire naturelle des Crustacés, comprenant l'anatomie, la physiologie et la classification de ces animaux*. Paris: Roret, 1834-40.

3 v. fold. tables. 22 cm., and atlas (32 p., 42 (i.e.44) plates). 26 cm. (Suites à Buffon). Tan library binding.
BM II p.510. Nissen ZBI 2838.

791

MOHR, ERNA, ed. *Säugetiere; eine Auswahl von 192 Einzeldarstellungen auf 192 Tafeln in 7-bis 8-farbigem Offsetdruck nach Originalen von W. Eigener, F. Murr, K. Grossmann, und H. Vogel*. Ausgewählt und bearbeitet von Erna Mohr, unter Mitarbeit von Theodor Haltenorth. [1. auflage]. Hamburg: Kronen-Verlag, [c1958].

15 p. 192 col. plates. 29 cm. In cloth box.
Nissen ZBI 2854.

792

MOQUIN-TANDON, CHRISTIAN HORACE BENÉDICT ALFRED (1804-1863). *Monographie de la famille des hirudinées*. Paris: Gabon et comp. [etc., etc.], 1827.

151, [1]p. VII plates (part col.) 26.5 x 22 cm. Paper boards.
BM III p.1346. Nissen ZBI 2882.

793

MOQUIN-TANDON, CHRISTIAN HORACE BÉNÉDICT ALFRED (1804-1863). *Monographie de la famille des hirudinées*. Nouvelle édition, rev. et augm., accompagnée d'un atlas de 14 planches gravées et coloriées. Paris: J.B. Baillière, 1846.

448 p. 22 cm. and atlas (23 p., xiv col. plates.) 27 cm. Quarter blue morocco/marbled boards, marbled endpapers.
BM III p.1346. Nissen ZBI 2882.

794

MORRIS, FRANCIS ORPEN (1810-1893). *A history of British butterflies*. London: Groombridge, 1857.

vi, 168, 29 p. [73] leaves of plates, illus. (some col.) 26 cm. Label mounted on p.[2] of cover: Arthur King, 10 De Crespigny Park, Denmark Hill. LSU copy imperfect: pp.167-168 bound in incorrect order. Manuscript writing on fly leaf. Half calf/marbled boards, marbled endpapers.
BM II p.1354 (earlier eds.) Nissen ZBI 2892.

795

MORRIS, FRANCIS ORPEN (1810-1893). *A history of British butterflies*. 6th edition. London: John C. Nimms, 1891.

viii, 184 p. 74 plates (72 col.) 26 cm. In ms. on fly leaf: A.H. Marshall Lat. grammar & composition. Class VI. ? July, 1891. Original green cloth binding, gilt decoration.
BM III p.1354 (earlier eds.) Nissen ZBI 2892.

796

MORRIS, FRANCIS ORPEN (1810-1893). *A history of British moths*. With an introduction by W. Egmont Kirby. 5th edition, with 132 hand-coloured plates, containing 1933 distinct specimens. London: J.C. Nimmo, 1896.

4 v. in 2. illus. 27 cm. Green cloth binding.
BM II p.1354. Nissen ZBI 290.

797

MÜLLER, ADOLF (1821-1910). *Thiere der Heimath; Deutschlands Säugethiere und Vögel, geschildert von Adolf und Karl Müller.* Mit original-Illustrationen in Farbendruck nach Zeichnungen und Aquarellen von C.F. Deiker und Adolf Müller. 2 Aufl. Kassel: T. Fischer, 1891-

v. illus., col. plates. 30 cm. LSU owns v.2. New cloth binding with original pictorial cover preserved.
BM III p.1364. Nissen VBI 655. Zimmer p.447.

*798

Là Natura d'ogni Genere delle Farfalle, ò siano cavallieri, che producano la seta ed' altri animali... [n.p.,n.p., ca.1750].

107 p. col. illus. 30 cm. Text in ms. with hand-drawn and hand-colored illustrations. Bibliographical information on 2 typed leaves laid in. Disbound in blue cloth portfolio in blue cloth slipcase, marbled paper linings. Portfolio by William A. Payne.

799

A Natural History of Insects. Illustrated with anecdotes and numerous engravings. Designed for youth. Boston: Carter & Hendee; [etc., etc.], 1830.

107 p. illus. 15 cm. In ms. on title page: Butler. Quarter brown calf, paper boards.
Gift of Mildred P. Harrington.

*800

The Natural history of quadrupeds and cetaceous animals, from the works of the best authors, ancient and modern. Bungay: Brightly, 1811.

2 v. 120 hand col. plates. 22 cm. Contemporary calf, spine restored with gilt ornamentation.

801

NEWSTEAD, ROBERT (1859-). *Monograph of the Coccidae of the British Isles.* London: Printed for the Ray Society, 1901-03.

2 v. illus., plates (part col.) 22.5 cm. (The Ray Society [Publications: v.78, 1900; v.80, 1902]). Original blue cloth binding with "R.S." on front cover.
BM IV p.1655.

802

Novitates Zoologicae, a Journal of Zoology. v. 1-42, 1894-1948. London: Printed by order of the Trustees, British Museum (Natural History) [etc.], 1894-1948.

42 v. illus., plates (part col., diagrs.) 28.5 cm. Founded and for many years edited by L.W. Rothschild. Superseded by British Museum (Natural History) Bulletin. Zoology. Half calf/marbled boards.
BM III p.1454. Nissen IVB 683.

803

OLIVI, GIUSEPPE (1769-1795). *Zoologia Adriatica, ossia catalogo ragionato degli animali del golfo e delle lagune di Venezia: preceduto da una dissertazione sulla storia fisica e naturale del golfo; e accompagnato da memorie, ed osservazioni di fisica storia naturale ed economia...* Bassano [Venezia: G. Remondini e f.', stampatori], MDCCXCII.

5 p.l., 334, xxxii p. IX fold. plates. 32 cm. LSU copy imperfect: leaf between p.30 and p.31 omitted in numbering of pages. Page 279 misnumbered [p.]179. Full contemporary calf, marbled endpapers.
BM III p.1469. Nissen ZBI 3010.

*804

OLIVIER, GUILLAUME ANTOINE (1756-1814). *Entomologie; ou, Histoire naturelle des insectes, avec leurs caractères génériques et spécifiques, leur description, leur synonymie, et leur figure enluminée. Coléoptères.* Paris: De l'imprimerie de Baudouin, 1789-1808.

8 v. col. plates. 32 cm. LSU copy v.1, 7-8 imperfect: v.1, pp.43-46 (1st group) wanting; v.7, plate 7 of No.6, Cetoine wanting, plate 8 of No.6, Cetoine duplicated in v.8; v.8, No.38, 41-44, 48-58, 60-65, 71-72, 74-76, 78-81 wanting. Vols.1-2: text; v.7-8: plates. Vol.7 contains plates for no.1, Lucane-No.34, Elaphre. Vol.8, contains plates for No.35, Carage-No.83, Charanson. Vol.7 (lettered on spine as v.3) has engraved title page with title: Histoire Naturelle des insectes. V.1, 2, 7 full calf/marbled endpapers. V.8, half calf/marbled boards.
BM III p.1469. Nissen ZBI 3012.

*805

OPPIANUS. *Oppiani poëtae cilicis De venatione Lib. IIII. De piscatv Lib. v. Cum interpretatione latina, commentariis, & indice rerum in vtroque opere memorabilium locupletissimo, confectis studio & opera Conradi Rittershysii... Qui & recensuit hos libros denuò, & Adr. Tutnebi editionem parisiensem sum trib. Mss. palatinis contulit: inde & var. lect. & scholia graeca excerpsit.* Lvgdvni Batavorvm: Ex officina Plantiniana, apud Franciscum Raphelengium, 1597.

1015 p. in various pagings. 17 cm. Printer's device on title page. The author of *De piscatu (Halieutica)* is Oppianus of

Anazarbus (or of Corycus) in Cilicia; the *De venatione (Cynegetica)* is by a native (of the same name?) of Apamea in Syria. Text in Greek and Latin on opposite pages. Manuscript notes on fly leaves. Clipping containing bibliographical information mounted on fly leaf. Full calf, spine rebacked; marbled endpapers.
BM III p.1473 (dif. ed.)

806

PARKER, WILLIAM KITCHEN (1823-1890). *A monograph on the structure and development of the shoulder-girdle and sternum in the Vertebrata.* Published for the Ray Society by Robert Hardwicke, 1868.
xi, [1], 237 p., 31 l. XXX col. plates. 37 cm. Each plate accompanied by leaf with explanatory letterpress. (The Ray Society [Publications, v.42, 1867]). Half-title lacking. Blue cloth binding.
BM IV p.1522. Nissen ZBI 3089.

807

PENNANT, THOMAS (1726-1798). *British Zoology.* A new edition. London: Printed for Wilkie and Robinson [etc.], 1812.
4 v. 294 plates. 22 cm. Added title page, engraved, in each volume. Tan cloth binding.
BM IV p.1543. Nissen VBI 710. Yale p.223.

*808

PENNANT, THOMAS (1726-1798). *History of quadrupeds.* London: Printed for B. White, 1781.
2 v. in 1. (xxiv, 566, [14]p.) illus., 52 plates. 26 cm. Preface signed: Thomas Pennant. Armorial bookplate of T. Beale. Contemporary calf binding.
Nissen ZBI 3108.

809

PERRY, GEORGE. *Conchology, or The natural history of shells: containing a new arrangement of the genera and species, illustrated by coloured engravings executed from the natural specimens, and including the latest discoveries.* London: W. Miller, [1811].
2 p.l., 4 p., 62 leaves. 61 hand col. plates. 42.5 cm. Bookplate of Charles Atwood Kofoid. In ms. on fly leaf: Henrietta Brown, 1850. Pckyns Manor, Hurstpierpoint Stewart Pollard, 1889.
BM III p.1551. Nissen ZBI 3134.

810

PITMAN, CHARLES ROBERT SENHOUSE (1890-). *A guide to the snakes of Uganda...* Kampala, Uganda: The Uganda Society, 1938.
2 p.l., xxl, 362 p. col. front., plates (part col.) 2 fold. maps. 24 x 19 cm. "Limited to five hundred copies of which this is number 253." Blue cloth binding.

811

PITMAN, CHARLES ROBERT SENHOUSE (1890-). *A guide to the snakes of Uganda.* Revised edition. Codicote: Wheldon & Wesley, 1974.
xxii, 290 p. [3] fold. leaves, [32] leaves of plates, illus. (some col.) 26 cm. Erratum slip inserted. Red cloth binding.

*812

PLAYFAIR, ROBERT LAMBERT, SIR (1828-1899). *The fishes of Zanzibar. Acanthopterygii,* by R. Lambert Playfair. *Pharyngognathi,* etc., by Albert C.L.G. Gunther. Kentfield, Calif.: N.K. Gregg, [c1971].
32, xiv, 153 p. illus. (part col.) 33 cm. (Hand-colored reprint series, no.1). Reprint of the 1866 edition, together with a new introduction. 350 copies printed. Blue cloth binding.

813

POE, EDGAR ALLAN (1809-1849). *The conchologist's first book: or, A system of testaceous malacology, arranged expressly for the use of schools, in which the animals, according to Cuvier, are given with the shells, a great number of new species added, and the whole brought up, as accurately as possible, to the present condition of the science. With illustrations of two hundred and fifteen shells...* Philadelphia: Published for the author, by Haswell, Barrington, and Haswell, 1839.
156 p. 12 plates. 17 cm. Stated by R.W. Griswold, in the *International Monthly Magazine,* Oct. 1850, to be a copy nearly verbatim, of the text-book of conchology by Captain Thomas Brown, printed in Glasgow in 1833. In ms. on fly leaf: Joseph Jones. Original printed boards.

814

POE, EDGAR ALLAN (1809-1849). *The conchologist's first book: a system of testaceous malacology, arranged expressly for the use of schools, in which the animals, according to Cuvier, are given with the shells, a great number of new species added, and the whole brought up, as accurately as possible, to the present condition of the science. With illustrations of two hundred and fifteen shells, presenting a correct type of each genus.* 2d edition. Philadelphia: Published for the author, by Haswell, Barrington, and Haswell, 1840, [c1839].

166 p. 12 plates. 18 cm. Stated by R.W. Griswold, in the *International Monthly Magazine*, Oct. 1850, to be a copy nearly verbatim, of the text-book of conchology by Captain Thomas Brown, printed in Glasgow in 1833. Original printed boards.

815

PRITCHARD, ANDREW (1804-1882). *The natural history of animalcules: containing descriptions of all the known species of Infusoria...* London: Whittaker and co., 1834.

2 p.l., 196 p. fold. front., 6 plates. 21 cm. Grey library binding.
BM IV p.1617. Nissen ZBI 3243.

816

RAFINESQUE, CONSTANTINE SAMUEL (1783-1840). *The complete writings of Constantine Smaltz Rafinesque, on recent & fossil conchology.* Edited by William G. Binney and George W. Tryon, jr. New York: Bailliere brothers; [etc., etc.], 1864.

96, 7 p. illus., 3 plates. 22.5 cm. In pamphlet binder.
BM IV p.1639.

817

Recherches Zoologiques pour Servir à l'histoire de la faune de l'Amérique centrale et du Mexique, pub. sous la direction de...Milne Edwards. Paris: Imprimerie impériale, 18- .

v. 37 x 28 cm. (Mission scientifique au Mexique et dans l'Amérique centrale). LSU owns pt.3, sect.1, text v.1, v.2 and atlas, v.1, v.2; pt.4; pt.7, v.2. Original printed paper wrappers.
BM II p.608.

*818

REEVE, LOVELL AUGUSTUS (1814-1865). *[Conchologia iconica; or, Monographs of the genera of shells. Including Latin and English descriptions. Continued by G.B. Sowerby].* n.p. [Vincent Brooks, imp., 1843?-1877?].

20?v. illus., col. plates. 29 cm. Title pages missing. LSU lacks vols.7, 19, & 20. Vols.16, 17 black, red library binding; others vols. unbound.
BM IV p.1663. Nissen ZBI 3331.

*819

REICHENBACH, ANTON BENEDICT (1807-1880). *Der Schmetterlingsfreund. Ausführliche Beschreibung der deutschen Schmetterlinge, ihrer Raupen und Puppen nebst fasslicher Anweisung, sie auf zweckmässige Weise zu fangen, zu erziehen, zu tödten, aufzuspannen, systematisch zu ordnen und aufzubewahren, und einem systematischen Verzeichnisse der bekanntesten europäischen Schmetterlinge.* Mit 118 Abbildungen auf 8 Tafeln. Leipzig: H. Hartung, 1852.

vi, 169 p. col. illus. 20 cm. Original green cloth binding.

820

REICHENBACH, HEINRICH GOTTLIEB LUDWIG (1793-1879). *Monographia pselaphorvm, dissertatio entomologica amplissimi philosophorvm ordinis...illvstris ictorvm ordinis concessv in avditorio ivridico D.I.* Lipsiae: Impressit I.B. Hirschfeld, [1816].

77, [3]p. II plates. 21 cm. Dissertation, Leipzig. Paperbound.

*821

REITTER, EWALD. *Beetles.* [Translation by Paul Hamlyn of *Der Kafer.*] New York: Putnam, [c1961].

205 p. illus. (part col.) 35 cm. Tan cloth binding.

822

ROSSMÄSSLER, EMIL ADOLF (1806-1867). *Iconographie der land-und Süfswasser-Mollusken, mit vorzüglicher berücksichtigung der europäischen noch nicht Abgebildeten Arten...* Dresden und Leipzig: Arnoldische Buchhandlung, 1835-1844.

Various pagings. 1 folded table, 60 col. plates. 24.5 cm. Issued in parts: Heft I-VIII (II Bd. I & II), IX-X (II Bd. III, IV), XI (II Bd. V), XII (II Bd. VI). These parts have been rearranged so paging is not in order but all parts are complete. Text and plates have been rebound and renumbered in manuscript. "Index to the new arrangement of the plates, new numbers and original numbers of the text" in 16 ms. pages inserted in rebinding. Half green morocco/ marbled boards, decorated endpapers.
Nissen ZBI 3484.

823

SANDVED, KJELL BLOCH. *Butterfly magic.* Photographs by Kjell B. Sandved; text by Michael G. Emsley. New York: Viking Press, 1975.

128 p. col. illus. 28 cm. Quarter cloth, paper boards, dustjacket.

*824

SAY, THOMAS (1787-1834). *American conchology; or, Descriptions of the shells of North America, illustrated by coloured figures from original drawings executed from nature.* New Harmony, Indiana: Printed at the School press, 1830-[34].

8 v. in 1. 60 col. plates. 22 cm.

——— *A glossary to Say's conchology.* New Harmony, Indiana: Printed by R. Beck & J. Bennett, 1832. (1 p.l., 25 p. 22 cm.) Full leather, spine restored.
BM IV p.1817. Nissen ZBI 3614.

*825

SAY, THOMAS (1787-1834). *American entomology, or Descriptions of the insects of North America. Illustrated by coloured figures from original drawings executed from nature.* [Philadelphia]: Philadelphia Museum, S.A. Mitchell, 1824-28.

3 v. 54 col. plates. 23.5 cm. Slip mounted on lining paper of v.1-2: Mary Addington, from Henry; v.3: Mary Addington. Engraved half-title in v.1. Plates drawn by T.R. Peale. Paper boards with spine labels and label on front cover of v.1. In box.
BM IV p.1817. Nissen ZBI 3612.

826

SAY, THOMAS (1787-1834). *American entomology. A description of the insects of North America...with illustrations drawn and colored after nature.* Edited by John L. Le Conte, with a memoir of the author by George Ord. New York: J.W. Bouton, 1869.

2 v. 54 col. plates. 24.5 cm. Green cloth binding, gilt edges.
Nissen ZBI 3613.

827

SAY, THOMAS (1787-1834). *The complete writings of Thomas Say, on the conchology of the United States.* Edited by W.G. Binney. New York, London: H. Baillière, [etc., etc.], 1858.

vi, 252 p. LXXV col. plates. 24 cm. Blue library binding.
BM IV p.1817. Nissen ZBI 3615.

*828

SAY, THOMAS (1787-1834). *The complete writings of Thomas Say on the entomology of North America.* Edited by John L. Le Conte. With a memoir of the author, by George Ord. New York: Baillière Brothers, 1859.

2 v. 54 (i.e.55) plates, (54 col.) 24 cm. Vol.2, p.551 incorrectly numbered 561. Engraved half-title *American Entomology* in v.1. Half red morocco, green cloth binding.
BM IV p.1817. Nissen ZBI 3613.

829

SCHMARDA, LUDWIG KARL (1819-1908). *Neue wirbellose thiere beobachtet und gesammelt auf einer reise um die erde 1853 bis 1857.* Von Ludwig K. Schmarda. 1.bd... Leipzig: W. Engelmann, 1859-61.

1 v. in 2. illus., XXXVII col. plates. 33 x 25.5 cm. Original green cloth binding.
BM IV p.1842. Nissen ZBI 3698.

830

SCHNACK, FRIEDRICH (1888-). *Joyaux ailés; un atlas des plus beaux papillons du monde.* [2. éd. française]. [Paris]: Hachette, [c1955].

138 p. col. illus., 39 col. plates. Map (on lining paper). 31 cm. Yellow cloth binding.

831

SETON, ERNEST THOMPSON (1860-1946). *Life-histories of northern animals; an account of the mammals of Manitoba.* With 68 maps and 560 drawings by the author. New York: C. Scribner's sons, 1909.

2 v. fronts., illus., plates, maps. 26.5 cm. Paged continuously. Title pages in red and black. Tan cloth binding.
Nissen ZBI 3819.

*832

SETON, ERNEST THOMPSON (1860-1946). *Wild animals at home...with over 150 sketches and photographs by the author.* Garden City, N.Y.: Doubleday, Page & company, 1922, [c1913].

xv, 226 p. illus. 21 cm. Marginal illustrations. The author's presentation copy to Geo. C. Devorshak, with his signed autograph inscription. Original grey cloth binding with bear claw illustration on front cover.
Nissen ZBI 3824.

*833

SIEBOLD, PHILIPP FRANZ VON (1796-1866). *Fauna japonica sive descriptio animalium, quae in itinere per Japoniam, jussu et auspiciis superiorum, qui summum in India Batava imperium tenent, suscepto, annis 1823-1830 collegit, notis, observationibus et adumbrationibus illustravit Ph. Fr. de Siebold. Conjunctis studiis C.J. Temminck et H. Schlegel pro vertebratis atque W. de Haan pro invertebratis elaborata.* Regis auspiciis edita. Lugduni Batavorum, apud auctorem; Amstelodami: apud J. Müller, 1833-50; [Tokyo: Shokubutsu bunken kanko-kai (Society for publication of botanical works), 1934].

4 v. plates (part col., part double), double map. 38 cm. Limited edition of 300 copies. Originally issued in parts; reprinted, with title pages, notes and indexes to plates in Japanese. In Latin (v.1) and French (v.2-4). Label mounted on p.[2] of cover of each vol.: Inoue Book-Company, Hongo, Tokyo. Half dark red morocco/cloth.
BM IV p.1923. Nissen ZBI 3848. Yale p.266.

834

SIMON, HILDA. *The splendor of iridescence; structural colors in the animal world*. New York: Dodd, Mead, [c1971].

268 p. illus. (part col.) 29 cm. Turquoise cloth binding.

835

SIMPSON, GEORGE BANCROFT (1844-1901). *Anatomy and physiology of Anodonta fluviatilis...* [n.p., 19?].

caption-title, pp.[169]-191. 13 plates. 24 cm. "Senate document no.38." "Thirty-fifth report on the [New York] State Museum." In pamphlet binder.

836

SMELLIE, WILLIAM (1749-1795). *The philosophy of natural history*. Dublin: William Porter, 1790.

2 v. 21 cm. In ms. on title pages: Michel Smith. Contemporary calf binding, mended.
BM IV p.1941.

837

SMITH, JAMES EDWARD, SIR (1759-1828). *The natural history of the rarer lepidopterous insects of Georgia. Including their systematic characters, the particulars of their several metamorphoses, and the plants on which they feed. Collected from the observations of Mr. John Abbot, many years resident in that country, by James Edward Smith...* London: Printed by T. Bensley, for J. Edwards [etc.], 1797.

2 v. CIV col. plates. 41.5 cm. Text in English and French. Paged continuously. Full red morocco, gilt decoration, gilt edges.
BM IV p.1946

*838

SMITH, MEREDITH J. *Marsupials of Australia*. Illustrations by Rosemary Woodford Ganf, foreword by J.H. Calaby. Melbourne; New York: Lansdowne Editions, 1980.

v. col. illus. 52 cm. "A limited edition of 1000 copies numbered and signed by the author and the artist this being Number 165." Vol.1, beige cloth binding, tooled leather illustration inserted on front cover.

839

SOAR, CHARLES DAVID. *The British Hydracarina*. By Chas. D. Soar and W. Williamson. London: The Ray Society, 1925-29.

3 v. plates (part col.) 22.5 cm. (The Ray Society [Publications: 110, 112, 115]). Original blue cloth binding with "R.S." on front cover.
BM IV p.1655.

840

SOWERBY, GEORGE BRETTINGHAM (1788-1854). *A catalogue of the shells contained in the collection of the late Earl of Tankerville: arranged according to the Lamarckian conchological system; together with an appendix containing descriptions of many new species...illustrated with several coloured plates*. London: Printed by E.J. Stirling for G.B. Sowerby, 1825.

vii, 92, xxxiv p., [9] leaves of plates, col. illus. 24 cm. Paper boards.
BM V p.1980.

841

SPIX, JOHANN BAPTIST VON (1781-1826). *Serpentum Brasiliensium species novae, ou Histoire naturelle des espèces nouvelles de serpens, recueillies et observées pendant le voyage dans l'intérieur du Brésil dans les années 1817, 1818, 1819, 1820, exécuté par ordre de Sa Majesté le Roi de Bavière, publiée par Jean de Spix, écrite d'après les notes du voyageur par Jean Wagler*. Monachii: Typis Franc. Seraph. Hübschmanni, 1824.

viii, 75 p. 26 col. plates. 36 cm. Errata sheet: p.[77]. Half green morocco/paper.
BM V p.1992. Nissen ZBI 3952.

*842

STAUDINGER, OTTO (1830-1900). *Catalog der Lepidopteren des palaearctischen Faunengebietes*. Von O. Staudinger und H. Rebel. 3. Aufl. des Cataloges der Lepidopteren des europäischen Faunengebietes. Berlin: R. Friedländer & sohn, 1901.

2 v. in 1. front. (port.) 24 cm. In ms. on p.[2] of cover: E.L. Meade-Waldo. Green cloth binding.
BM V p.2005.

843

STEPHENSON, THOMAS ALAN. *The British sea anemones.* London: The Ray Society, 1928-1935.

2 v. illus., plates (part col.) 22.5 cm. (The Ray Society [Publications, v.113, 121]). Bookplate of Charles Oldham. Original blue cloth binding with "R.S." in gilt on front cover, gilt edges.
BM VIII p.1247. Nissen ZBI 3996.

844

SWAINSON, WILLIAM (1789-1855). *Exotic conchology.* [Princeton, N.J.: Published by Van Nostrand for the Delaware Museum of Natural History, 1968].

xxiv, 47 p. 48 col. plates. 30 cm. Includes facsimile title pages of the 1821, 1834, and 1841 editions. An edition of 2,000 copies. Reproduction of the 1841 edition with excerpts from the 1821 and 1834 editions. Red cloth/marbled endpapers.
BM V p.2054. Nissen ZBI 4049 (orig. ed.)

845

SWAINSON, WILLIAM (1789-1855). *On the habits and instincts of animals.* London: Printed for Longman, Orme, Brown, Green, & Longmans [etc.], 1840.

vi, 375 p. illus. 17 cm. (The Cabinet cyclopaedia. Conducted by D. Lardner. Natural history. [v.116]). Half green calf/marbled boards, spine taped.
BM V p.2055. Yale p.282.

846

SWAINSON, WILLIAM (1789-1855). *On the history and natural arrangement of insects.* By William Swainson and W.E. Shuckard. London: Printed for Longman, Orme, Brown, Green, & Longmans [etc.], 1840.

4 p.l., 406 p. illus. 17 cm.(The Cabinet Cyclopaedia. Conducted by D. Lardner...Natural history. [v.104]). Half black morocco/marbled boards.
BM IV p.2055.

847

SWAINSON, WILLIAM (1789-1855). *On the natural history and classification of fishes, amphibians, and reptiles...* London: Printed for Longman, Orme, Brown, Green & Longmans [etc.], 1838-39.

2 v. illus. 17.5 cm. (The Cabinet Cyclopaedia. Conducted by...D. Lardner...Natural History. [v.119, 120]). Added title page engraved: *The natural history of fishes...* Half black morocco/marbled boards.
BM V p.2054. Nissen ZBI 4051.

848

SWAINSON, WILLIAM (1789-1855). *On the natural history and classification of quadrupeds.* London: Printed for Longman, Brown, Green, and Longmans [etc.], 1835.

viii, 397, [1]p. illus. 16.5 cm. (The Cabinet cyclopaedia) Engraved title page. Half green morocco/marbled boards, spine taped.

849

SWAINSON, WILLIAM (1789-1855). *A treatise on malacology; or, Shells and shell fish.* London: Printed for Longman, Orme, Brown, Green & Longmans [etc.], 1840.

viii, 419 p. illus. 16.5 cm. Engraved title page. (The Cabinet cyclopaedia. Conducted by...D. Lardner... Natural history. [v.122]). Half green morocco/marbled boards, spine taped.
BM V p.2054.

850

SWAMMERDAM, JAN (1637-1680). *Bybel der natuure...of historie der insecten, tot zeekere zoorten gebracht: door voorbeelden, ontleedkundige onderzoekingen van veelerhande kleine gediertens, als ook door kunstige kopere plaaten opgeheldert. Verrykt met ontelbaare waarnemingen van nooit ontdekte zeldzaamheden in de natuur. Alles in de Hollandsche, des auteurs moedertaale, beschreven...* Leyden: I. Severinus, 1737-38.

2 v. illus. 39 cm. Paged continuously. Added title page in Latin. Dutch and Latin in parallel columns.

——— *De Tafereelen en derzelver korte uitleggingen, waar door de Bybel der Natuure van Joan Swammerdam wert opgeneldert.* [Leyden: I. Severinus, 1738]. (1 v. illus. 39 cm.) Called v.3 on spine. Vellum binding.
BM V p.2055. Nissen ZBI 4054.

851

TENNENT, JAMES EMERSON, SIR, BART. (1804-1869). *Sketches of the natural history of Ceylon; with narratives and anecdotes illustrative of the habits and instincts of the mammalia, birds, reptiles, fishes, insects, & c. including a monograph of the elephant...* London: Longman, Green, Longman, and Roberts, 1861.

xxiii, [1], 500 p. front., illus., plates. 19 cm. In ms. on fly leaf: Perceval Maxwell. Original green cloth binding.
BM V p.2084. Nissen ZBI 4095. Yale p.286.

852

TERVUREN, BELGIUM. *Musée Royal de l'Afrique Centrale. Zoologie.* Sér.1, t.1- , Nov. 1898- ; Sér.2, t.1- ,

Sept. 1907-; Sér.3, section 1, t.1-, Nov. 1903-; section 2, t.1- , Juil. 1909-; section 3, t.1-, Juin 1925-; Sér.4, t.1-Nov. 1905-; Sér.5, t.1- , 19—: Bruxelles.

v. illus. (part col.) 34-38 cm. LSU set incomplete: lacks scattered numbers. Black library binding, some numbers unbound.

853

THAYER, GERALD HANDERSON (1883-). *Concealing-coloration in the animal kingdom; an exposition of the laws of disguise through color and pattern.* Being a summary of Abbott H. Thayer's discoveries, by Gerald H. Thayer, with an introductory essay by A.H. Thayer, illustrated by Abbott H. Thayer, Gerald H. Thayer, Richard S. Meryman and others and with photographs. New York: The Macmillan co., 1909.

xix, 260 p. front. plates (part. col.) 28.5 cm. Original dark green cloth binding, spine mended.
BM V p.2088. Nissen VBI 930. Yale p.287. Zimmer p.630.

854

THEOBALD, FREDERICK VINCENT (1868-). *A monograph of the Culicidae, or mosquitoes.* Mainly compiled from the collections received at the British Museum from various parts of the world in connection with the investigation into the cause of malaria conducted by the Colonial Office and the Royal Society. London: Printed by order of the Trustees, 1901-10.

5 v. illus., plates, photos., fold. diagr. and atlas (viii p., XXXVII col. plates, photos.) 22 cm. LSU owns v.2, 3, and atlas. Brown cloth binding.
BM VIII p.1298. Nissen ZBI 4108.

*855

TINKER, SPENCER WILKIE. *Hawaiian fishes; a handbook of the fishes among the islands of the central Pacific Ocean.* Illustrated by Gordon S.C. Chun and Y. Oda. Honolulu, Hawaii: Tongg Publishing Company, 1944.

404 p. illus., 8 col. plates. 24 cm. Illustrated paper boards.

856

TOPSELL, EDWARD (1572-1625?). *The history of four-footed beasts and serpents and insects.* With a new introduction by Willy Ley. New York: Da Capo Press, 1967.

3 v. (1130 p.) illus. 32 cm. "An unabridged republication of the 1658 edition published in London... Reproduced from a copy in the rare book collection of the Library of the American Museum of Natural History." Quarter imitation leather/paper boards.
BM V p.2122 (orig. ed.) Nissen ZBI 4147.

857

TORTORI, EGISTO. *Genesi organizzazione e metamorfosi degli Infusori.* Opera postuma di Egisto Tortori. Firenze: S. Landi, 1895.

196 p. plates (part col.) 29 cm. Printed boards.
BM IV p.2126. Nissen ZBI 409.

*858

TREMEAU DE ROCHEBRUNE, ALPHONSE (1834-). *Faune de la Sénégambie.* Tome 1. Paris: O. Doin, 1883-1887.

8 pts. in 3 v. col. illus., fold. map, port. 29 cm. No more published. Originally issued in 5 fasc., an introduction and table, an atlas, and supplement. Original covers and title pages for each part included. Part nos. arbitrarily assigned. Stamped on covers of pts.2, 3, and 7: Bibliotheque Adrien Dollfus. Pts.3 and 7 are author's autograph presentation copies. Green cloth binding.
BM IV p.1715. Nissen ZBI 3447.

859

TRYON, GEORGE WASHINGTON (1838-1888). *Structural and systematic conchology; an introduction to the study of the Mollusca...* Philadelphia: The author, 1882-84.

3 v. front. (fold. map), 140 plates (incl. front. of v.3). 23 cm. LSU owns only v.3. From the library of L.S. Frierson, conchologist. Original cloth binding.
BM V p.2146. Nissen ZBI 4175.

860

TURTON, WILLIAM (1762-1835). *Conchylia insularum Britannicarum. The shells of the British islands systematically arranged.* London: M.A. Nattali, [etc., etc., 1822].

xivii, [1], 279, [1]p. 20 col. plates. 25 x 20 cm. Descriptions of the species in Latin and English. Reissued as *Bivalve shells of the British islands*, London, 1830. In ms. on fly leaf struck out: Caroline Baker, 1832. In ms. on fly leaf: W.J. Corbett Winder. Original brown cloth binding.
BM V p.2155. Nissen ZBI 4189.

861

UNITED STATES. WAR DEPT. *Report of the Secretary of War, communicating, in compliance with a resolution of the Senate of February 2, 1857, information respecting the purchase of camels for the purposes of military transportation.* Washington: A.O.P. Nicholson, printer, 1857.

238 p. illus., fold. plates. 23 cm. (34th Cong., 3d sess. Senate. Ex. doc. 62). Jefferson Davis, Secretary of War. "The Zemboureks, or the dromedary field artillery of the Persian Army. By Colonel F. Colombari... Extracted from the *'Spectateur Militaire'*, 1853. Translated by Brevet Major Henry C. Wayne, U.S. Army, 1854.": pp.[201]-237. Full burnt orange morocco binding, gilt decorated spine. Binding by William A. Payne.

862

VEJDOVSKÝ, FRANTISEK (1849-1939). *System und Morphologie der Oligochaeten, bearb. im Auftrage des Comité's für naturhistorische Landesdurchforschung Böhmens. Veröffentlicht durch Subvention der Akademie der Wissenschaften in Wien.* Prag: F. Rivnác, 1884.

166 p. 16 col. plates. 40 cm. Printed boards.
BM V p.2199. Nissen ZBI 4241.

*863

VIALLANES, HENRI. 1, thèse: *Recherches sur l'histologie des insectes et sur les phénomènes histologiques qui accompagnent le développement postembryonnaire de ces animaux.* 2, thèse: *Propositions données par la faculté par H. Viallanes.* Paris: G. Masson, 1883.

2 p.l., 348, 2 l., 12 p. 19 leaves of plates, illus. 25 cm. Thesis: *Faculté des sciences de Paris.* This copy is the author's presentation copy to M. Duval, with his signed autograph inscription. In ms. on fly leaf: Warren Whitcomb, Jr. Quarter brown morocco, marbled boards.
BM I p.2210.

864

WESTWOOD, JOHN OBADIAH (1805-1893). *An introduction to the modern classification of insects; founded on the natural habits and corresponding organisation of the different families.* London: Longman, Orme, Brown, Green, and Longmans, 1839-40.

2 v. col. front. (v.1), illus. 21.5 cm. Half brown calf/marbled boards.
BM V p.2302. Nissen ZBI 4380.

865

WIEDERSHEIM, ROBERT ERNST EDUARD (1848-1923). *Die Anatomie der Gymnophionen.* Jena: Gustav Fischer, 1879.

vii, 101 p. 9 colored plates. 30 cm. Bookplate of Hans Theodore Rust. Half cloth/marbled boards.
BM V p.2316. Nissen ZBI 4403.

*866

WILKES, BENJAMIN. *One hundred and twenty copperplates of English moths, butterflies; representing their changes into the caterpillar, chrysalis, and fly-states, and the plants, flowers, and fruits whereon they feed. Coloured with great exactness from the subjects themselves. With a natural history of the moths and butterflies, describing the method of managing, preserving, and feeding them. To which is added, an index of the insects and plants, adapted to Linnaeus's system.* London: Printed for R.N. Rose, 1824.

viii, [22], 63, [4]p. hand col. plates. 32 cm. Earlier editions issued under title: *The English moths and butterflies.* In ms. on p.[2] of cover: From Dr. Quesnel to Alfred Denzilve Esquire of Jersey by whom presented to Thomas Sorrell, 1867. In ms. on fly leaf: Christiana Pope, Alveston Cottage, Sutton-Surrey, August 4th 1880. Further ms. notes. Half green morocco/cloth.
BM V p.2322 (1773 ed.) Nissen ZBI 4410a.

*867

WILSON, JAMES (1795-1856). *Illustrations of zoology; being representations of new, rare, or remarkable subjects of the animal kingdom, drawn and coloured after nature, with historical and descriptive details.* Edinburgh: W. Blackwood; London: T. Cadell, 1831.

ii, 4, [141]p. 36 col. plates. 42 cm. On spine: v.1. Half green morocco/marbled boards, marbled endpapers.
BM V p.2333. Nissen VBI 999. Yale p.313.

*868

WOLF, JOSEF (1820-1899). *The life and habits of wild animals.* Illustrated by designs by Joseph Wolf, engraved by J.W. & Edward Whymper. With descriptive letterpress by Daniel Giraud Elliot. London: Alexander Macmillan & co., 1874.

5 p.l., 72 p. 20 plates. 48 cm. Full red morocco, gilt ornamentation, marbled endpapers.
BM V p.2349.

*869

WOLF, JOSEF (1820-1899). *Zoological sketches. Made for the Zoological Society of London, from animals in their vivarium, in the Regent's Park.* Edited, with notes, by Philip Lutley Sclater. London: H. Graves & Co., printsellers to Her Majesty [Vinton & Son, printers], 1861-67.

2 v. 100 mounted col. plates. 62 cm. Added title page for each vol., illustrated in colors. Imprint date on wrapper of

pt.1: 1856. On wrapper of pt.1: Edited, with notes, by D.W. Mitchell. On title page of v.2: Second series. Original wrapper of pt.1 included. Each vol. in portfolio. Original red portfolio.
BM V p.2349. Nissen VBI 1012.

870

WOOD, WILLIAM (1774-1857). *Index testaceologicus; an illustrated catalogue of British and foreign shells, containing about 2800 figures accurately coloured after nature...* A new and entirely revised edition, with ancient and modern appellations, synonyms, localities, etc. etc. by Sylvanus Hanley... London: Willis and Sotheran, 1856.

xx, 234 p. col. plates. 25 cm. In ms. on title page: S.N. Martin. Half dark green morocco/marbled boards, marbled endpapers.
BM V p.2353. Nissen ZBI 4459 (earlier eds.)

871

WOODWARD, SAMUEL PECKWORTH (1821-1865). *A manual of the Mollusca; or, Rudimentary treatise of recent and fossil shells...* Illustrated by A.N. Waterhouse and Joseph Wilson Lowry. London: John Weale, 1851.

158, 12 p. front., plates. 19 cm. In ms. on title page: Professor Chilton, University of Louisiana New Orleans. Original green cloth binding with label on front cover.
BM V p.2359. Nissen ZBI 4472.

872

WOODWARD, SAMUEL PECKWORTH (1821-1865). *A manual of the Mollusca; or, A rudimentary treatise of recent and fossil shells...* Illustrated by A.N. Waterhouse and Joseph Wilson Lowry. London: John Weale, 1851-56.

3 v. in 1. fronts. (v.1, 3) illus., 24 plates. 18.5 cm. (Weale's rudimentary series, no.72). Paged continuously. Green library binding.
BM V p.2359. Nissen ZBI 4472.

873

WOODWARD, SAMUEL PECKWORTH (1821-1865). *A manual of the Mollusca: a treatise of recent and fossil shells.* 2d edition, with an appendix of recent and fossil conchological discoveries to the present time, by Ralph Tate. With numerous illustrations, by A.N. Waterhouse and J.W. Lowry. London: Virtue & co., 1868.

xiv, 542, 85 p. front., illus., plates, fold. map. 18 cm. Original green cloth binding.
BM V p.2359. Nissen ZBI 4472.

*874

YARRELL, WILLIAM (1784-1856). *A history of British fishes.* Illustrated by nearly 400 woodcuts. London: J. Van Voorst, 1836.

2 v. illus. 23 cm. Signed in ms. on lining paper of v.1-2: Burleigh James. Clipping containing bibliographical information for this title mounted on lining paper of v.1. Label mounted on lining paper of each vol. and sup.1: The Wawayanda Company, Sporting and specialty booksellers, Warwick, N.Y., U.S.A.

——— *Supplement* [1]-[2]. Illustrated with woodcuts. London: J. Van Voorst, 1839-60. (pts. in illus. 23 cm.) *Supplement* 2 , 1860, edited by Sir John Richardson. LSU owns only the 1839 supplement (2 pts. in 1 v. illus. 23 cm.) Original green cloth binding.
BM V p.2372. Nissen ZBI 442.

875

Zoological Journal. v.1-5 (nos.1-20), March 1824-1834. London, 1824-34.

5 v. plates (part col.) 22.5 cm. Half green morocco/cloth, marbled endpapers.

——— *Supplement.* no.1-5, 1832/34. [London, 1832-34]. (5 nos. 22.5 cm.) LSU lacks supplement.
BM V p.2396.

876

ZOOLOGICAL SOCIETY OF LONDON. *Transactions.* v.1-; 1835- . London: Printed for the Society, 1835— .

v. plates (part col.) 31 cm. LSU lacks v.4. Suspended publication 1917-25. Bindings vary.
BM V p.2397. Nissen ZBI 4788.

PART VII

Ornithology

Title page from *Nederlandsche Vogelen* by Cornelius Nozeman. Amsterdam, 1770-1829.

* 877

ADAMS, HENRY GARDINER (1811 or 12-1881). *Humming-birds, described and illustrated: with an introductory sketch of their structure, plumage, haunts, habits, etc.* With 8 coloured plates and several wood engravings. London: Groombridge and sons, 1856.

xxiii, [29]-70 p., 1 l. 8 leaves of plates. col. illus. 19 cm. Armorial bookplate of John Davies Knatchbull Lloyd. Original green cloth binding.
Nissen VBI 3. Yale p.3. Zimmer p.2.

878

ALBIN, ELEAZAR (fl. 1713-1759). *A natural history of singing birds; and particularly that species of them most commonly bred in Britain. To which are added, figures of the cock, hen, and egg of each species, exactly copied from nature, and elegantly engraven on copper. Together with the figure, description, and use of the day-net, and the manner of catching small birds of all kinds. By a lover of birds.* Edinburgh: Printed for J. Dickson, 1776.

vii, 120 p. illus. front. 17 cm. Book plate of T.B. Grierson. In ms. on p.[2] of cover: T.B. Grierson, by his father, 1825. Full contemporary calf.
Anker 4. BM I p.25 (1737, 1779 eds.) Nissen VBI 17. Yale p.5 (1779 ed.) Zimmer p.3 (1737 ed.)

879

ALFERAKI, SERGĪEĪ NIKOLAEVICH (1850-1918). *The geese of Europe and Asia; being the description of most of the old world species.* With twenty-four coloured plates by F.W. Frohawk and frontispiece by Dr. P.P. Sushkin. London: R. Ward, ltd., 1905.

ix, 198 p., 1 l. col. front., 24 col. plates. 34.5 x 28 cm. "A translation of my *Gusi rossii* published in Russia in 1904." —Preface. Green cloth binding.
Anker 92. BM VI p.13. Nissen VBI 83. Zimmer p.6.

* 880

ALI, SALIM A. *The book of Indian birds.* With a map, 188 plates in colour (depicting 196 species), 3 in a line and 24 in half-tone. 3d edition, revised and enlarged. Bombay: Bombay Natural History Society, [1944].

xliii, 438 p. illus. (part col.) 19 cm. Stamped on fly leaf: Robert E. Barron, III, M.D. Green cloth binding.
Nissen VBI 20 (1942 ed.) Yale p.6.

881

ALLEN, ELSA GUERDRUM. *The history of American ornithology before Audubon.* Philadelphia: American Philosophical Society, 1951.

387-591 p. illus., ports. 30 cm. (Transactions of the American Philosophical Society, new ser., v.41, pt.3). Green cloth binding.

882

ALLEN, ROBERT PORTER. *Birds of the Caribbean.* New York: Viking Press, 1961.

256 p. illus. (col. plates), map (on lining papers) 29 cm. (A Studio book). Original green cloth binding, gilt decoration.

* 883

Amazon Parrots. Paintings by Elizabeth Butterworth. Text by Rosemary Low. London: Rodolphe d'Erlanger/The Basilisk Press, c1983.

178 p. illus., 27 leaves of plates, maps. 46 cm. Edition limited to 515 signed and numbered copies. LSU owns copy no.54. Accompanied by hand-coloured etching made specifically for the book by Elizabeth Butterworth, no.13 of 70. Red silk binding by Smith and Settle, Bindley, Yorkshire. Black cloth Solander box.

884

ARCHER, GEOFFREY FRANCIS, SIR (1882-). *The birds of British Somaliland and the Gulf of Aden; their life histories, breeding habits, and eggs.* By Sir Geoffrey Archer and Eva M. Godman. [With coloured plates by Archibald Thorburn and Henrik Grönvold]. Edinburgh: Oliver and Boyd, 1937-61.

4 v. col. fronts., illus., plates (part col.) port., 3 fold. maps. 29 cm. Paged continuously. Vols.1-2 published by Gurney and Jackson of London. Green cloth binding, gilt ornamentation.
Nissen VBI 34. Yale p.10.

* 885

ASIATIC SOCIETY, CALCUTTA. MUSEUM. *Catalogue of the birds in the Museum Asiatic Society.* By Edward Blyth. Published by order of the society. Calcutta: J. Thomas, 1849.

xxxiv, 403 p. 23 cm. Imprint date, 1852?, in ms. on title page. Manuscript notes on fly leaf. Stamp of the Royal Geographical Society, London on verso of title page and p.403. Label mounted on cover: R. Friedländer & sohn, Berlin. Label mounted on p.[2] of cover: Williams & Norgate. Label partially covered by bookplate of Royal Geographical Society. "Library of the Royal Geographical Society regulations" (1 leaf) mounted on p.[3] of cover. Quarter green cloth/paper boards.
BM I p.180. Zimmer p.62.

* 886

AUDEBERT, JEAN BAPTISTE (1759-1800). *Oiseaux dorés ou a reflets métalliques.* Par J.B. Audebert et L.P. Vieillot. Paris: Desray, 1802.

2 v. 190 hand col. plates. 36 cm. Quarter brown morocco/marbled boards, marbled endpapers.

Anker 14. BM I p.71. Nissen VBI 47. Yale p.13. Zimmer p.17.

887

AUDOUIN, JEAN VICTOR (1797-1841). *Audouin's Explication sommaire des planches d'oiseaux de l'Égypte et de la Syrie, publieé par Jules-César Savigny.* Edited by Alfred Newton. London, [Cambridge]: Printed by C.J. Clay and son], 1883.

vii, 139, [1]p. 21.5 cm. Reprint of Explication which originally appeared in the 2d edition of Jules-César Savigny's *Description de l'Égypte*, v.23, published in 1828. Original pagination ([302]-456) also given. At head of title: The Willughby Society. Green library binding.

Anker 15. BM I p.71. Zimmer p.18.

888

AUDUBON, JOHN JAMES (1785-1851). *The Audubon folio.* Text by George Dock, Jr. New York: H.N. Abrams, [c1964].

32 p. illus. 36 cm. and portfolio (30 col. plates). 44 cm. Pictorial paper wrappers.

889

AUDUBON, JOHN JAMES (1785-1851). *Audubon in Louisiana.* New Orleans: Louisiana State Museum, Friends of the Cabildo, 1966.

[40]p., chiefly illus. 23 x 31 cm. "Reproductions of *The birds of America* elephant folio from the Louisiana State Museum." Paperbound.

890

AUDUBON, JOHN JAMES (1785-1851). *Audubon watercolors and drawings.* [Exhibited at] Munson-Williams-Proctor Institute, Utica, New York, and The Pierpont Morgan Library, New York, New York. Foreword by Frederick B. Adams, Jr. [Utica, N.Y.: Widtman Press, 1965].

57 p. 95 plates. 27 cm. Paperbound.

* 891

AUDUBON, JOHN JAMES (1785-1851). *Bird biographies.* Selected and edited by Alice Ford. New York: Macmillan, 1957.

xv, 282 p. col. plates. 26 cm. Selections from the author's *Ornithological biography*, the text to his folio, *The birds of America*. Green cloth binding, dustjacket.

892

AUDUBON, JOHN JAMES (1785-1851). *The birds of America.* From original drawings by John James Audubon... London: Published by the author, 1827-38.

4 v. 435 col. plates. 100 cm.Originally issued in 87 parts. Imprint dates: v.1, 1827-30; v.2, 1831-34; v.3, 1834-35; v.4, 1835-38, June 20. Plate 64 drawn from nature by Lucy Audubon. Plates 1-2, 6-7 engraved by W.H. Lizars, retouched by R. Havell, junr.; plates 8-9 engraved by W.H. Lizars; plates 3-5, 101-105, 108, 110 engraved, printed and coloured by R. Havell, junr.; plates 10-100, 106-107, 109, 111-435 engraved, printed and coloured by R. Havell. Plates 32, 49 dated 1828; plates 2, 7, 62 dated 1829; plates 81, 83, 87, 89, 93 dated 1830; plates 106-110,112-115 dated 1831; plates 131-140, 143-155 dated 1832; plates 156-177, 179-182, 184-185 dated 1833; plates 136-197, 199, 202-235 dated 1834; plates 236-287, 289-290 dated 1835; plates 286, 288, 291-350 dated 1836; plates 351-400 dated 1837; plates 401-435 dated 1838. Plate 254 marked "CCLVI." Bookplates of the Duke of Northumberland. "The plates were published without any text, to avoid the necessity of furnishing copies gratis to the public libraries in England, agreeably to the law of copyright." From Sabin, A dictionary of books relating to America, v.1, p.315.

Text to accompany the plates was published in 5 vols., rl. 8 vo., Edinburgh, 1831-39, under the title *Ornithological biography, or, An account of the habits of the birds of the United States of America...* Later editions, of text and plates combined, with alterations, were published in 7 and 8 vols., rl. 8 vo., under the title *The Birds of America, from drawings made in the United States and their territories.*

LSU copy purchased by the LSU Foundation with a grant from the Crown Zellerbach Foundation. Full calf, gilt decoration.

Anker 17. BM I p.71. Nissen VBI 49. Yale p.13. Zimmer pp.18-19.

893

AUDUBON, JOHN JAMES (1785-1851). *The birds of America, from drawings made in the United States and their territories.* New York: J.J. Audubon; Philadelphia: J.B. Chevalier, 1840-44.

7 v. illus., 500 col. plates. 27 cm. Vols.6-7 have imprint: New York, Philadelphia: J.J. Audubon, 1843-44. Stamped on each plate in v.1-7, c.2: Bradford Library. Stamped in v.1-7, c.2: Bound for Bernard Quaritch, ltd., by Sangorski and Sutcliffe. Bookplate of M.H. Bartlett, Vermejo Park, N.M. Half green morocco/marbled boards, marbled endpapers.

BM I p.71. Nissen VBI 51. Yale p.13. Zimmer p.22.

894

AUDUBON, JOHN JAMES (1785-1851). *The birds of America, from drawings made in the United States and their territories.* New York: V.G. Audubon, 1859.

7 v. illus., 500 col. plates. 27 cm. Issued in 100 parts. The majority of the plates were originally published in folio, without text, London, 1827-38. The first edition of the text was published in octavo, Edinburgh, 1831-39, under title: *Ornithological biography.* In all volumes in ms. on fly leaf: W.E. Thrall. In ms. on fly leaf and title page of all vols.: R.T. Bouri. Label on p.[2] of cover of all vols.: W.E. Thrall. Full brown morocco binding.
Yale p.13.

895

AUDUBON, JOHN JAMES (1785-1851). *The birds of America, from drawings made in the United States and their territories...* Re-issued by J.W. Audubon. New York: Roe Lockwood & Son, Publishers, 1861.

7 v. in 5. illus. 26 cm. Each volume has separate title page. Embossed on title page of v.1, 3, 5-7: Jose M. de Cardenas y Rodriguez. Originally presented to the Friends of the LSU Library by Chancellor and Mrs. Murrill (cover letter laid in). Presented by the Friends to the McIlhenny Collection. Half brown calf/marbled boards, gilt decorated spine.
Anker 18. Nissen VBI 50. Yale p.13. Zimmer p.24.

896

AUDUBON, JOHN JAMES (1785-1851). *The birds of America...with an introduction and descriptive text by William Vogt.* New York: The Macmillan company, 1937.

xxvi p., 500 col. plates on 250 leaves. front. (port.) 32 cm. "This edition...is limited to two thousand five hundred copies," "Of the...500 plates, the first 435 were originally published by Audubon in London, during the years 1827-38...a set believed to be the finest in uncutstate in America...has...been followed in making the present reproductions. The final sixty-five subjects were painted at a later date... They were first reproduced in the octavo *Birds of America* published by Audubon in New York (and J.B. Chevalier in Philadelphia), 1840-1844. The present plates 436-500 are reproduced from that book." The colored plates are preceded by a reproduction of the title page of one of the original portfolios dated 1827-1830, not included in the paging. Quarter green cloth/marbled boards.

*897

AUDUBON, JOHN JAMES (1785-1851). *Birds of America.* Introduction and descriptive captions by Ludlow Griscom. Popular edition. New York: Macmillan, 1950.

320 p. (pp.33-320 col. plates) 21 cm. In ms. on fly leaf of copy 5: J.S. Conner, Carlton Hotel, Berkeley, California, Aug. 12, 1950. Quarter green cloth/pictorial boards, dust-jacket.

898

AUDUBON, JOHN JAMES (1785-1851). *The birds of America.* With a foreword and descriptive captions by William Vogt. New York: The Macmillan company, 1957 [c1937].

xxvi p. incl. front. (port.) 435 col. plates on 218 leaves. 31.5 cm. "Published November 1937. Reprinted 1941 ...1957." LSU has another copy, third printing 1961. The colored plates are preceded by a reproduction of the title page of one of the original portfolios dated 1827-30, not included in the paging. "The 435 plates...during the years 1827-1938...a set believed to be the finest in uncut state in America...has...been followed in making the present reproductions." Green cloth binding.

*899

AUDUBON, JOHN JAMES (1785-1851). *Audubon's Birds of America.* Edited by Roger Tory Peterson and Virginia Marie Peterson. New York: Abbeville Press, 1981.

[190]p. [501]p. of plates, col. illus., port. 39 cm. Limited to 2500 copies handbound in leather and signed by the authors. LSU owns copy no.850. Brown calf, gilt decoration, marbled endpapers.

900

AUDUBON, JOHN JAMES (1785-1851). *Birds of America; fifty selections with commentaries by Roger Tory Peterson.* [New York]: Macmillan, [19—].

50 colored plates in portfolio. 33 cm. Beige paperboard portfolio.

*901

AUDUBON, JOHN JAMES (1785-1851). *My style of drawing birds.* Introduction by Michael Zinman. [n.p.]: Published by the Overland Press for the Haydn Foundation, 1979.

26 p. illus. 30 cm. Consists of two essays: "My style of drawing birds," published in M. Audubon's *Audubon and His Journals*, 1897; and "Method of Drawing Birds," published in the *Edinburgh Journal of Science*, v.8, 1828. Green cloth binding.

902

AUDUBON, JOHN JAMES (1785-1851). *The original water-color paintings by John James Audubon for "The*

Birds of America," reproduced in color for the first time from the collection at the New York Historical Society. Introduction by Marshall B. Davidson. [Original edition]. New York: American Heritage Pub. Co.; distributed to booksellers by Houghton Mifflin Co., Boston, 1966.

2 v. illus. (part col.), facsim., 431 col. plates (part fold.), ports. (1 col.) 35 cm. M.B. Davidson edited the descriptive captions, which include many quotations from Audubon's *Ornithological biography*. Brown cloth, decorative endpapers, slipcase.

*903

AUDUBON, JOHN JAMES (1785-1851). *Ornithological biography, or An account of the habits of the birds of the United States of America; accompanied by descriptions of the objects represented in the work entitled "The Birds of America," and interspersed with delineations of American scenery and manners.* Edinburgh: A. Black; [etc., etc.], 1831-1849 [i.e.1839].

5 v. illus. 27 cm. Vol.1 is the author's presentation copy to Dr. Hooker, with his signed autograph inscription. First edition. Later editions of text and plates combined, with alterations, were published in octavo, under the title: *The birds of America, from drawings made in the United States and their territories.*

The folio edition of *The Birds of America* was published without text; v.1 of the present work furnishes text for plates I-C; v.2 for plates CI-CC; v.3 for plates CCI-CCC; v.4 for plates CCCI-CCCLXXXVII; v.5 for plates CCCLXXXVIII-CCCCXXXV, and in addition contains "Descriptions of species found in North America, but not figured in the *Birds of America*: pp.[305]-336, and *Appendix*: comprising additional observations on the habits, geographical distribution, and anatomical structure of the birds described in this work: together with corrections of errors relative to the species": pp.[337]-646. Half red morocco/marbled boards, marbled endpapers.

Anker 18. BM I p.71. Yale p.13. Zimmer p.20.

904

AUDUBON, JOHN JAMES (1785-1851). *A synopsis of the birds of North America.* Edinburgh: A. and C. Black, 1839.

xii, 359 p. 23 cm. "The figures and descriptions contained in the works entitled *The Birds of America*, and *Ornithological biography*... having been issued in the miscellaneous manner which was thought best adapted to the occasion...seemed to require a systematic index, in which the nomenclature should be corrected, and the species arranged agreeably to my present views. This Synopsis, then, will afford a methodical catalogue of all the species...described...in the works above named." — Preface. Stamped on fly leaf: Ex Libris Louis Bennett Bishop.

BM I p.71. Zimmer pp.21-22.

905

BAILEY, FLORENCE AUGUSTA (MERRIAM), MRS. (1863-). *Birds of New Mexico.* With contributions by the late Wells Woodbridge Cooke. Illustrated with colored plates by Allan Brooks, plates and text figures by the late Louis Agassiz Fuertes and many other cuts from drawings, photographs, and maps; based mainly on field work of the Bureau of biological survey, United States Department of Agriculture. [Santa Fe, N.M.]: Published by the New Mexico Department of Game and Fish in cooperation with the State Game Protective association and the Bureau of Biological Survey, 1928.

xxiv, 807 p. col. front., illus., plates (part col.), maps. 25.5 cm. Green library binding.

Nissen VBI 56. Yale p.17.

906

BAILEY, FLORENCE AUGUSTA (MERRIAM), MRS. (1863-). *Birds through an opera glass...* Boston and New York: Houghton, Mifflin and company, 1891.

xiii, 223 p. illus. 17.5 cm. (The Riverside library for young people, no.3). "Many of the articles herein contained were published in the Audubon magazine in 1886. These have been revised and largely rewritten." Blue cloth binding.

Yale p.17. Zimmer p.30.

907

BAILEY, HAROLD HARRIS (1879-). *The birds of Florida; a popular and scientific account of the 425 species and subspecies of birds that are now, and that have been found within the state and its adjacent waters; with special reference to their relation to agriculture...illustrated with 76 full page four-color plates, figuring over 480 birds.* By George M. Sutton; and with an outline map of the state showing areas; and a topographical drawing of a bird... Limited edition. Baltimore: Priv. pub. for the author by the Williams & Wilkins company, 1925.

xxi p., 1 l., 146 p. 1 illus., 76 col. plates, fold. map. 31.5 cm. Brown cloth binding.

Nissen VBI 59. Yale p.17.

908

BAIRD, SPENCER FULLERTON (1823-1887). *Birds.* By Spencer F. Baird with the co-operation of John Cassin and George N. Lawrence. Washington, D.C., 1858.

lvi, 1005 p. 28 cm. (United States. War Department. Reports of exploration and surveys, to ascertain the most practicable and economical route for a railroad from the Mississippi River to the Pacific Ocean. Vol.9). Original brown cloth binding, spine missing.
BM V pp.2176-7. Zimmer pp.646-7.

909

BAIRD, SPENCER FULLERTON (1823-1887). *The birds of North America; the descriptions of species based chiefly on the collections in the Museum of the Smithsonian Institution.* By Spencer F. Baird with the cooperation of John Cassin and George N. Lawrence. With an atlas of one hundred plates. Philadelphia: J.B. Lippincott & co., 1860.

3 p.l., lvi, 1005 p. and atlas of C col. plates. 30 cm. "The present work is, in part, a reprint of the General report on North American birds presented to the Department of war, and published in October, 1858, as one of the series of *Reports of explorations and surveys of a railroad route to the Pacific ocean*. In this volume, however, will be found many important additions and collections." Original brown cloth binding.
Anker 24. BM I p.88. Nissen VBI 62. Yale p.18. Zimmer p.33.

*910

BAIRD, SPENCER FULLERTON (1823-1887). *A history of North American birds.* By S.F. Baird, T.M. Brewer, and R. Ridgway. *Land birds,* illustrated by 64 plates and 593 woodcuts. Boston: Little, Brown, and company, 1874.

3 v. illus., plates. 27.5 cm. LSU v.1-3, c.1 has bookplate: Ex libris E. Robert Wiese. In ms. on half title of all vols., c.1: Francis Beach White. In ms. on fly leaf of v.2, v.3, c.1: E. Robert Wiese, Christmas, 1901. Copy 1, original green cloth binding; copies 2 & 3, original blue cloth binding.
BM I p.88. Nissen VBI 63. Yale p.18. Zimmer p.34.

*911

BAKER, EDWARD CHARLES STUART (1864-1944). *Birds.* 2d edition. London: Taylor and Francis, 1922-30.

8 v. fronts. (v.1-2, 5-6; part col.), illus., fold. map, plates (part col.) 24 cm. (The Fauna of British India, including Ceylon and Burma). The first edition appeared in four volumes, 1889-98, v.1-2 being by E.W. Oates, v.3-4 by W.T. Blanford. Stamped on title page and fly leaf of v.3-6, c.2: Royal Army Medical College library; on title page and fly leaf of v.7-8: R.A.M. College Library. Manuscript note on fly leaf of v.7-8: Zoology, general. LSU copy imperfect: plate 1 wanting; v.2, plates 1-7 wanting. Vols.1-2, half brown morocco/cloth; v.3-8, red cloth binding.
Nissen VBI 68. Yale p.19.

*912

BAKER, EDWARD CHARLES STUART (1864-1944). *The game-birds of India, Burma and Ceylon.* 2d edition. Bombay?: Published by the Bombay Natural History Society; London: J. Bale, Sons & Danielsson, 1921-30.

3 v. 2 col. maps, 78 plates (60 col.) 28 cm. Each vol. has also special title page. Vols.2-3 without ed. note. Coloured plates by H. Grönvold, G.E. Lodge, and J.G. Keulemans. "Reprinted from the *Bombay Natural History Society's Journal*; with corrections & additions." Half green morocco/cloth.
Nissen VBI 67. Yale p.19. Zimmer p.36.

913

BAKER, EDWARD CHARLES STUART (1864-1944). *The Indian ducks and their allies.* With 30 colored plates by H. Grönvold, G.E. Lodge and J.G. Keulemans. Bombay: Bombay Natural History Society, 1908.

xi, 292 p. 30 col. plates. 28 cm. "Reprinted from the *Bombay Natural History Society's Journal*: with corrections & additions." "This edition limited to 1200 copies, after which the lithographs will be erased. Half green morocco/cloth, marbled endpapers.
BM VI p.50. Nissen VBI 65.

*914

BAKER, HAROLD ROBERT (1869-   ). *The birds of southern India, including Madras, Malabar, Travancore, Cochin, Coorg and Mysore.* By H.R. Baker and Chas. M. Inglis. Madras: Printed by the Superintendent, Govt. Press, 1930.

xxxiii, 504 p. [22] leaves of plates, illus. (some col.) 24 cm. Stamped on title page: Táru, no.3643/111, Mori Gate, Delhi-110006, India. Additional illus. mounted on p. x, 351, 436, 463, 465, 485 and verso of plates 2, 6, 18-22. LSU copy imperfect: front. (plate 1) wanting. Printed cloth binding.

915

BANNERMAN, DAVID ARMITAGE (1886-   ). *Birds of the Atlantic islands.* Illustrated in colour by D.M. Reid-Henry. Edinburgh: Oliver & Boyd, [1963-68].

4 v. illus., maps, col. plates, ports. 28 cm. Vols.2-4 illustrated by D.A. Bannerman and W.M. Bannerman. Vol.3 illustrated by D.M. Reid-Henry and G. Lodge; v.4 illustrated by D.M. Reid-Henry and P.A. Clancey. Blue cloth binding.

916

BANNERMAN, DAVID ARMITAGE (1886-   ). *The birds of West and Equatorial Africa.* With a foreword by Sir Alan Burns. Edinburgh: Oliver and Boyd,

[1953].

2 v. (xiii, 1526 p.) illus. (part col.) 24cm. "Based on [the author's]...*Birds of tropical West Africa.*" Red cloth binding. Nissen VBI 3: 73a.

*917

BANNON, LOIS ELMER. *Handbook of Audubon prints.* By Lois Elmer Bannon and Taylor Clark. Gretna, La: Pelican Publishing Co., 1980.

122 p. illus. 22 cm. This copy is the authors' presentation copy to Jack McIlhenny, with their signed autograph inscription. Brown cloth binding, dustjacket.

*918

BARTLETT, EDWARD (d. 1908). *A monograph of the weaver-birds, Ploceidae, and arboreal and terrestrial finches, Fringillidae...* Maidstone: The author, 1888-89.

201 p., 31 leaves of plates. 27 cm. Cover title. Various pagings. Originally issued in 5 pts. Cover of pt.1 bound in. List of plates in ms.; plates numbered in ms. In ms. on fly leaf: Frances L. Bunyard. "Prospectus" ([5] leaves) inserted. Includes 3 clippings containing bibliographical information mounted on leaf with ms. notes pertaining to the publication of this work. "Obituary: Mr. Edward Bartlett" (pp.43-44) from the Maidstone Museum Report, 1907, inserted. Includes 2 newspaper clippings of the obituary of the author. Half red morocco/cloth binding.
BM I p.104. Nissen VBI 77. Yale p.22. Zimmer p.41.

*919

BECHSTEIN, JOHANN MATTHAUS (1757-1822). *The natural history of cage birds; their management, habits, food, diseases, treatment, breeding, and the methods of catching them.* A new edition. London: Groombridge, [1837?].

vi, 311 p. col. front., illus., col. plates. 18 cm. Stamped on p.[2] of cover: Robert E. Barron, III, MD. LSU copy imperfect: front. wanting. Red cloth binding, gilt ornamentation, gilt edges.
Yale p.24.

*920

BEEBE, CHARLES WILLIAM (1877-1962). *A monograph of the pheasants.* London: Published under the auspices of the New York Zoological Society by Witherby & co., 1918-22.

4 v. col. fronts., plates (part col.) maps. 41.5 cm. Dark red cloth binding.
Anker 31. Nissen VBI 84. Yale p.24. Zimmer p.49.

921

BEEBE, CHARLES WILLIAM (1877-1962). *Pheasants, their lives and homes.* 1st edition. Published under the auspices of the New York Zoological Society. Garden City, N.Y.: Doubleday, Page & company, 1926.

2 v. col. fronts., plates (part col.) "In its first form, William Beebe's *Monograph* was published...in four folio volumes at a cost of two hundred and fifty dollars for each set... The present volumes include all but the technical descriptions... Much of it has been rewritten and brought up to date." Original green cloth binding with gilt medallion.
Anker 31. Nissen VBI 85. Yale p.24.

*922

BEEBE, CHARLES WILLIAM (1877-1962). *Pheasants, their lives and homes.* Garden City, N.Y.: Published under the auspices of the New York Zoological Society by Doubleday, Doran, 1931 [c1926].

2 v. illus. (part col.), map. 27 cm. "In its first form, William Beebe's *Monograph* was published...in four folio volumes at a cost of two hundred and fifty dollars for each set... The present volumes include all but the technical descriptions... Much of it has been rewritten and brought up to date." Original green cloth binding with gilt medallion.
Anker 31. Nissen VBI 85. Yale p.24.

923

BELEM, BRAZIL. *Museu Paraense Emilio Goeldi.* Die Vogelwelt des Amazonenstromes,... Veroffenlicht auf Anordnung von Jose Paes de Carvalho; Zeichnungen von Ernst Lohse. Zürich: Polygraphischen Institutes, 1900-1906.

3 v. in 1. 48 col. plates. 31 cm. In portfolio, original covers retained.
BM IV p.1882. Nissen VBI 843. Yale p.257. Zimmer p.562.

*924

BELON, PIERRE (1517?-1564). *L'histoire de la nature des oyseaux: avec leurs descriptions, & naïfs portraicts retirez du naturel; escrite en sept livres...* Paris: On les vend en la grand salle du Palais, en la boutique de G. Corrozet, pres la chambre des consultations, 1555.

14 p.l., 381 p. illus. 33 cm. Colophon: Imprime a Paris par Benoist Preuost, demeurant en la rue Frementai, pres le cloz Bruneau, a l'enseigne de l'estoille d'or. Manuscript notes on title page and throughout the text. LSU copy imperfect: p.174 incorrectly numbered 176; pp.377-378 wanting. Half brown calf/marbled boards.
BM I p.131. Nissen VBI 86. Yale p.26. Zimmer p.52.

925

BENOIT, LUIGI. *Ornitologia Siciliana, o sia catalogo ragionato degli uccelli che si trovana in Sicilia.* Messina: G. Fiumara, 1840.

viii, 231 p. 22 cm. Full marbled paper over boards.
BM I p.136. Zimmer p.54.

926

[BERWICK, E.J.H., ed. *Birds of Borneo.* Kuching]: Published by Borneo Literature Bureau in association with Longmans of Malaya ltd., c1963.

6 charts in portfolio. 40 x 55 cm. Unbound.

*927

BEWICK, THOMAS (1753-1828). *Figures of British land birds, engraved on wood by T. Bewick. To which are added, a few foreign birds with their vulgar and scientific names.* Vol.1. Newcastle upon Tyne: Printed by S. Hodgson for R. Beilby and T. Bewick, 1800.

2 p.l., 134 plates. 24 cm. No more published. All plates numbered except the last. In ms. on fly leaf: William Guidry Clark. Devys 1844. Full calf, figures of birds tooled and painted on front and back boards.
BM I p.158. Yale p.29. Zimmer pp.58-59.

*928

BEWICK, THOMAS (1753-1828). *A history of British birds.* Newcastle: Printed by C.H. Cook, for R.E. Bewick; sold by him, Longman, London, 1832.

2 v. illus. 23 cm. Quarter calf/marbled boards.
BM I p.158 (other eds.) Yale p.29. Zimmer pp.58-59.

929

BEWICK, THOMAS (1753-1828). *A natural history of foreign birds.* Thirty-four engravings on wood. Alnwick, Eng.: [n.p., 1814?].

36 p. illus. 14 cm. Ex libris George Allison Armour. Full green morocco, gilt decoration. Binding by Zaehnsdorf.

*930

BEWICK, THOMAS (1753-1828). *A natural history of water birds.* Thirty-four engravings on wood. [n.p.]: Alnwick, [1814?].

36 p. illus. 14 cm. Stamped on p.[2] of cover: Bound by Zaehnsdorf. Ex libris George Allison Armour. Full green morocco, gilt decoration.

*931

BIAGGI, VIRGILIO. *Las aves de Puerto Rico.* Illustraciones por Lucila Madruga de Piferrer y Christine Boyce. 2d edition revised. [Rio Piedras]: Editorial Universitaria, Universidad de Puerto Rico, 1974.

xii, 373 p. illus., col. plates. 25 cm. Captions for illustrations and plates in Spanish, English, and Latin. Card mounted on p.[2] of cover: Francisco Gonzales Veve. Red cloth binding.

*932

[*Bird Notes.* n.p.: n.p., 1910-1916].

11 pamphlets in 1 v. illus. (some col.), maps. 24 cm. Label mounted on p.[2] of cover: Plantation Book Shop, Natchez, Miss. Contents: Cooke, W.W. — *The migratory movements of birds in relation to the weather,* 1910; McAtee, W.L. — *Our grosbeaks and their value to agriculture,* 1911; McAtee, W.L., and Beal, F.E.L. — *Some common game, aquatic, and rapacious birds in relation to man,* 1912; *Fifty common birds of farm and orchard,* 1913; Cooke, W.W. — *Our shorebirds and their future,* 1914; Kalmbach, E.R. — *Birds in relation to the alfalfa weevil,* 1914; Cooke, W.W. — *Distribution and migration of North American rails and their allies,* 1914; Beal, F.E.L., McAtee, W.L., and Kalmbach, E.R. — *Common birds of southeastern United States in relation to agriculture,* 1916; Beal, F.E.L. — *Some common birds useful to the farmer,* 1915; Cooke, W.W. — *Bird migration,* 1915; Cooke, W.W. — *Distribution and migration of North American gulls and their allies,* 1915. Red cloth binding.

933

Distribution and migration of North American gulls and their Birds of an Indian garden. By Thomas Bainbrigge Fletcher and C.M. Inglis. 2nd edition, revised and enlarged. Calcutta: Thacker, Spink, 1936.

xii, 201 p. illus., plates (part col.) 26 cm. "All the coloured plates, except three, and most of the text of these articles originally appeared in the *Agricultural Journal of India* during the years 1919-1924 under the title of 'Some common Indian birds.'" — Publishers' note. Green cloth binding, gilt ornamentation.
Nissen VBI 323. Yale p.96.

934

Birds of Cyprus. By David A. Bannerman and W. Mary Bannerman. Illustrated in colour by D.M. Reid-Henry and Roland Green. Edinburgh: Oliver and Boyd, [c1958].

lxix, 384 p. illus., 31 plates (part col.) fold. col. map, tables. 28 cm. Green cloth binding.

*935

Birds of Nepal; with reference to Kashmir and Sikkim. By Robert L. Fleming, Sr., Robert L. Fleming, Jr., Lain Singh Bangdel; foreword by Elvis J. Stahr; illustrated by Hem Poudyal...[et al.]. Kathmandu: Flemings, 1976.

349 p. col. illus., maps (on lining papers). 19 cm. Red cloth binding, dustjacket.

*936

The Birds of Zimbabwe. Text by Michael P. Stuart Irwin. Plates by Peter Fogarty. Salisbury, Zimbabwe: Quest, 1981.

xvi, 464 p., [25]p. of plates. illus. (some col.), maps. 24 cm. Limited edition of 140 signed copies, 125 of which are numbered and 15 lettered. LSU owns copy no.60. With author's autograph on title page and fly leaf. Issued in slipcase. Full green calf binding and slipcase, map on lining papers. Printed and bound by Mardons of Bulawayo, Zimbabwe.

*937

BLAAUW, FRANS ERNST (1860-). *A monograph of the cranes.* Illustrated by 22 coloured plates (the greater number drawn under the immediate superintendence of the late Dr. G.F. Westerman) by Heinrich Leutemann and J.G. Keulemans. Leiden; London: E.J. Brill; R.H. Porter, 1897.

viii, 64, [2]p., 1 l. illus., 18 (i.e. 22) col. mounted plates. 48 cm. Original green pictorial cloth binding.
BM I p.168. Yale p.31. Zimmer p.59.

*938

BLAKE, EMMET REID (1908-). *Manual of neotropical birds.* Chicago: University of Chicago Press, 1977- .

v. illus., maps. 26 cm. Brown cloth binding, dustjacket.

*939

BLAKSTON, W.A. *The illustrated book of canaries and cage-birds, British and foreign.* By W.A. Blakston, W. Swaysland and August F. Wiener. London: Cassell, [188-?].

viii, 448 p. [56] leaves of plates. illus. (some col.) 29 cm. Half red morocco/cloth, marbled endpapers, marbled edges.
BM I p.171. Nissen VBI 108. Yale p.32.

*940

BOLTON, JAMES (fl. 1775-1795). *Harmonia ruralis; or, an essay towards a natural history of British song birds. Illustrated with figures the size of life, of the birds, male and female, in their most natural attitudes; their nests and eggs, food, favourite plants, shrubs, trees, c.&c. faithfully drawn, engraved, and coloured after nature.* By the author, on forty copper-plates. Stannary, near Halifax [Eng.]: Printed for and sold by the author; sold also by B. and J. White, in London, 1794-96.

2 v. in 1 (viii, 82 leaves). 80 hand col. plates. 34 cm. Bound by H.G. Bohn, London. Full calf binding, marbled endpapers, marbled edges.
BM I p.192. Nissen VBI 115. Yale p.34. Zimmer p.64.

*941

BONAPARTE, CHARLES LUCIEN JULES LAURENT, PRINCE DE CANINO (1803-1857). *American ornithology; or, The natural history of birds inhabiting the United States, not given by Wilson.* With figures drawn, engraved, and coloured from nature. Philadelphia: Carey, Lea & Carey, 1825-33.

4 v. 27 hand col. plates. 38 cm. In ms. on fly leaf of all volumes: Paschal P. Pope. Half red morocco/marbled boards.
Anker 47. BM I p.194. Nissen VBI 116. Yale p.34. Zimmer pp.64-65.

*942

BONAPARTE, CHARLES LUCIEN JULES LAURENT, PRINCE DE CANINO (1803-1857). *A geographical and comparative list of the birds of Europe and North America.* London: J. Van Voorst, 1838.

vii, 67 p. 23 cm. Stamped on fly leaf: Roderick D.M. Chisholm and Erchless Circulating Library. Blue cloth binding.
BM I p.194. Yale p.34. Zimmer p.67.

943

BONAPARTE, CHARLES LUCIEN JULES LAURENT, PRINCE DE CANINO (1803-1857). [7 article reprints, all unbound]

_______ *Monographie des Laniens.* Extrait de la *Revue et magasin de zoologie*, no.7, 1853. (8 p. 20.5 cm.)
_______ *Notes sur les Larides.* Extrait de la *Revue et magasin de zoologie*, no.11, 1854. (13 p. 22.5 cm.)
_______ *Notes sur les Tangaras et leurs affinités.* Descriptions d'espèces nouvelles. Extrait de *Revue et magasin de zoologie*, no.3, Mars 1851. (51 p. 23 cm.)
_______ *Notes sur le genre Moquinus nouvelle forme intermédiaire aux Turnides, aux Laniides et aux Muscicapides; sur le nouveau genre Myiagrien schwaneria et sur le catalogue des oiseaux d'Europe et d'Alegérie.* Extrait de la *Revue et magasin de zoologie*, no.2, 1857. (15 p., 1 plate.

22 cm.)

——— *Tableau des oiseaux de proie*. Extrait de la *Revue et magasin de zoologie*, no.8, 1854. (16 p. 22 cm.)

——— *Tableau des oiseaux-Mouches*. Extrait de la *Revue et magasin de zoologie*, no.5, 1854. (12 p. 22 cm.)

——— *Tableau des perroquets*. Extrait de la *Revue et magasin de zoologie*, no.3, 1854. (16 p. 22 cm.)

944

BONHOTE, JOHN LEWIS JAMES (1875-1922). *Birds of Britain*. With 100 illustrations in colour selected by H.E. Dresser from his *Birds of Europe*. London: A. and C. Black, 1907.

x, 404 p., 1 l. incl. plates. 100 col. plates (incl. front.) 23 cm. Each plate accompanied by guard sheet with descriptive letterpress. Original brown binding with gilt pictorial cloth.
Anker 50. Nissen VBI 120. Yale p.35. Zimmer p.78.

945

BONHOTE, JOHN LEWIS JAMES (1875-1922). *Birds of Britain and their eggs, with eighty-two full-page illustrations in colour*. [3d edition]. London: A. & C. Black, [1927].

vi, 404 p. 82 plates. illus. front. 23 cm. Half red morocco/cloth, marbled endpapers, with bird medallions in gilt on front cover.
Anker 50. Nissen VBI 120. Zimmer pp.79-80.

946

The Book of birds; birds of town and country, the warblers and American game birds, with 331 color portraits of North American birds and 94 illustrations in black and white... Washington, D.C.: The National Geographic Society, 1925.

4 p.l., 215 p. illus. (part col., incl. maps). 25.5 cm. "Color plates from paintings from life by Louis Agassiz Fuertes." "Articles published in the National geographic magazine during the last ten years." Published in part in 1914 under title: *Common birds of town and country*. In ms. on p.[2] and [3] of cover: Van Deusen, H.M. Tan cloth binding.
Yale p.202.

*947

The Book of birds; common birds of town and country and American game birds. By Henry W. Henshaw. Illustrated in natural colors with 250 paintings by Louis Agassiz Fuertes. With chapters on "Encouraging birds around the home," by F.H. Kennard; "The mysteries of bird migration," by Wells W. Cooke; and "How birds can take their own portraits," by George Shiras, and 45 illustrations and 13 charts in black and white. Washington: [c1921].

viii, 195 p. illus. (part col.), maps. 26 cm. "Published in the National geographic magazine during the last six years." Published in part in 1914 under title: *Common birds of town and country*. Tan cloth binding.
Yale p.202.

*948

The Book of birds, the first work presenting in full color all the major species of the United States and Canada. Edited by Gilbert Grosvenor and Alexander Wetmore with 950 color portraits by Major Allan Brooks. Washington, D.C.: National Geographic Society, [c1937].

2 v. illus. (part col., incl. maps). 26.5 cm. Green cloth binding.
Yale p.202.

*949

BOOTH, EDWARD THOMAS (1840-1890). *Rough notes on the birds observed during twenty-five years' shooting and collecting in the British Islands*. With plates from drawings by E. Neale, taken from specimens in the author's possession. London: Published by R.H. Porter and Messrs. Dulau, 1881-1887.

3 v. illus., 2 col. maps, 114 col. plates. 44 cm. Vol.2 has armorial bookplate of Kylemore Castle, with motto: Vincit veritas. Half brown morocco/cloth, marbled endpapers.
Anker 51. BM I p.199. Nissen VBI 121. Yale p.35. Zimmer pp.79-80.

950

BORRER, WILLIAM. *The birds of Sussex*. London: R.H. Porter, 1891.

xviii, 385 p. fold. col. map, col. plate. 23 cm. Blue cloth binding with birds in gilt on spine.
BM I p.203. Nissen VBI 123a. Yale p.36. Zimmer p.82.

951

BRABOURNE, WYNDHAM WENTWORTH KNATCHBULL-HUGESSEN, 3D BARON (1885-1915). *The birds of South America*. By Lord Brabourne and Charles Chubb. London: R.H. Porter [etc., 1912]-17.

2 v. in 1. 38 col. plates, fold. map. 29 cm. Vol.II (Plates), issued in 6 parts, 1915-17. "The work was to have comprised 16 volumes with 400 hand coloured plates, but owing to the death of Lord Brabourne at Neuve-Chapelle in 1915 only vol.I of the text has been published. The 38 plates already finished by Mr. Grönvold, with notes on most of the species by H. Kirke Swann, have been issued as vol.II.

Publisher's note, v.2. Half red morocco/cloth, marbled endpapers; original covers bound in back.
Anker 56. BM VII p.572. Nissen VBI 129. Yale p.38.

952

BRASHER, REX (1869-1960). *Birds & trees of North America.* Comprising 875 full-color plates: 1094 species and sub-species of birds, 383 species of trees. With an explanatory text by the artist, edited [rev.] and annotated by Lisa McGaw. New York: Rowman and Littlefield, 1961-62.

4 v. col. plates. 33 x 43 cm. Quarter black imitation leather/cloth.
Gift of Miss Mabel Brasher.

*953

BRASHER, REX (1869-1960). *Treasury of bird paintings.* Comprising 162 full-color plates. With a text by the artist. Edited and annotated by Lisa McGaw. New York: Rowman and Littlefield, 1967.

4 p.l., 24 p. 744 col. plates. illus., ports. 32 x 42 cm. Quarter green cloth/pictorial boards.

*954

BREE, CHARLES ROBERT (1811-1886). *A history of the birds of Europe, not observed in the British Isles.* London: Groombridge, 1860-63.

4 v. col. illus. 25 cm. Rebound, parts of original red cloth with gilt decorative binding retained.
Anker 59. BM I p.228. Nissen VBI 136. Yale p.39. Zimmer pp.87-88.

*955

BREHM, ALFRED EDMUND (1829-1884). *Cassell's book of birds.* From the text of Dr. Brehm by Thomas Rymer Jones. With upwards of 400 engravings, and a series of coloured plates. London, New York: Cassell, Petter and Galpin, [1869-73?].

4 v. illus. (part col.) 26 cm. Translation of Abt. 2 of *Illustrirtes Thierleben.* LSU copy imperfect: v.1-2, c.1; 3-4 col. plates wanting. Original green pictorial cloth binding.
BM I p.228. Nissen VBI 139 a. Yale p.39. Zimmer p.88.

*956

BREHM, ALFRED EDMUND (1829-1884). *Ornithology; or, The science of birds.* From the text of Dr. Brehm. With two hundred and twelve illustrations by Theodore Jasper. Columbus: J.H. Studer, 1878.

156 p. illus., 37 plates. 39 cm. LSU copy 1 imperfect: plate 9 wanting. Copy 1, half calf/marbled boards; c.2, half calf/cloth, marbled endpapers.
Yale p.39.

957

BRISSON, MATHURIN JACQUES (1723-1806). *Ornithologie; ou, Méthode contenant la division des oiseaux en ordres, sections, genres, especes & leurs variétés. A laquelle on a joint une description exacte de chaque espece, avec les citations des auteurs qui en ont traité, les noms qu'ils leur ont donnés, ceux que leur ont donnés les différentes nations, & les noms vulgaires. Ouvrage enrichi de figures in taille douce...* Paris: Chez C.J.-B. Bauche, 1760.

6 v. 255 fold. plates. 26 cm. French and Latin in parallel columns.

——— *Supplément...* [Paris, 1760]. (146, xxii p., 1 l. 6 fold. plates. 26 cm.) Bound with v.6 of main work. French and Latin. Full calf, marbled endpapers.
Anker 69. BM I p.237. Nissen VBI 145. Yale p.40. Zimmer p.94.

958

British birds: their haunts and habits. London: Printed for private distribution only, 1868.

2 p.l., 282 p. 23 cm. Label on p.[2] of cover: Jan Peet Boekverkoojier, Amsterdam. Full brown morocco binding, gilt edges, marbled endpapers.

959

BRITISH MUSEUM (NATURAL HISTORY). DEPT. OF ZOOLOGY. *Catalogue of the birds of the tropical islands of the Pacific Ocean, in the collection of the British Museum.* By George Robert Gray. London: Printed by order of the Trustees, 1859.

2 p.l., 72 p. 21.5 cm. "Sion College Library" stamped on title page. Paper wrappers.
BM I p.246. Yale p.41.

960

BRITISH MUSEUM (NATURAL HISTORY). DEPT. OF ZOOLOGY. *Catalogue of the genera and subgenera of birds contained in the British Museum.* By George Robert Gray. London: Printed by order of the Trustees, 1855.

192 p. 18 cm. In ms. on fly leaf: Presented (with great pleasure) to Mr. Leonhard Stejneger by P.L. Sclater, 10th Feby. 1890. Half calf/marbled boards, spine taped.
BM I p.245. Yale p.41.

961

BRITISH MUSEUM (NATURAL HISTORY). DEPT. OF ZOOLOGY. *List of the specimens of birds in the collection of the British Museum.* By G.R. Gray. London: Printed by order of the Trustees, 1844-68.

5 pts. in 3 v. illus. 18 cm. Gray's name does not appear on title page of pt.II, sec. I; pt.III; and pt.III, sec.I. "FB" stamped on cover of vol. containing pt.III, sec.I; pt.IV; and pt.III, sec.II. Manuscript notes throughout text of pt.III, sec.I; pt.IV; and pt.III, sec.II. Ex libris privatis Fr. Billaud stamped on title page of each part. Part III, sec.I; pt.IV; and pt.III, sec.II bound in 1 vol. with pt.III, sec.II following pt.IV. Pt.III, sec.I; pt.IV; and pt.III, sec.II interleaved with blank leaves. Pt.2, sec.I, pt.III unbound. Pt.3. sec.I, sec.II, pt.IV in 1 v., tan cloth binding.
BM I p.245. Yale p.41.
Gift of George Lowery.

*962

BUFFON, GEORGES LOUIS LECLERC, COMTE DE (1707-1788). *The history of singing birds: containing an exact description of their habits & customs, & their manner of constructing their nests, their times of incubation, with the peculiar excellencies of their several songs, the method of rearing them in cages & the preparation and choice of their food, also the disorders they are subject to with the mode of treatment, including the history & management of canary birds. Translated from the French of the Count de Buffon; the whole ornamented with copper plates from drawings after nature.* Edinburgh: Silvester Doig Royal Exchange, 1791.

4 p.l., 192 p. [22] leaves of plates, illus. 17 cm. Translation from the author's *Histoire naturelle des oiseaux.* In ms. on fly leaf: Wyatt Rawson, March 6th, 1858. Full calf binding.
Anker 76-77 (French ed.) Nissen VBI 158 (French ed.)

963

BULLER, WALTER LAWRY, SIR (1838-1906). *Buller's birds of New Zealand.* A new edition of Sir Walter Lawry Buller's *A history of the birds of New Zealand.* Reproducing in six-colour offset the 48 stone-plate lithographs by J.G. Keulemans, from the 2nd edition, 1888. Now edited & brought up to date by E.G. Turbott. Christchurch: Whitcombe & Tombs, 1967.

xviii, 261 p. illus., 48 col. plates. 38 cm. Quarter blue morocco/cloth, marbled endpapers, silver gilt ornamentation.
Anker 85 (orig. ed.) BM I p.284 (orig. ed.) Nissen VBI 163 (orig. ed.) Yale p.48 (orig. ed.) Zimmer p.114 (orig.ed.)

*964

BULLER, WALTER LAWRY, SIR (1838-1906). *A history of the birds of New Zealand.* London: J. Van Voorst, 1873.

xxiii, 384 p. illus., 35 col. plates. 32 cm. Stamped on title page: Als Dublette ausgeschieden. Label on lining paper: Wien Gebunden bei F. Krauss, Bürgerspital. Half brown morocco/marbled boards, marbled endpapers.
Anker 85. BM I p.284. Nissen VBI 163. Yale p.48. Zimmer p.115 (1888 ed.)

965

BULLER, WALTER LAWRY, SIR (1838-1906). *A history of the birds of New Zealand.* 2d edition. London: Published by the Author, 1888.

2 v. in 1. illus., 50 plates (48 col.) 38 cm. Issued in parts. Coloured plates by Keulemans. Green cloth binding.
Anker 85. BM I p.284. Nissen VBI 163. Yale p.48.

*966

BURGESS, THORNTON WALDO (1874-1965). *The Burgess bird book for children.* With illustrations in color by Louis Agassiz Fuertes. Boston: Little, Brown, 1923 [c1919].

xvi, 1 l., 353 p. col. front., col. plates. 21 cm. Bookplate of Harold Frederick Moon, Jr. Blue cloth pictorial binding.
Nissen VBI 165. Yale p.48.

*967

BURLEIGH, THOMAS DEARBORN (1895-). *Georgia birds.* With reproductions of original paintings by George Miksch Sutton. [1st ed.]. Norman: University of Oklahoma Press, [c1958].

xxix, 746 p. illus., col. plates, port., maps. 27 cm. Copy 2 has Sutton's autograph. Blue cloth binding, illustrated endpapers.

968

BURMEISTER, HERMANN (1807-1892). *A manual of entomology.* Translated from the German of Dr. Hermann Burmeister. By W.E. Shuckard, with additions by the author, and original notes and plates by the translator. London: E. Churton, 1836.

1 p.l., xii, 654 p. col. front., 32 plates (7 col.) 23 cm. Green library binding.
BM I p.289. Nissen ZBI 763.

969

BURROUGHS, JOHN (1837-1921). *John James Audubon.* Boston: Small, Maynard & company, c1902.

4 p.l., xvii p., 1 l., 144 p. front. (port.) 14.5 cm. (Added

title page: The Beacon biographies of eminent Americans, edited by M.A. De W. Howe). Black library binding.
Yale p.14.

970

BUTLER, ARTHUR GARDINER (1844-1925). *Birds of Great Britain and Ireland, order Passeres.* Illustrated by H. Grönvold and F.W. Frohawk. Hull: Brumby & Clarke, [1896-97?].

2 v. col. illus. 30 cm. Blue cloth binding.
Anker 88. BM I p.293. Yale p.50. Zimmer pp.120-21.

*971

BUTLER, ARTHUR GARDINER (1844-1925). *Birds of Great Britain and Ireland, order Passeres.* Illustrated by H. Grönvold and F.W. Frohawk. London: Caxton Pub. Co., [1907-08].

2 v. illus., 115 col. plates. 30 cm. Half green morocco/cloth binding.
Anker 88. BM I p.293. Nissen VBI 168. Yale p.50. Zimmer pp.120-21.

972

BUTLER, ARTHUR GARDINER (1844-1925). *British birds with their nests and eggs.* Illustrated by F.W. Frohawk. London: Brumby & Clarke, [1896].

6 v. illus. (part col.) 32 cm. Armorial bookplate with motto: Peradventure. Yellow cloth binding, gilt decoration, gilt edges, decorated endpapers.
Anker 88. BM I p.293. Nissen VBI 167. Yale p.50. Zimmer p.119.

973

BUTLER, ARTHUR GARDINER (1844-1925). *Foreign finches in captivity.* 2d edition. Illustrated by F.W. Frohawk. Hull, London: Brumby and Clarke, limited, 1899.

4 p.l., viii, 317, [1]p. col. front., 58 col. plates. Copy 1, 27.5 cm.; c.2, 28 cm. Copy 1, bookplate of Ronald A. Wright. Copy 2, Ex libris Rudolph Valentino. In ms. on fly leaf: Geoffrey Hope Morley, 1910. Copy 1, quarter maroon morocco/cloth; c.2, red cloth binding.
Anker 89. BM I p.293. Nissen VBI 169. Yale p.50. Zimmer p.120.

974

CABANIS, JEAN LOUIS (1816-1906). *Museum Heineanum.* Verzeichniss der ornithologischen Sammlung des Oberamtmann Ferdinand Heine, auf Gut St. Burchard vor Halberstadt. Mit kritischen Anmerkungen und Beschreibung der neuen Arten, systematisch bearbeitet von Jean Cabanis. Halberstadt: In commission bei R. Frantz, 1850-63.

4 v. in 2. 23 cm. Vols.2-4 are by Jean Cabanis and Ferdinand Heine, the younger. Manuscript note on slip attached to p.[4] of cover of v.2-4. Vol.2 is the author's autograph presentation copy. Vol.2 is Heine's presentation copy to A. von Pelzeln. LSU v.2 imperfect: pp.49-175 wanting; v.3, pp.83-102 wanting. LSU lacks v.4. Half brown calf/marbled boards, original wrappers retained.
BM I p.295. Zimmer pp.121-22.
Gift of George Lowery.

*975

CASTAÑEDA, PORFIRIO G. *A portfolio of Philippine birds.* 30 full color plates by Porfirio G. Castañeda; introductions by Hilario S. Francia and Discoro S. Rabor. Manila: Ayala Museum, [1977?].

3 p.l. 30 leaves of plates, col. illus. 44 cm. "A publication of Filipinas Foundation, inc." Full red morocco binding.

*976

CHAPIN, JAMES PAUL (1889-). *The birds of the Belgian Congo.* New York, 1932-54.

4 v. illus. (part col.), maps (1 fold. col.) 25 cm. (American Museum of Natural History, Bulletin, v.65, 75, 75 A-B). Bookplate in vols.1-2: Ex libris Howard H. Davis. Part 1, section A issued also as thesis, Columbia University. Blue cloth binding.

*977

CHAPMAN, FRANK MICHLER (1864-1945). *Bird-life; a guide to the study of our common birds.* With seventy-five full-page plates and numerous text drawings by Ernest Seton Thompson. New York: D. Appleton and Company, 1897.

xii, 269 p. front., illus. 19.5 cm. In ms. on fly leaf: A.H. Acton, June 4th, 1897. Clipping pertaining to the author tipped in. Original green pictorial cloth binding.
Nissen VBI 183. Yale p.57. Zimmer p.130.

*978

CHENU, JEAN CHARLES (1808-1879). *Encyclopédie d'histoire naturelle; ou, Traité complet de cette science d'après les travaux des naturalistes les plus éminents de tous les pays et de toutes les époques: Buffon, Daubenton, Lacépède, G. Cuvier, F. Cuvier, Geoffroy Saint-Hilaire, Latreille, De Jussieu, Brongniart, etc., etc.* Ouvrage resumant les observations des auteurs anciens et comprenant toutes les decouvertes modernes jusqu'a nos jours. Oiseaux, avec la collaboration de M. Des Murs. Paris: Morescq, [1852-54].

6 v. illus., plates. 28 cm. v.1, plate 24 incorrectly numbered 23. LSU owns only v.1. Half green morocco/marbled boards.
BM I p.341. Nissen VBI 193. Yale p.58. Zimmer p.131.

*979

The Children's picture-book of birds. Illustrated with sixty-one engravings by W. Harvey. New York: Harper, 1861.

xii, 276 p. illus. 18 cm. Label mounted on p.[2] of cover: Sold by William W. Swayne, bookseller & stationer, 210 Fulton Street, Brooklyn. Stamped on p.274: Library of the Young Women's Christian Association, Brooklyn, N.Y. Lavender cloth binding, gilt decoration.

980

CHUBB, CHARLES. *The birds of British Guiana, based on the collection of Frederick Vavasour McConnell, Camfield Place, Hatfield, Herts.* By Charles Chubb with a preface by Mrs. F.V. McConnell... London: B. Quaritch, 1916-21.

2 v. illus., col. plates, port., fold. map. 27 cm. Red cloth binding, original wrappers retained.
Anker 100. Nissen VBI 198. Yale p.60. Zimmer p.132.
Gift of E.W. Mudge, Jr.

*981

CLANCEY, PHILLIP ALEXANDER. *Gamebirds of Southern Africa; being a guide to all the major sporting birds of Africa south of the Cunene, Okavango and Zambezi Rivers.* With 12 colour plates, 35 line drawings and 10 maps. 1st edition. Cape Town: Purnell, c1967.

xviii, 224 p. col. front., illus., col. plates, maps. 26 cm. Label mounted on lining paper: Collector's edition of which this book is no.70. With author's autograph. Full blue morocco binding.

*982

CORTI, ULRICH ARNOLD (1904-). *Passereaux.* Illustrations de Walter Linsenmaier. Version française par Paul Géroudet. Zurich: Éditions Silva, c1956.

130 p. 60 mounted col. plates. 30 cm. (His *Les Oiseaux nicheurs d'Europe*). Dark red cloth binding.

983

CORY, CHARLES BARNEY (1857-1921). *The birds of Haiti and San Domingo.* Boston: Estes & Lauriat, 1885.

[5]-198 p. illus. (part col.), map (front.) 30 cm. Half brown morocco/red cloth, marbled endpapers.
BM I p.387. Nissen VBI 204. Yale p.66. Zimmer p.138.

*984

CORY, CHARLES BARNEY (1857-1921). *The birds of the West Indies. Including all species known to occur in the Bahama Islands, the Greater Antilles, the Caymans, and the Lesser Antilles, excepting the islands of Tobago and Trinidad.* Boston: Estes & Lauriat, 1889.

1 p.l., 324 p. illus., 2 maps. 27 cm. "Most of the matter contained in the present work appeared in the *Auk* for 1886, 1887, and 1888." — Introduction. Quarter calf/red cloth binding.
BM I p.387. Yale p.66. Zimmer p.139.

*985

COTTON, JOHN. *Beautiful birds described.* Edited from the manuscript of John Cotton, by Robert Tyas. With thirty-six illustrations in colours by James Andrews. London: Houlston and Wright, 1868.

3 v. illus., 36 col. plates. 17 cm. Armorial bookplate in each vol.: James Augustus Hewlett. Vol.1 has label mounted on lining paper: Bound by Burn & Co., Kirby St., E.C. Red cloth binding, gilt decoration.
Nissen VBI 207 (1854-56 ed.) Yale p.67 (1854-56 ed.) Zimmer p.142.

986

COUES, ELLIOTT (1842-1899). *Key to North American birds; containing a concise account of every species of living and fossil bird at present known from the continent north of the Mexican and United States boundary.* Illustrated by 6 steel plates and upwards of 250 woodcuts. Salem, [Mass.]: Naturalists' agency; New York: Dodd and Mead [etc., etc.], 1872.

4 p.l., 361 p. illus., 6 plates (incl. front.) 28.5 cm. In ms. on half title: M.A. Harris, 1874. Red library binding.
BM I p.393. Nissen VBI 208. Yale p.67. Zimmer p.143.

*987

COUES, ELLIOTT (1842-1899). *Key to North American birds. Containing a concise account of every species of living and fossil bird at present known from the continent north of the Mexican and United States boundary, inclusive of Greenland.* 2nd edition, revised to date, and entirely rewritten: with which are incorporated *General ornithology*...and *Field ornithology*...by

Elliott Coues. Boston: Estes and Lauriat, 1884.
xxx, 863 p. col. front., illus. 25.5 cm. LSU copy imperfect: front. wanting. Half brown morocco/cloth, marbled endpapers.
BM I p.393. Nissen VBI 208. Yale p.67. Zimmer p.143.

*988

The Country Life book of birds of prey. [By] Gareth Parry and Rory Putman. Feltham, Eng.: Country Life Books; London, New York: Hamlyn, c1979.
120 p. illus., 25 plates. 36 cm. Blue cloth binding, dust-jacket.

*989

COWARD, THOMAS ALFRED (1867-1933). *The birds of the British Isles.* Rev. by A.W. Boyd. 3d series, comprising their migration and habits and observations on our rarer visitants. With 68 accurately coloured illustrations by Archibald Thorburn and others, reproduced from Lord Lilford's work, *Coloured figures of the birds of the British Islands.* [4th edition]. London, New York: F. Warne, [1950].
286 p. 127 plates (part col.) front. 18 cm. (The Wayside and woodland series). Stamped on p.[2] of cover: Robert E. Barron III, M.D. Tan cloth binding, pictorial endpapers.
Anker 109.

*990

COWARD, THOMAS ALFRED (1867-1933). *The birds of the British Isles and their eggs.* Rev. by A.W. Boyd. 1st series, comprising the families Corvidae to Phoenicopteridae. With 252 accurately coloured illustrations by Archibald Thorburn and others, reproduced from Lord Lilford's work, *Coloured figures of the birds of the British Islands,* and 73 photographic illustrations by Richard Kearton and others. [7th ed.] London, New York: F. Warne, [1950].
400 p. plates (part col.) 18 cm. (The Wayside and woodland series). Stamped on p.[2] of cover: Robert E. Barron III, M.D. Tan cloth binding, pictorial endpapers.
Anker 108. Nissen VBI 209 (1920-26 ed.) Yale p.68 (1934-39 ed.) Zimmer p.151 (1920 ed.)

*991

COWARD, THOMAS ALFRED (1867-1933). *The birds of the British Isles and their eggs.* Rev. by A.W. Boyd. 2d series, comprising the families Anatidae to Phasianidae. With 203 accurately coloured illustrations by Archibald Thorburn and others, reproduced from Lord Lilford's work, *Coloured figures of the birds of the British Islands,* and 66 photographic illustrations by E.L. Turner, R. Kearton and others. [8th edition]. London, New York: F. Warne, [1953].
384 p. 157 plates (part col.) front. 18 cm. (The Wayside and woodland series). "Eighth edition 1950...reprinted 1953." Stamped on p.[2] of cover: Robert E. Barron III, M.D. Tan cloth binding, pictorial endpapers.
Anker 109.

992

Curiosities of ornithology. With beautifully colored illustrations, from drawings by T.W. Wood, and other eminent artists. London: Groombridge and sons, [18 ?].
64 p. illus. plates. 22 cm. Tan cloth binding with gilt and black decoration, gilt edges.
Yale p.316.

993

DAGLISH, ERIC FITCH (1892-1966). *Birds of the British Isles.* Described & engraved by Eric Fitch. London: J.M. Dent, [1948].
xviii, 222 p. illus. (part col.) 28 cm. Limited edition of 150 copies. Blue cloth binding.
Yale p.70.

*994

DAWSON, JOHN WILLIAM, SIR (1820-1899). *Handbook of zoology, with examples from Canadian species, recent and fossil. Part I., Invertebrata.* With 275 illustrations. Montreal: Dawson Brothers, 1870.
v, [1]p., 1 l., 264 p. illus. 17 cm. No more published. Original green cloth binding.
BM I p.430 (1886 ed.)

995

DAWSON, WILLIAM LEON (1873-). *The birds of California; a complete, scientific and popular account of the 580 species and subspecies of birds found in the state.* Illustrated by 16 photogravures, 32 full-page duotone plates and more than 1100 half-tone cuts of birds in life, nests, eggs, and favorite haunts, from photographs, chiefly by Donald R. Dickey, Wright M. Pierce, William L. Finley and the author, together with 44 drawings in the text and a series of 48 full-page color plates, chiefly by Major Allan Brooks. Booklover's edition. San Diego, Los Angeles [etc.]: South Moulton company, 1923.

4 v. fronts. (2 col.), illus., plates (part col.) 28 cm. "This edition contains selections of the various types of full page illustrations employed in the larger format and its circulation is limited to 1000 copies." Paged continuously. Parts of plates accompanied by leaves with descriptive letterpress. Sold only by subscription. Green cloth binding.
Nissen VBI 225. Yale p.73. Zimmer p.162.

996

DAWSON, WILLIAM LEON (1873-). *The birds of Washington: a complete, scientific and popular account of 372 species of birds found in the state.* By William Leon Dawson assisted by John Hooper Bowles. Illustrated by more than 300 original half-tones of birds in life, nests, eggs, and favorite haunts, from photographs by the author and others, together with 40 drawings in the text and a series of full-page color plates by Allan Brooks. Author's edition. Sold only by subscription. Seattle: The Occidental Publishing co., [c1909].

2 v. fronts., illus., plates (part col.) 33 cm. Paged continuously. "Of this work in all its editions 1250 copies have been printed and the plates destroyed. Of the Author's edition 250 sets have been printed and bound of which this copy is no.177 and 237." Red cloth binding.
Nissen VBI 224. Yale p.73. Zimmer pp.161/2.

997

DELACOUR, JEAN THÉODORE (1890-). *Les Oiseaux de l'Indochine française.* Par J. Delacour & P. Jabouille. [Aurillac, Impr. du Cantal républicain, 1931].

4 v. illus., LXVII col. plates, maps. 28.5 cm. At head of title: Exposition Coloniale Internationale, Paris, 1931. Indochine Française. Vol.1, 3, half ivory cloth/marbled boards; v.2, 4, tan cloth binding.
Nissen VBI 228. Yale p.74.

*998

DELACOUR, JEAN THÉODORE (1890-). *The pheasants of the world.* Illustrated with thirty-two plates by J.C. Harrison, and twenty-one maps & diagrams. London: Country Life; Salt Lake City: Allen Pub. Co., 1964 [c1957].

351 p. plates (part col.), maps. 29 cm. Title page in red and black. Beige cloth binding.
Nissen VBI 229 (1951 ed.) Yale p.74 (1951 ed.)

999

DELACOUR, JEAN THÉODORE (1890-). *The waterfowl of the world.* With sixteen plates in colour by Peter Scott and thirty-three distributive maps. London: Country Life, 1954-64.

4 v. illus., col. plates, maps. 26 cm. Blue cloth binding.
Yale p.74.

* 1000

DE SCHAUENSEE, RODOLPHE MEYER (1901-). *The birds of Colombia, and adjacent areas of South and Central America.* Illustrated by Earl L. Poole and George Miksch Sutton. Narberth, Pa.: Livingston Pub. Co., [c1964].

xvi, 427 p. illus., 20 plates (12 col.) 24 cm. Maps on lining-papers. Published for the Academy of Natural Sciences of Philadelphia. Green cloth binding.

* 1001

DE SCHAUENSEE, RODOLPHE MEYER (1901-). *A guide to the Birds of Venezuela.* By Rodolphe Meyer de Schauensee and William H. Phelps, Jr.; 53 color and black and white plates with facing page notes by Guy Tudor and H. Wayne Trimm, John Gwynne, and Kathleen D. Phelps. Line drawings by Michel Kleinbaum. Princeton, N.J.: Princeton University Press, c1978.

xxii, 424 p. [53] leaves of plates. illus. 24 cm. Blue cloth binding, dustjacket.

1002

DESCOURTILZ, JEAN THÉODORE (d.1855). *Beija-flores do Brasil; pintados e descritos pelo Th. Descourtilz.* Tradução de Carlos Drummond de Andrade. Estudo crítico por Oliverio M. de O. Pinto. Rio de Janeiro: [Biblioteca Nacional], 1960.

50, 35 p. 23 col. plates. 43 cm. Title page in red and black. French and Portuguese. Facsimile reproduction of a ms. in the Biblioteca nacional, Rio de Janeiro. Original title page reads: *Oiseaux-mouches orthorynques du Brésil*...Rio de Janeiro, 1831. Blue library binding.

* 1003

DESCOURTILZ, JEAN THÉODORE (d.1855). *Oiseaux remarquables du Brésil.* Rio de Janeiro: Imprimerie lithographique de Heaton & Rensburg, [1843].

1 leaf, 32 plates (30 col.) 51 cm. Leaf (title page) preceding plates is the upper cover of a part issue, numbered 11. Imprint date in ms. on title page. Manuscript notes on title page and plates 1-7. Plates 8 and 16 duplicated, one in color and one in black and white at end. Bookplate, with motto: Ex multis non multos. Green cloth binding/marbled endpapers.

* 1004

DESCOURTILZ, JEAN THÉODORE (d.1855). *Ornithologie brésilienne; ou, Histoire des oiseaux du Brésil, remarquables par leur plumage, leur chant ou leurs habitudes.* Rio de Janeiro: T. Reeves, [1854-56].

42 p. 48 col. plates. 64 cm. Imprint date in ms. on title page. Ex libris W. Godfrey Spears. Green cloth binding.
BM I p.443. Nissen VBI 236. Yale p.76. Zimmer p.166.

1005

DESCOURTILZ, JEAN THÉODORE (d.1855). *Pageantry of tropical birds in their natural surroundings.* Commentary by João Moojen. Amsterdam, Rio de Janeiro: Colibris Editora, 1960.

23 p. 60 plates (part col.) 44 cm. Translation of *Oiseaux brillans et remarquables du Brésil.* Quarter green cloth/beige cloth binding.
Nissen VBI 235 (orig. ed.)

* 1006

DHARMAKUMARSINHJI, RAOL SHRI (1917-). *Birds of Saurashtra, India, with additional notes on the birds of Kutch and Gujerat.* [Dil Bahar, 195-].

liii, 561 p. illus., plates (part col.), maps. 30 cm. Stamped on title page: Taru, No.3643/111, Mori Gate, Delhi-110006, India. Red cloth binding.
Yale p.77.

1007

DRESSER, HENRY EELES (1838-1915). *A manual of palaearctic birds.* London: The author, 1902-03.

2 v. in 1 (vii, 922 p.) plates (part col.) 23 cm. Green library binding.
BM VI p.279. Yale p.82. Zimmer pp.179-80.

* 1008

DRESSER, HENRY EELES (1838-1915). *A monograph of the Meropidae, or family of the bee-eaters.* London: The author, 1884-1886.

xix, [1], 144 [i.e.146]p. 34 col. plates. 40 cm. Plates by J.G. Keulemans. Originally issued in 5 parts. Original red cloth binding, gilt ornamentation.
BM I p.479. Nissen VBI 269. Yale p.82. Zimmer p.178.

* 1009

DU BUS DE GISIGNIES, BERNARD AMÉ LÉONARD, VICOMTE (1808-1874). *Esquisses ornithologiques; descriptions et figures d'oiseaux nouveaux ou peu connus.* Bruxelles: A. Vandale, [1845-48].

livr. 37 col. plates. 40 cm. Plates 1-12; 15-37 lithographed by G. Severeyns; plates 13-14 by J. Dekeghel. Parts in original printed wrappers laid in protective covers.
Yale p.83.

1010

DUFFIELD, THOMAS, comp. *Protected native birds of South Australia.* Introduction and descriptions by Alfred Geo. Edquist. Lithographed by Alfred Vaughan. Drawings by C. Wall. Adelaide; R.E.E. Rogers, Gov't. Ptr., 1910.

30 p. 12 col. illus. 25 cm. (South Australia. Department of Intelligence. Special bulletin). Original printed paper wrappers.
Gift of J. Dally.

* 1011

DUGMORE, ARTHUR RADCLYFFE (1870-1955). *Bird homes; the nests, eggs, and breeding habits of the land birds breeding in the eastern United States; with hints on the rearing and photographing of young birds.* Illustrated with photographs from nature by the author. New York: Doubleday, Page, 1905.

xviii, 183 p., [80] leaves of plates, illus. (some col.) 26 cm. (The Nature library, 3). Stamped on title page and p.31: Zion College. Tan cloth binding.
Yale p.83. Zimmer p.184.

* 1012

DU PONT, JOHN ELEUTHÈRE. *Philippine birds.* With color illustrations by George Sandström and John R. Peirce. Greenville: Delaware Museum of Natural History, [c1971].

x, 480 p. 85 col. plates. 28 cm. (Delaware Museum of Natural History. Monograph series, no.2). Red cloth binding.

1013

EATON, ELON HOWARD (1866-). *Birds of New York.* Albany: University of New York, 1910-14.

2 v. illus., 106 col. plates, maps (1 fold.), diagrs. 30 x 24.5 cm. (New York State Museum, Memoir 12). In ms. on fly leaf of v.2: To Hon. Thomas W. White. Stamped on fly leaf of v.2: With compliments of Henry D. Patton, Member of Assembly. Green cloth binding.
BM VI p.291. Nissen VBI 109. Zimmer pp.190/91.

* 1014

EDWARDS, ERNEST PRESTON (1919-). *A field guide to the birds of Mexico, including all birds occur-*

ring from the northern border of Mexico to the southern border of Nicaragua. Illustrated by Murrell Butler...[et al.]; Spanish descriptions by Miguel Alvarez del Toro, Ernest P. Edwards. 1st edition. Sweet Briar, Va.: E.P. Edwards, 1972.

vi, 300, A79 p. illus., 24 col. plates. 23 cm. Spiral bound, paper covers.

* 1015

EDWARDS, ERNEST PRESTON (1919-). *Finding birds in Mexico.* Illustrated by Edward Murrell Butler, Ernest P. Edwards, and Frederick K. Hilton. 2d edition, revised and enlarged. Sweet Briar, Va., [c1968].

xxi, 282 p. illus. (part col.), maps. 23 cm. Green cloth binding, dustjacket.

* 1016

EDWARDS, ERNEST PRESTON (1919-). *1976 supplement to finding birds in Mexico (1968): bird-finding information brought up-to date by Ernest Preston Edwards.* Sweet Briar, Va.: Edwards, c1976.

135 p. maps. 23 cm. Paperbound.

* 1017

EDWARDS, GEORGE (1694-1773). *A natural history of uncommon birds, and of some other rare and undescribed animals, quadrupedes, reptiles, fishes, insects, &c., exhibited in two hundred and ten copperplates from designs copied immediately from nature, and curiously coloured after life. With a full and accurate description of each figure. To which is added a brief and general idea of drawing and painting in water-colours; with instructions for etching on copper with aqua fortis: likewise some thoughts on the passage of birds; and additions to many of the subjects described in this work.* London: Printed for the author, at the College of Physicians, [1743]-64.

7 v. col.front. (v.1), 362 col. plates (incl. map). 30 cm. L.S.U. v.1 imperfect: col. front. wanting. Issued in parts. Added title page for v.1 and title page for v.2-4 reads, with minor changes: *A natural history of birds. Most of which have not been figur'd or describ'd...* Vols.5-7 have title: *Gleanings of natural history, exhibiting figures of quadrupeds, birds, insects, plants, &c...* With added title page: *Glanures d'histoire naturelle...* [1.]-2 ptie. traduit de l'anglois par J. Du Plessis, 3 ptie. traduit par Edmond Barker. 1758-64. Vols.5-7 have text in English and French in parallel columns. With this (v.1-4) is bound: His *Histoire naturelle d'oiseaux peu communs*, 1751 [i.e. 1745]-51. Full calf, marbled endpapers.

Anker 124-126. BM I p.510. Nissen VBI 286, 287. Yale p.86. Zimmer pp.192-202.

* 1018

[Egg sales at J.C. Stevens, 38 King Street, Covent Garden, London, 1902-5; dispersal of the collections of Mr. Philip Crowley, Dr. A.C. Stark, Mr. R.J. Ussher, Mr. Edward Bidwell, Mr. Heatley Noble. London: n.p., 1902-1905].

12 pamphlets in 1 v. 23 cm. Manuscript notes on fly leaves and throughout catalogs. Armorial bookplate of Thomas Parkin. Blue cloth binding.

1019

ELLIOT, DANIEL GIRAUD (1835-1915). *A classification and synopsis of the Trochilidae.* Washington: Smithsonian Institution, 1879.

xii, 277 p. illus. 32 cm. (Smithsonian contributions to knowledge; vol. XXIII, art.5). Green cloth binding, original covers bound in.

BM II p.522. Zimmer p.207.

* 1020

ELLIOT, DANIEL GIRAUD (1835-1915). *The gallinaceous game birds of North America, including the partridges, grouse, ptarmigan, and wild turkeys; with accounts of their dispersion, habits, nesting, etc., and full descriptions of the plumage of both adult and young, together with their popular and scientific names.* With forty-six plates. London: Suckling, 1897.

xviii, 220 p. 1 l. 46 plates (incl. front.) col. chart. 27 cm.

——— *Another copy.* New York: F.P. Harper, 1897. 100 copies on large paper for America and England. LSU owns no.20, London; no.21, New York. Signed by the author. Title page in red and black. Ivory pictorial cloth binding.

BM II p.522. Nissen VBI 299. Yale pp.88-89. Zimmer p.209.

* 1021

ELLIOT, DANIEL GIRAUD (1835-1915). *A monograph of the Bucerotidae, or family of the hornbills.* [London]: Published for the subscribers by the author, [printed by Taylor and Francis], 1882.

xxxii, [117]p. 60 plates (part col.) 39 cm. Originally issued in 10 pts., 1877-82. Original covers for pts.1-10 bound at end. "Anorrhinus austeni (plate 44) mentioned in contents, is not figured, as no specimen was available." Fifty-seven plates are in color and were lithographed by J.G. Keulemans; 3 plates are not in color and were lithographed by J. Smit. Half green morocco/cloth, marbled endpapers.

BM I p.522. Nissen VBI 297. Yale p.88. Zimmer p.207.

* 1022

ELLIOT, DANIEL GIRAUD (1835-1915). *A monograph of the Phasianidae or family of the pheasants.* New

York: Published by the author, 1872.

2 v. 81 plates (79 col.) 61 cm. Title page for v.2 bound in v.1. Originally issued in 6 parts, 1870-72. Plates not bound according to list of plates accompanying volumes. Full red morocco/marbled boards, gilt ornamentation, gilt edges.
Anker 130. BM II p.522. Nissen VBI 295. Yale p.88. Zimmer p.206.

* 1023

ELLIOT, DANIEL GIRAUD (1835-1915). *A monograph of the Tetraoninae; or, family of the grouse.* New York: The author, 1865.

5 pts. in 4. 27 col. plates. 62 cm. Title page and 9 preliminary leaves bound at end of pts.4-5. Published in 5 pts., 1864-65. Each pt. has cover title. Each plate accompanied by from one to three pages of descriptive letterpress. Portrait inserted in pt.2: T. Laurentz. In green cloth portfolio.
Anker 128. BM I p.522. Nissen VBI 293. Yale p.88. Zimmer p.205.

1024

ELLIOT, DANIEL GIRAUD (1835-1915). *The wild fowl of the United States and British possessions: or, The swan, geese, ducks, and mergansers of North America. With accounts of their habits, nesting, migrations, and dispersions, together with descriptions of the adults and young, and keys for the ready identification of the species. A book for the sportsman, and for those desirous of knowing how to distinguish these web-footed birds and to learn their ways in their native wilds.* London: Suckling, 1898.

xxii, 316 p., 64 leaves of plates. illus., port. 27 cm. Each plate accompanied by guard sheet with descriptive letterpress. "100 copies on large paper for America and England." LSU owns copies no.28 and 29. With author's autograph. White cloth binding.
BM II p.522. Nissen VBI 300. Yale p.89. Zimmer p.209.
Copy 2, gift of Dr. & Mrs. A. Brooks Cronan.

* 1025

ENNION, ERIC ARNOLD ROBERTS (d. 1981). *The living birds of Eric Ennion.* Introduction and commentary by John Busby. London: Victor Gollancz Ltd., 1982.

128 p. illus. (some col.) 29 cm. Brown cloth binding, illustrated dustjacket.

1026

Exotic birds: parrots, birds of paradise, toucans. [London]: Andre Deutsch, [c1963].

[38]p. 16 col. plates. 41 cm. "16 colour plates [of which 14 are] by Jacques Barraband and [2 by] Auguste. Text after François Levaillant, translated by Eric Mosbacher." Text adapted and plates reproduced from Le Vaillant's *Histoire naturelle des oiseaux* and his *Les perroquets.* Illustrated paper wrappers.

1027

FATIO, VICTOR (1838-1906). *Les oiseaux de la Suisse ("Catalogue des oiseaux de la Suisse de V. Fatio et Th. Studer") élaboré par ordre du Département fédéral de l'intérieur (inspection des forêts, chasse et pêche) par G. von Burg avec le concours de nombreux observateurs de tous les cantons...* Berne et Genève: 1889-19- .

v. in . maps (part fold.) 24-25 cm. Paged continuously. Title varies. Imprint varies. LSU owns 16 vols. in 6. Half green imitation leather/marbled boards.
BM II p.557. Yale p.279 (Ger. ed.) Zimmer pp.607-8.

* 1028

FINCH-DAVIES, CLAUDE GIBNEY (1875-1920). *The birds of prey of southern Africa.* Colour plates by Claude Gibney Finch-Davies. Text by Alan Kemp. Johannesburg: Winchester Press, c1980.

339 p. 140 col. illus., port. 30 cm. Plates reproduced from the original paintings in the Transvaal Museum. Limited edition of 1726 copies. LSU has one of 26 special presentation volumes: Volume R, Troy H. Middleton Library. Quarter brown morocco/marbled boards, brown cloth slipcase.

* 1029

FINCH-DAVIES, CLAUDE GIBNEY (1875-1920). *The birds of southern Africa.* Colour plates by Claude Gibney Finch-Davies. Text by Alan Kemp. Johannesburg: Winchester Press, c1982.

488 p. 176 col. illus., port. 30 cm. Plates reproduced from the original paintings in the Transvaal Museum. Limited edition of 3026 copies. LSU owns volume R, one of 26 special presentation copies, A to Z. Half tan morocco/marbled boards. Tan cloth slipcase with marbled edges.

1030

FINN, FRANK (1868-1932). *The birds of Calcutta.* 3d edition. Calcutta: Thacker, Spink & Co., 1917.

vi, 166 p. illus. 19 cm. Quarter red cloth/pictorial boards.

* 1031

FINN, FRANK (1868-1932). *How to know the Indian waders.* 2d edition, revised. Calcutta: Thacker, Spink, 1920.

xi, 200 p. plates. 19 cm. Stamped on title page: Taru, no. 3643/111, Mori Gate, Delhi—110006, India. Paper wrappers.

1032

FLOWER, STANLEY SMYTH. *The principal species of birds protected by law in Egypt, giving their English, French, Arabic, and scientific names, their local status, their approximate size, and concise notes on their coloration, for purposes of identification.* By Captain S.S. Flower and Mr. M.J. Nicoll. Cairo: Government press, 1918.

iv, 8 p. VIII col. plates. 27 cm. At head of title: Ministry of Agriculture, Egypt. "The substance of the following was issued as a Circular, no.84, dated May 5, 1917." Caption title. Printed paper covers.
BM VI p.293.

* 1033

FORBUSH, EDWARD HOWE (1858-1929). *Birds of Massachusetts and other New England states.* Illustrated with colored plates from drawings by Louis Agassiz Fuertes and figures and cuts from drawings and photographs by the author and others. Issued by authority of the legislature. [Norwood, Mass.: Printed by Berwick and Smith company], 1925-29.

3 v. front. (v.3), illus. (incl. map), plates (part col.) 25.5 cm. Vols.1-2 reprinted August, 1929. In ms. on fly leaf of v.1: to Tabitha Wilson, From: Billy Cain, June 10-1930. Original green cloth binding.
Nissen VBI 328. Yale p.98.

* 1034

FORBUSH, EDWARD HOWE (1858-1929). *A natural history of American birds of eastern and central North America.* With ninety-six full color illustrations by Louis Agassiz Fuertes, Allan Brooks and Roger Tory Peterson. Revised and abridged with the addition of more than one hundred species by John Bichard May. New York: Bramhall House, c1955.

xxv, 553 p. [49] leaves of plates, illus. (some col.) 30 cm. Reprint of the 1939 edition, published by Houghton, Mifflin, Boston, under the title: *Natural history of the birds of eastern and central North America.* An abridgement and revision of the author's *Birds of Massachusetts and other New England states* issued in three volumes in 1925 and subsequent years. Autograph of Roger Tory Peterson on title page. Green cloth binding, dustjacket.
Yale p.98 (1939 ed.)

1035

FORD, ALICE ELIZABETH (1906-). *John James Audubon.* [1st ed.] Norman: University of Oklahoma Press, [c1964].

xiv, 458 p. 24 cm. Grey cloth binding.

1036

FORSHAW, JOSEPH MICHAEL. *Australian parrots.* With five paintings by John C. Yrizarry. [Melbourne]: Lansdowne Press, [c1969].

xiv, 306 p. illus. (part col.) maps. 32 cm. Errata sheet inserted. Green cloth binding.

1037

FORSHAW, JOSEPH MICHAEL. *Parrots of the world.* Illustrated by William T. Cooper. Garden City, N.Y.: Doubleday, 1973.

584 p. illus. (part col.), maps. 40 cm. Purple cloth binding.

1038

FORSTER, GEORG (1754-1794). *Vögel der Südsee.* 23 Gouachen u. Aquarelle nach Zeichnungen Georg Forsters, entstanden während s. Weltumsegelung 1772-1775. Herausgegeben und kommentiert von Gerhard Steiner und Ludwig Baege. Leipzig: Insel-Verlag, 1971.

79 p. 23 col. plates. 37 cm. Blue cloth binding.

* 1039

FRASCONI, ANTONIO. *Birds from my homeland; ten hand-colored woodcuts.* With notes from W.H. Hudson's *Birds of La Plata.* [New York], 1958.

[30]p. (on double leaves) col. illus. 21 x 29 cm. "200 numbered copies... This copy...signed by the artist, is numbered 60." Illustrated paper boards, uncut.

1040

FRIDERICH, C.G. *Naturgeschichte der Vögel Europas.* 6., dem gegenwärtigen Stand der ornithologischen Wissenschaft entsprechend vermehrte und verbesserte. Auflage Neu bearbeitet von Alexander Bau. Stuttgart: E. Schweizerbart'sche Verlagsbuchhandlung, G.m.b.H., 1923 [c1922].

[d]-n, lxxvii, 884 p. illus., 53 col. plates. 30 cm. Printed paper over boards.
Nissen VBI 331. Zimmer p.232.

1041

FRIEDMANN, HERBERT (1900-). *The honey-guides.* Washington: Smithsonian Institution, 1955.

vii, 292 p. illus. (part col.) 24 cm. (United States National Museum. Bulletin, 208). Printed paper wrappers.

1042

FRIES, WALDEMAR H. (1889-). *The double elephant folio: the story of Audubon's Birds of America. . .* Chicago: American Library Association, 1973 [i.e.1974].

xxii, 501 p. illus. 27 cm. Author's autograph on title page. Quarter imitation leather/painted cloth, decorative endpapers.

1043

FUERTES, LOUIS AGASSIZ (1874-1927). *Artist and naturalist in Ethiopia.* By Louis Agassiz Fuertes and Wilfred Hudson Osgood. Illustrations painted from life by Louis Agassiz Fuertes and reproduced in this volume by special permission of Field Museum of Natural History. Garden City, N.Y.: Doubleday, Doran & company, inc., 1936.

xi, [3], 249 p. col. front., illus. (map), col. plates. 26 x 21.5 cm. Original black cloth binding.
Gift from Piccadilly Cafeteria Library by T.H. Hamilton.

* 1044

GALLAGHER, MICHAEL. *The birds of Oman.* By Michael Gallagher and Martin W. Woodcock. Foreword by His Majesty Sultan Qaboos Bin Said, Sultan of Oman. New York: Quartet Books, 1980.

310 p., [2] leaves of plates, illus. (some col.), col. maps. 45 cm. Limited edition of 500 copies. LSU owns copy no.72. With authors' autographs. Color print of the sooty falcon autographed by Martin Woodcock laid in. Full green morocco, gilt ornamentation.

1045

The Game birds of California. By Joseph Grinnell, Harold Child Bryant, and Tracy Irwin Storer. Berkeley: University of California Press, 1918.

2 p.l., iii-x, 642 p. illus., 16 col. plates (incl. front.), fold. table. 26.5 cm. Blue cloth binding, spine mended, gilt medallion.
Anker 190. Zimmer p.275.

1046

GEROUDET, PAUL. *Water-birds with webbed feet.* Translated by Phyllis Barclay-Smith. With 48 illustrations (24 in colour) by Robert Hainard and 59 line drawings. London: Blandford Press, [c1965].

314 p. illus. 21 cm. Translation of *Les palmipèdes.* Blue cloth binding.

* 1047

GESNER, KONRAD (1516-1565). Conradi Gesneri... *historiae animalium Lib. I. de quadrupedibus viviparis. Opvs philosophis, medicis, grammaticis, philologis, poetis & omnibus rerum linguarumque variarum studiosis...* Tigvri: Apvd C. Froschovervm, 1551.

[38], 1104, [11]p. illus. 40 cm. Full black morocco, red edges.
BM II p.668. Nissen ZBI 1549.

1048

GIGLIOLI, ENRICO HILLYER (1845-1909). *Avifauna italica: elenco delle specie di uccellistazionarie o di passaggio in Italia... Per servire alla inchiesta ornitologica.* Firenze: Successori Le Monnier, 1886.

623 p. 23 cm. Author's autograph presentation copy to Dr. Leonhard Stejneger. Original printed wrappers in portfolio.
BM II p.675. Nissen VBI 353. Yale p.108. Zimmer p.242.

* 1049

GODFREY, W. EARL. *The birds of Canada.* Colour illustration by John A. Crosby, line drawings by S.D. MacDonald. Ottawa: National Museum of Natural Sciences, c1979.

428 p. illus. (some col.), maps. 29 cm. (National Museum of Canada. Bulletin, no.203; Biological series, no.73). Grey cloth binding, map on endpapers, dustjacket.

* 1050

GOODWIN, DEREK. *Crows of the world.* Illustrations by Robert Gillmor. London: British Museum (Natural History), 1976.

vi, 354 p., 3 leaves of plates. illus. (some col.), maps. 29 cm. (British Museum (Natural History), Publication, no.771). Green cloth binding, dustjacket.

1051

GORDON, WILLIAM JOHN. *Our country's birds and how to know them; a guide to all the birds of Great Britain.* With an illustration in colour of every species and many original diagrams by G. Willis and R.E. Holding. London: Simpkin, Marshall, Hamilton, Kent, [1901].

vii, 152 p. illus. col. plates. 20 cm. Original green pictorial binding.
Nissen VBI: 366. Yale p.110. Zimmer p.249.

* 1052

GOSSE, PHILIP HENRY (1810-1888). *The birds of Jamaica.* By Philip Henry Gosse; assisted by Richard Hill, esq., of Spanish-town. London: J. Van Voorst, 1847.

x, 447 p. 20 cm. Plates were issued in 1849 with title: *Illustrations of the birds of Jamaica.* Errata slip inserted. Black cloth binding.
BM II p.697. Nissen VBI 367. Yale p.111. Zimmer p.250.

1053

GOULD, JOHN (1804-1881). *Birds of Asia.* Illustrated from the lithographs of John Gould. Text by A. Rutgers. New York: Taplinger, [1969, c1968].

321 p. 160 col. plates. 26 cm. Grey cloth binding.

1054

GOULD, JOHN (1804-1881). *Birds of Australia.* Text by Abram Rutgers. London: Methuen, [c1967].

5 p.l., 321 p. col. illus. 26 cm. Reproduction of 160 plates from the author's *The birds of Australia* with new updated text. Blue cloth binding.

1055

GOULD, JOHN (1804-1881). *Birds of Europe.* Text by A. Rutgers. London: Methuen, [c1966].

321 p. col. illus. 25 cm. Reproduction of 160 plates from Gould's *The birds of Europe* and *The Birds of Great Britain.* With new updated text. Red cloth binding.

* 1056

GOULD, JOHN (1804-1881). *The birds of Great Britain.* London: Printed by Taylor and Francis, published by the author, 1873.

5 v. col. plates. 56 cm. 367 plates (colored by hand) with introduction and accompanying letterpress; issued in 25 parts, 1862-73. Armorial bookplate in v.1-5. Full black morocco, gilt decoration, gilt edges.
BM II p.702. Nissen VBI 372. Yale p.114. Zimmer pp.261-62.

* 1057

GOULD, JOHN (1804-1881). *A century of birds from the Himalaya Mountains.* Plates drawn from nature and on stone by E. Gould. Text by N.A. Vigors. London: Published by the author, 1832.

6 p.l., 72 leaves. 80 col. plates. 56 cm. Most of the plates accompanied by leaves with descriptive letterpress. LSU copy imperfect: plates 4, 5, 8, 11, 13, 18-21, 25, 27, 29, 31, 34, 37-42, 45, 46, 48-50, 55, 56, 59, 64, 66-70, 72, and 73 wanting. Full green morocco, gilt decoration, marbled boards.
Anker 168. BM II p.701. Nissen VBI 374. Yale p.112. Zimmer p.251.

* 1058

GOULD, JOHN (1804-1881). *Humming birds (Trochilidae).* London: 1849-1887.

2 v. 240 col. plates. 57 cm. Each plate accompanied by descriptive letterpress. Title page in each vol. appears to be done by hand. Plates are signed: J. Gould and H.C. Richter, del. et lith.; Hullmandel & Walton, imp. Issued in parts, 1849-61. Incomplete set of *A monograph of the Trochilidae....* Full blue morocco, gilt ornamentation including inside front and back covers, moire fly leaves.
Anker 177. BM II p.701. Nissen VBI 380. Yale p.113. Zimmer pp.258, 263.

1059

GOULD, JOHN (1804-1881). *An introduction to the Trochilidae, or family of humming-birds.* London: Printed by Taylor and Francis, 1861.

4 p.l., iv, 216 p. 22 cm. Red library binding.
BM II p.701. Nissen VBI 380. Yale p.114. Zimmer p.261.

* 1060

GOULD, JOHN (1804-1881). *John Gould's birds.* New York: A & W publishers, c1980.

239 p. illus. (some col.) 35 cm. Red paper over boards, illustrated dustjacket.

* 1061

GOULD, JOHN (1804-1881). *A monograph of the Odontophorinae, or partridges of America.* London: Published by the author, printed by R. and J.E. Taylor, 1850.

4 p.l., [11]-23 p., 33 leaves. 32 col. plates. 57 cm. Each plate accompanied by leaf with descriptive letterpress. A few of the leaves are printed on both sides. Half black morocco/cloth, marbled endpapers.
Anker 176. BM II p.701. Nissen VBI 376. Yale p.114. Zimmer p.257.

* 1062

GOULD, JOHN (1804-1881). *A monograph of the Ramphastidae, or family of toucans.* London: Published by the author, 1834.

3 p.l., [12]p., 36 leaves, [4]p. 34 plates (33 col.) 57 cm. Bookplate: Egremont. Full green morocco, gilt decoration, gilt edges. Bound by Hering.

Anker 170. BM II p.701. Nissen VBI 378. Yale p.112. Zimmer pp.259-60.

* 1063

GOULD, JOHN (1804-1881). *A monograph of the Trogonidae, or family of trogons.* [2d edition]. London: Published by the author, printed by Taylor and Francis, 1875.

xx, [98]p. 47 col. plates. 57 cm. Each plate accompanied by leaf with descriptive letterpress. Full black morocco, gilt ornamentation, gilt edges, marbled endpapers.

Anker 171. BM II p.701. Nissen VBI 381. Yale p.112. Zimmer p.261.

* 1064

GOULD, JOHN (1804-1881). *Oiseaux d'Asie.* Selectionné d'après John Gould, *The birds of Asia* [Par] A. Rutgers. Gorssel: S.A. Editions 'Littera Scripta Manet', c1966-1970.

2 v. col. illus. 26 cm. (Le paradis des oiseaux en couleurs, v.5-6). Grey cloth binding; dustjacket.

* 1065

GOULD, JOHN (1804-1881). *Oiseaux d'Australie.* Sélectionné de John Gould, *The birds of Australia* [Par] A. Rutgers; [traduction par Georgette Swaenepoel]. Gorssel: S.A. Editions 'Littera Scripta Manet', c1966-1967.

2 v. col. illus. 26 cm. (Le paradis des oiseaux en couleurs, v.3-4). Translation of *Vogels van Australië.* Orange cloth binding, dustjacket.

* 1066

GOULD, JOHN (1804-1881). *Oiseaux d'Europe.* Sélectionné des oeuvres de John Gould, *The birds of Europe et The birds of Great Britain* [par] A. Rutgers; [traduction par Georgette Swaenepoel]. Gorssel: S.A. Editions 'Littera Scripta Manet', c1965.

2 v. col. illus. 26 cm. (Le paradis des oiseaux en couleurs; v.1-2). Translation of *Vogels van Europa.* Green cloth binding, dustjacket.

* 1067

GOULD, JOHN (1804-1881). *Oiseaux, de l'Amerique du Sud.* Sélectionné des oeuvres de John Gould, *Monograph of the Odontophorinae, Monograph of the Trogonidae, Monograph of the Ramphastidae* [Par] A. Rutgers. Gorssel: S.A. Editions 'Littera Scripta Manet', c1966-1970.

2 v. col. illus. 26 cm. (Le paradis des oiseaux en couleurs, v.9-10). Red cloth binding, dustjacket.

* 1068

GOULD, JOHN (1804-1881). *Oiseaux de la Nouvelle-Guinée.* Sélectionné d'après John Gould, *The birds of New Guinea.* Gorssel: S.A. Editions 'Littera Scripta Manet', c1966-1970.

2 v. col. illus. 26 cm. (Le paradis des oiseaux en couleurs, v.7-8). Blue cloth binding, dustjacket.

* 1069

GOULD, JOHN (1804-1881). *Tropical birds.* From plates by John Gould. With an introduction and notes on the plates by Sacheverell Sitwell. London, New York: B.T. Batsford, [1948].

[4], 12 p. illus., 16 col. plates. 24 cm. (Batsford Colour books). Yellow paper over boards, dustjacket.

Yale p.114.

1070

GRÄSSNER, FÜRCHTEGOTT. *Die Vögel Deutschlands und ihre Eier; eine vollständige Naturgeschichte sämmtlicher Vögel Deutschlands und der benachbarten Länder mit besonderer Berücksichtigung ihrer Fortpflanzung. 2. sehr vermehrte und gänzlich umgearbeitete Aufl. des früher erschienenen Werkes: Die Eier der Vögel Deutschlands von Naumann und Buhle.* Neue billige Ausg. Halle: G.C. Knapp, 1865.

viii, 215 p. 10 col. plates. illus. 34 cm. Half green morocco/marbled boards, marbled endpapers; original wrappers bound in back.

Nissen VBI 669. Zimmer p.266.

* 1071

GRAVES, GEORGE (fl. 1777-1834). *British ornithology: being the history, with a coloured representation, of every known [sic] species of British birds.* 2d edition. London: Printed for the author, by W. and S. Graves; and sold by Sherwood, Neely, and Jones, 1821.

3 v. 144 col. plates. 24 cm. Vol. 3 without edition statement. Armorial bookplate of Richard Hobson in each vol. Slip mounted on lining paper of each vol.: Mr. William Morris. No.9, case 1. Presentation inscription in all three volumes. Full calf binding, gilt spine.

BM II p.709. Nissen VBI 386. Yale p.115. Zimmer p.266.

* 1072

GRAY, GEORGE ROBERT (1808-1872). *A fasciculus of the birds of China.* [London: Printed by Taylor and Francis, 1871].

8 p. 12 col. plates. 35 cm. The "plates were designed and placed on stone...William Swainson."—Introduction. Half red morocco/cloth.
BM II p.711. Nissen VBI 389. Yale p.116. Zimmer p.271.

1073

GRAY, GEORGE ROBERT (1808-1872). *The genera of birds: comprising their generic characters, a notice of the habits of each genus, and an extensive list of species referred to their several genera.* Illustrated by David William Mitchell...1844-1849. London: Longman, Brown, Green, and Longmans, 1849.

3 v. 334 plates (part fold., 185 col.) 39 cm. Originally issued in parts, 1844-49. Half green morocco, marbled boards, marbled endpapers.
BM II p.711.

1074

GREENE, WILLIAM THOMAS. *Parrots in captivity.* With notes on several species by the Hon. and Rev. F.G. Dutton. London: G. Bell and sons, 1884-87.

3 v. col. fronts., col. plates. 25.5 cm. On cover of v.2: 1890. LSU owns only v.1-2. Green cloth binding, gilt ornamentation, bindings do not match.
BM II p.730. Nissen VBI 393. Yale p.117. Zimmer p.274.

* 1075

GREENEWALT, CRAWFORD H. (1902-). *Hummingbirds.* With a foreword by Dean Amadon. Garden City, N.Y.: Published for the American Museum of Natural History by Doubleday, c[1960].

1 p.l., [xvi], 250, [xvii]-xxi p. illus. (part mounted col.), maps, diagrs. 31 cm. "This edition...limited to five hundred copies, signed by the author, of which this is number 276." Copy 2 is no.195. Full green calf, photograph inserted on front cover.

* 1076

GREENEWALT, CRAWFORD H. (1902-). *Hummingbirds.* [n.p., n.p.], c1977.

5 v. ([9]p., 148 leaves of plates) col. illus. 31 cm. "The prints have been reproduced photographically from the original transparencies." Photographs mounted on boards in portfolio.

1077

HACHISUKA, MASAUJI (1903-1953). *The dodo and kindred birds: or, the extinct birds of the Mascarene Islands.* London: Witherby, 1953.

xvi, 250 p. illus. 29 cm. Grey cloth binding.
Yale pp.120-21.

1078

HARGITT, EDWARD (1835-1895). *Catalogue of the Picariae in the collection of the British Museum. Scansores, containing the family Picidae.* London: Printed by order of the Trustees, sold by Longmans & co. [etc.], 1890.

3 v. illus., XV col. plates. 23 cm. (Added t.p.: Catalogue of the birds in the British Museum. vol. xviix [in 3 v.]) Interleaved and annotated in manuscript. Half green morocco/ marbled boards.
BM II p.786. Yale p.123. Zimmer p.285.

* 1079

HARRISON, JOHN CYRIL (1898-). *The birds of prey of the British Islands.* Illustrated by J.C. Harrison. Descriptive text by David Evans. Foreword by Aylmer Tryon. Kingston Deverill, Wiltshire: Fine Bird Books, 1980.

[29]p. [20] leaves of plates. col. illus. 52 cm. "This edition is limited to two hundred and seventy-five numbered copies each signed by the artist." LSU owns copy no.147. Half maroon morocco/beige cloth, illustrated endpapers; beige cloth slipcase.

1080

HARRISON,THOMAS PERRIN (1897-), ed. *The first water colors of North American birds.* Austin: University of Texas Press, [1964].

59 p. col. illus. 23 cm. At head of title: John White and Edward Topsell. Beige cloth binding, dustjacket.

* 1081

HARVEY, NORMAN BRUCE. *A portfolio of New Zealand birds.* Rutland, Vt.: C.E. Tuttle Co., 1971, c1970.

60 p. illus., 25 col. plates. 37 cm. Beige cloth binding.

1082

HEINROTH, OSKAR (1871-1945). *Die Vögel Mitteleuropas.* In allen Lebens-und Entwicklungsstufen photographisch aufgenommen und in ihrem Seelenleben bei der Aufzucht vom Ei ab beobachtet von Oskar und Magdalena Heinroth. [Nachdruck. Leipzig]: Edition Leipzig, [c1966-68].

4 v. with illus. 30 cm. Reprint of the Berlin, H. Bermühler 1926-28 edition. Grey cloth binding.

1083

HEWITSON, WILLIAM CHAPMAN (1806-1878). *Coloured illustrations of the eggs of British birds, accompanied with descriptions of the eggs, nests, etc.* London: John Von Voorst, 1846.

2 v. col. plates. 22 cm. Armorial bookplate of Samuel Holker Haslam. In ms. on fly leaf: Presented to ——— by J. Atkinson Esq. 14.9.10. Stamped between lines of presentation: Cronulla School of Arts and Literary Institute. Half brown calf/marbled boards, spine mended.

BM II p.839. Nissen 442. Yale p.132 (3rd ed.) Zimmer pp.302-03.

* 1084

HOLDEN, GEORGE HENRY (1848-1914). *Canaries and cage-birds.* New York, Boston: G.H. Holden, [c1883].

2 p.l., 364, viii p. col. front., illus., col. plates. 26.5 cm. In ms. on fly leaf: Emma T. Currier, May 23, 1884. Original green cloth binding, gilt ornamentation.

Yale p.134.

1085

HORSBRUGH, BOYD ROBERT (1871-1916). *The game-birds & water-fowl of South Africa.* With coloured plates by C.G. Davies. London: Witherby & Co., 1912.

xii, 159 p. 65 (i.e. 67) col. plates. 26 cm. Quarter green morocco/cloth binding.

Anker 211. BM VI p.484. Nissen VBI 450. Yale p.137. Zimmer p.301.

* 1086

HOUSSE, ÉMILE (1883-). *Les Oiseaux du Chili.* Paris: Masson, 1948.

393 p. illus. 26 cm. "Le texte de cet ouvrage a paru dans les Annales des sciences naturelles, zoölogie. 10, série, tome XX, 1937: 11, série, tomes II, 1929[i.e. 1939] III, 1941, IV, 1942." Pts.2-3 have running title: "Les oiseaux de proie du Chili;" pts.4-5: "Les oiseaux des Andes." Green cloth binding.

1087

HOWARD, HENRY ELIOT (1873-1940). *An introduction to the study of bird behavior.* Cambridge: The University Press, 1929.

xi, [1], 135, [1]p. front., illus., X plates, 2 plans. 32.5 x 26.5 cm. Blue cloth binding.

Nissen VBI 454a. Yale p.138.

1088

HUDSON, WILLIAM HENRY (1841-1922). *Birds of La Plata.* By W.H. Hudson, with twenty-two coloured illustrations by H. Grönvold. London [etc.]: J.M. Dent & Sons; New York: E.P. Dutton & Co., 1920.

2 v. col. front., col. plates. 25 cm. "There have been printed of this edition 1500 copies for England and 1500 copies for United States of America." "The matter contained in this work is taken from the two volumes of the Argentine ornithology, published in 1888-89... [It forms] a companion work to *The Naturalist in La Plata.*" — Introduction. Green cloth binding.

Anker 216. BM VI p.491. Nissen VBI 460. Yale p.139. Zimmer p.312.

* 1089

HUME, ALLAN OCTAVIAN (1829-1912). *The game birds of India, Burmah, and Ceylon.* [By] Hume and Marshall. Calcutta: Published by A.O. Hume and C.H.T. Marshall, 1879-81.

3 v. 153 col. plates. 27 cm. Dates 1878, 1879, and 1880, respectively, are on spine of vols. LSU copy imperfect: plates, v.1, p.81; v.2, p.223, 235, 253, 261; v.3, p.47, 289, 305, and 373 wanting. Manuscript note on title page of v.1-2: G.F. Dawson from T.M. Franklen, 1902; v.3: G.F. Dawson from T. Mansel Franklen, Oct.1902. Original green cloth binding, gilt ornamentation.

BM II p.891. Nissen VBI 463. Yale p.140. Zimmer p.314.

1090

Humming-birds. By Mary and Elizabeth Kirby. London: T. Nelson and Sons, 1874.

102 p. col. illus. 15 cm. Illustration mounted on front cover. Original blue cloth binding.

1091

INGRAM, COLLINGWOOD (1880-). *The birds of the Riviera; being an account of the avifauna of the Côte d'Azur from the Esterel Mountains to the Italian frontier. With six plates by various artists and text figures by the author.* London: H.F.&G. Witherby, 1926.

xv, 155 p. front., illus., 5 plates. 22.5 cm. Presentation copy from the author with original sketch on fly leaf. Original red cloth binding.

1092

INTERNATIONAL ORNITHOLOGICAL CONGRESS. *Proceedings.* lst.—, 1884-   . Various publishers, 1884-   .

v. illus. 23-24 cm. Title varies slightly. 1954 issued as *Experientia. Supplementum* 3. Italian, English, French, and German. Red library binding.

BM I p.375. Yale 144.

1093

IREDALE, TOM (1880-   ). *Birds of New Guinea.* Illustrated with thirty-five plates in colour, figuring 347 birds by Lilian Medland. Melbourne: Georgian House, 1956.

2 v. plates. 28 cm. "An Australian Society publication." Beige linen binding.

Yale p.145.

1094

IREDALE, TOM (1880-   ). *Birds of paradise and bower birds.* With coloured illustrations of every species by Lilian Medland. Melbourne: Georgian House, [1950].

xii, 239 p. 33 col. plates, fold. map. 28 cm. Quarter black morocco/cloth binding.

Yale p.145.

* 1095

JÄGERSKIÖLD, LEONARD (1867-   ). *Nordens fäglar, av L.A. Jägerskiöld och Gustaf Kolthoff. Under medverkan av Rud. Söderberg. Med 170 tavlor av Olof Gylling. Andra upplagan.* Stockholm: A. Bonnier, [1926].

2 v. illus., 170 col. plates. 37 cm. Green cloth binding, marbled endpapers.

Anker 221. Nissen VBI 468. Yale p.161 (1898 ed.)

* 1096

JARDINE, WILLIAM, SIR, BART. (1800-1874). *The natural history of humming-birds.* Illustrated by thirty-five plates, coloured, and numerous woodcuts... Edinburgh: W.H. Lizars, [etc., etc.], 1833.

2 v. 2 front. (port.), illus., 64 col. plates. 18 cm. (The Natural History Library. Ornithology, Vol.I-II). Half black calf/marbled boards.

Anker 223. BM II p.927. Nissen VBI 471. Yale p.147. Zimmer p.324.

1097

JARDINE, WILLIAM, SIR, BART. (1800-1874), ed. *The naturalist's library.* Conducted by Sir William Jardine. *Ornithology.* Edinburgh: W.H. Lizars, 1833-43.

14 v. fronts. (ports.), illus., plates (part col.) 18 cm. The volume numbering differs in various editions of this work. LSU owns only vol.5 & 6. Full green morocco.

Anker 224. BM III p.176. Nissen VBI 471. Yale p.147. Zimmer pp.325-32.

Gift of Dr. and Mrs. A. Brooks Cronan.

1098

JEWETT, STANLEY GORDON (1885-   ). *Birds of Washington State.* [By] Stanley G. Jewett [and others]. Published in co-operation with the U.S. Dept. of the Interior, Fish and Wildlife Service, Washington, D.C. Seattle: University of Washington Press, 1953.

xxxii, 767 p. illus. (part col.), maps (1 fold. col. in pocket). 25 cm. First author's initial corrected by label. Red cloth binding.

Yale p.149.

1099

JOHNSON, ALFREDO WILLIAM (1897-   ). *The birds of Chile and adjacent regions of Argentina, Bolivia, and Peru.* With 100 colour-plates by J.D. Goodall. Buenos Aires: Platt Estab. Graficos, 1965-   .

v. illus. (part col.) maps. Stamped on fly leaf: Morris D. Williams. Quarter calf/tan cloth binding, red spine labels. Bound by Pedro A. Vadillo Sanchez, Lima, Peru.

Gift of Morris D. Williams.

* 1100

JONES, HENRY (1838-1921). *The bird paintings of Henry Jones.* [Text by] Bruce Campbell; foreword by the Duke of Edinburgh; preface by Lord Zuckerman. London: Folio Fine Editions with the Zoological Society of London, c1976.

[60]p., [24] leaves of plates, col. illus. 40 x 47 cm. This edition limited to five hundred copies. LSU owns copy no.220. Issued in slipcase. Half green calf, marbled endpapers.

* 1101

JONSTONUS, JOANNES (1603-1675). *Histoire naturelle et raisonnée des différens oiseaux qui habitent le globe...* Tr. du latin de Jonston, considérablement augm.,

& mise à la portée d'un chacun. De laquelle on a fait précéder l'histoire particulière des oiseaux de la ménagerie du roi, peints d'après nature par le célèbre Robert, & gravés par lui-même. Le tout orné de quatre-vingt-cinq planches... & divisée en deux parties, dont la première traite des oiseaux de la ménagerie du roi; la seconde, est l'ouvrage même de Jonston. Pour servir de suite à l'histoire des insectes & plantes de Mademoiselle de Merian... Paris: L.C. Desnos, 1773-74.

2 v.in 1. 85 plates. 50 cm. In ms. on title page: John Gebhard Jun. In ms. on verso of 2nd preliminary leaf: Mr. J. Albert Lintner, Will please accept these volumes describing the Natural History of Surinam, by Marie Sybilla Merian, as a slight token of the esteem and friendship of his Humble servant, John Gebhard, Jun. Schoharie, Jany. 1, 1872. The inscription and binder's title incorrectly link this work with Merian's. Contemporary full calf, restored.
Anker 234-235, 237. BM III p.942 (Latin ed.) Nissen VBI 481 (Latin ed.) Yale p.152 (Latin ed.) Zimmer p.341 (Latin ed.)

* 1102

KARMALI, JOHN (1917-). *Birds of Africa: a bird photographer in East Africa.* Foreword by Roger Tory Peterson. London: Collins, 1980.

191 p. illus. (some col.), maps (some col.) 33 cm. Black cloth binding, dustjacket.

* 1103

KEELER, CHARLES AUGUSTUS (1871-1937). *Evolution of the colors of North American Land Birds.* San Francisco: California Academy of Sciences, 1893.

xii, 361 p. 19 plates (part col.) 24 cm. (California Academy of Sciences. Occasional papers, III). Label mounted on p.[2] of cover: 396. Ex libris C. William Beebe. Manuscript notes on added title page. Half black morocco/cloth, spine taped.
Anker 241. BM II p.964. Nissen VBI 130. Zimmer p.346.

1104

KJAERBØLLING, NIELS (1806-1871). *Skandinaviens fugle; med saerligt hensyn til Danmark og de nordlige bilande.* 2. fuldstaendigt omarb. udg. Ved Jonas Collin. Kjøbenhavn: A. Jørgensen, 1875-79.

6 p.l., [vii]-li p., 1 l., 838 , [2]p. 20 cm. and atlas of 106 col. plates. (40.5 x 23.5 cm.) First edition published 1847-52, under title: *Danmarks fugle.* LSU lacks atlas. Quarter brown calf/marbled boards.
Anker 251. BM II p.988. Zimmer p.352.

* 1105

KNOWLTON, FRANK HALL (1860-1926). *Birds of the world; a popular account.* With a chapter on the anatomy of birds, by Frederic A. Lucas; the whole edited by Robert Ridgway. With 16 colored plates and 236 illustrations. New York: H. Holt, 1909.

xiii, 873 p. illus. (part col.) 26 cm. (American nature series. Group I. Natural History). Green cloth binding.
BM VII p.574. Nissen VBI 512. Yale p.160. Zimmer p.358.

1106

KURODA, NAGAMICHI (1889-). *Birds of the island of Java.* Tokyo: 1933-36.

2 v. (xv, 794 p.) 34 col. plates, fold. maps. 38 cm. Brown cloth binding.
Nissen VBI 528. Yale p.162.

1107

LACK, DAVID LAMBERT. *The life of the robin.* Illustrated by Robert Gillmor. 4th edition. London: H.F. & G. Witherby, 1965.

239 p. illus. 23 cm. Blue cloth binding, dustjacket.

1108

LACORDAIRE, CLAUDE FRANÇOIS LÉON (1801-1875). *Atlas des oiseaux d'Europe.* [n.p., 1842?-50?].

3 v. 324 colored plates. 24 x 33 cm. "Le petit pingouin," by Magaud d'Aubusson ([2] p.) from *Le Naturaliste,* 15 Juillet, 1890; 3 p. in ms.: 2 p. pertaining to Lacordaire and his Atlas, 1 p. relating to the penguin; and a photo. of the author with manuscript notes on verso inserted in v.1. Quarter red morocco, gilt decoration. This is an original album of watercolor illustrations, apparently the only existing copy.

* 1109

LAISHLEY, RICHARD. *A popular history of British birds' eggs.* London: Lovell Reeve, 1858.

xi, 313 p. 20 leaves of plates, col. illus. 17 cm. In ms. on fly leaf: A.T. Barraud, February 1867; Broadland, Lawrence Road, West Norwood. Original blue cloth binding, gilt ornamentation, gilt edges.
BM III p.1047. Zimmer p.367.

* 1110

LANSDOWNE, JAMES FENWICK. *Birds of the eastern forest.* Paintings by J. Fenwick Lansdowne. Text by John A. Livingston. Boston: Houghton Mifflin, 1968-1970.

2 v. col. plates. 34 cm. Blue cloth binding.

1111

LANSDOWNE, JAMES FENWICK. *Birds of the west coast*. Paintings, drawings, and text by J.F. Lansdowne; foreword by S. Dillon Ripley. Toronto: M.F. Feheley, c1976- .

v. col. illus. 37 cm. LSU owns v.1 ([176]p., 53 plates). Each plate accompanied by descriptive letterpress. Beige cloth binding, dustjacket.

1112

LATHAM, JOHN (1740-1837). *A general synopsis of birds*. London: Printed for B. White, 1781-85.

3 v. CVI plates. 26 cm. Vols.2-3 have imprint: London, Printed for Leigh & Sotheby.

——— *Supplement to the General synopsis of birds*. London: Printed for Leigh & Sotheby, 1787-1801. (2 v. in 1. CVII-CXL plates. 26 cm.) Contemporary calf binding, restored.

Anker 277-78. BM III p.1063. Nissen VBI 532. Yale p.164. Zimmer pp.371-72.

1113

LATHAM, JOHN (1740-1837). *Index ornithologicus, sive Systema ornithologiae; complectens avium divisionem in classes, ordines, genera, species, ipsarumque varietates: adjectis synonymis, locis, descriptionibus, &c. Studio et opera Joannis Latham*. Londini: sumptibus authoris, 1790.

2 v. 26 cm. Paged continuously. "Supplementum indicis ornithologici, sive, systematis ornithologiae: (lxxiv p. at end of v.2) has special title page. Also issued as a part of *A General synopsis of birds*. Full contemporary calf binding, restored.

Anker 277-79. Nissen VBI 532. Yale p.164. Zimmer pp.372-73.

*1114

LATOUCHE, JOHN DAVID DIGUES DE. *A handbook of the birds of eastern China (Chihli, Shantung, Kiangsu, Anhwei, Kiangsi, Chekiang, Fohkien, and Kwangtung provinces)*. London: Taylor and Francis, 1925- .

v. illus., fold. col. maps, plates. 24 cm. Issued in parts. Yale closes out at 1934. Title page for v.2 in v.2, pt.6. LSU owns v.1, pts.1-4; v.2, pts.1, 3-6. Unbound in portfolios.

Yale p.165.

*1115

LEAR, EDWARD (1812-1888). *Illustrations of the family of Psittacidae, or parrots; the greater part of them species hitherto unfigured, containing forty-two lithographic plates, drawn from life, and on stone*. London: E. Lear, 1832.

4 p.l., 42 col. plates. 58 cm. Each plate has number in ms. In portfolio.

Anker 283. BM III p.1072. Nissen VBI 536. Yale p.166. Zimmer p.380.

1116

LEBERMAN, ROBERT C. *The birds of the Ligonier Valley*. Color plates by H. Jon Janosik; line drawings by Carol H. Rudy. Pittsburgh: Carnegie Museum of Natural History, 1976.

67 p. illus. (some col.) 28 cm. (Carnegie Museum of Natural History Special publication, no.3). Paperbound, illustrated cover.

1117

LEGGE, WILLIAM VINCENT. *A history of the birds of Ceylon*. London: Published by the author, 1880.

1237 p. illus., col. plates, col. map. 33 cm. Black cloth binding.

Anker 284. BM III p.1081. Nissen VBI 135. Zimmer p.382.

*1118

LEMAIRE, C.L. *Histoire naturelle des oiseaux, exotiques*. Ouvrage orné de figures peintes d'après nature par Pauquet. Et gravées sur acier. Paris: Pauquet, 1836.

156 p. 80 col. plates. 23 cm. Errata p.[148]. Full black morocco, gilt decoration.

BM III p.1086. Nissen VBI 540. Yale p.167. Zimmer p.383.

1119

LE MOINE, JAMES MACPHERSON, SIR (1825-1912). *Ornithologie du Canada*. Quelques groupes d'après la nomenclature du Smithsonian institution, de Washington. Québec: Imprimé par E.R. Fréchette, 1860-61.

2 v. 17.5 cm. Paged continuously. Title of v.2 differs slightly: *Les oiseaux du Canada*. Label on p.[2] of cover: Librarie de J.T. Brousseau, No.7, Rue Baade, Vis-a-Vis le Presbytere. Quebec. Vol.1, unbound; v.2, original beige cloth binding.

BM III p.1088.

*1120

LESSON, RENÉ PRIMEVÈRE (1794-1849). *Histoire naturelle des colibris suivie d'un supplément à l'Histoire naturelle des oiseaux-mouches; ouvrage orné de planches dessinées et gravées par les meilleurs artistes*. Et dédié

a M. le Baron Cuvier. Paris: A. Bertrand, [1847?].
x, 196 p. 66 col. plates. 24 cm. Originally published in thirteen parts. Full red morocco, gilt decoration, gilt edges.
Anker 293 BM III p.1096. Nissen VBI 548. Yale p.169 (1830-31 ed.) Zimmer p.388.

* 1121

LESSON, RENÉ PRIMÈVERE (1794-1849). *Histoire naturelle des oiseaux de paradis et des épimaques; ouvrage orné de planches, dessinées et gravées par les meilleurs artistes*. Paris: A. Bertrand, [1835].
1 p.l, [v]-vij, 34 p., 1 l., 248 p. 40 (i.e. 43) col. plates (3 fold.) 31 cm. Preface dated 1835. Originally published in parts, 1833-35. LSU copy imperfect: plates 5, 7, 25, 25 bis, 26, 29, and 33 wanting. Quarter green morocco, gilt spine.
Anker 296. BM III p.1096. Nissen VBI 550. Yale p.169. Zimmer p.390.

1122

LESSON, RENÉ PRIMÈVERE (1794-1849). *Manuel d'ornithologie, ou description des genres et des principales espèces d'oiseaux*. Paris: Roret, 1828.
2 v. illus. 14 cm. Paper boards.
Anker 289-90. BM III p.1095. Nissen VBI 546. Zimmer p.386.

* 1123

LE VAILLANT, FRANÇOIS (1753-1824). *Histoire naturelle des perroquets*. Paris: Levrault, Schoell et C.E., 1804-05.
2 v. 139 (i.e.145) hand col. plates. 52 cm. Full tooled calf, marbled endpapers.
Anker 302. BM III p.1100. Nissen VBI 558. Yale p.170. Zimmer p.392.

* 1124

LE VAILLANT, FRANÇOIS (1753-1824). *Histoire naturelle d'une partie d'oiseaux nouveaux et rares de l'Amérique et des Indes*. Ouvrage destiné par l'auteur à faire partie de son *Ornithologie d'Afrique*. Tome premier. Paris: J.E.G. Dufour, 1801.
iii, 112 p. 49 col. plates. 53 cm. No more published. Quarter red morocco/marbled boards, marbled endpapers.
Anker 300-301. BM III p.1100. Nissen VBI 557. Yale p.170. Zimmer p.392.

* 1125

LILFORD, THOMAS LITTLETON POWYS, BARON. *Coloured figures of the birds of the British Islands*. 2d edition. London: R.H. Porter, 1891-1897.
7 v. 421 col. plates, port. 27 cm. Armorial bookplate of Adam Rivers Steele in each volume. Bound by R.H. Porter. Full green morocco, gilt decoration, marbled endpapers.
Anker 308. BM IV p.1606. Nissen VBI 563. Yale p.172. Zimmer p.399.

1126

LILFORD, THOMAS LITTLETON POWYS, BARON. *Notes on European ornithology. Being four papers extracted from the "Ibis" for 1860, 1862, 1865, & 1866*. London: Printed by Taylor and Francis, 1867.
107 p. 2 col. plates. 23 cm. Inscribed by author. Original red cloth binding, uncut.
Yale p.172.

1127

LINCOLN, FREDERICK CHARLES. *The migration of American birds*. Illustrated by Louis Agassiz Fuertes. 1st edition. New York: Doubleday, Doran & company, inc., 1939.
xii p., 1 l., 189 p. illus. (maps), XII col. plates. 26 cm. Tan library binding.
Yale p.172.

* 1128

LIVINGSTON, JOHN A. *Birds of the northern forest* [By] J. F. Lansdowne with John A. Livingston. Boston: Houghton Mifflin, 1966.
247 p. 56 plates. illus. (part col.) 35 cm. Each plate accompanied by descriptive letterpress. Blue cloth binding, dustjacket.

* 1129

LOCKWOOD, GEOFFREY. *Geoff Lockwood's Garden birds of Southern Africa*. Written and illustrated by Geoffrey Lockwood. Subscribers edition. Johannesburg: Winchester Press, 1981.
ix, 180 p. illus. (some col.), maps. 30 cm. Brown cloth binding, illustrated endpapers, dustjacket.

* 1130

LONG, WILLIAM JOSEPH (1867-1952). *Fowls of the air*. By William J. Long; illustrated by Charles Copeland. Boston and London: Ginn & co., The Athenaeum Press, [c1901].
xi, 310 p. front., plates. 20 cm. Marginal illustrations. Label on p.[2] of cover: Stanstead Wesleyan College, Stanstead Que. Presented for excellence in ______. With name in ms.: Dorothy Flint I, Academy, June 1911. Green cloth binding, gilt decoration.

* 1131

LOUISIANA. DEPT. OF CONSERVATION. *The birds of Louisiana.* [Baton Rouge: Ramires Jones Print. Co.], 1931.

598 p. illus. (part col.), map. 24 cm. (La. Dept. of Conservation, Bulletin no.20). Half blue morocco/marbled boards.

1132

LOWERY, GEORGE HINES (1913-). *Louisiana birds.* Illustrated by Robert E. Tucker. [1st edition, Baton Rouge]: Published for the Louisiana Wild Life and Fisheries Commission by Louisiana State University Press, [c1955].

xxix, 556 p. illus. (part col.), maps, tables. 23 cm. Author's autograph on p.[i]. Green cloth binding, pictorial endpapers.
Yale p.176.

* 1133

MAC CLEMENT, WILLIAM THOMAS (1861-). *The new Canadian bird book for school and home.* With life-like illustrations, concise sketches of Canadian birds for Canadian schools. Toronto: Dominion Book Co., 1914.

xvi, 324 p. [60] leaves of plates, illus. (some col.) 24 cm. "Errata" slip tipped in. In ms. on fly leaf: Jean MacInnes. Red cloth binding.

* 1134

MAC DONALD, MALCOLM (1901-). *Birds in my Indian garden.* Illustrated with 98 photographs by Christina Loke. New York: Knopf, 1961[c1960].

192 p. illus., col. front. 34 cm. Beige cloth binding.

1135

MAC DONALD, MALCOLM (1901-). *Birds in the sun.* Photographs by Christina Loke. Bombay: D. B. Taraporevala Sons, [c1962].

128 p. illus. (part col.) 29 cm. Green cloth binding.

1136

MAC GILLIVRAY, WILLIAM (1796-1852). *A history of British birds, indigenous and migratory: including their organization, habits, and relations; remarks on classification and nomenclature; an account of the principal organs of birds, and observations relative to practical ornithology.* Illustrated by numerous engravings. London: Printed for Scott, Webster and Geary, 1837-52.

5 v. illus., XXIX plates. 22 cm. Vol.1, c.2 bears the signature of John J. Audubon on the title page, p.97 and p.237; vol.2, c.2 is the author's autograph presentation copy to John J. Audubon. Vol.1, c.3 and v.2, c.3 are the author's autograph presentation copies to Th. Durham Weir Espe. Copy 2, half green morocco/green cloth, marbled endpapers; copy 3, half tan calf/green cloth, marbled edges.
BM III p.1209. Nissen VBI 584. Yale p.179. Zimmer p.409.

* 1137

MALHERBE, ALFRED (d. 1866). *Monographie des picidées; ou, Histoire naturelle des picidés, picumninés, yuncinés ou torcols; comprenant dans la première partie, l'origine, mythologique, les moeurs, les migrations, l'anatomie, la physiologie, la répartition géographique, les divers systèmes de classification de ces oiseaux grimpeurs zygodactyles, ainsi qu'un dictionnaire alphabétique des auteurs et des ouvrages cités, par abréviation; dans la deuxième partie, la synonymie, la description en latin et en français, l'histoire de chaque espèce, ainsi qu'un dictionnaire alphabétique et synonymique latin de toutes les espèces.* Metz: Typographie de J. Verronnais, imprimeur de la Sociéte d'histoire naturelle de la Moselle, 1861-62.

4 v. illus., 121 (i.e.123) col. plates. 56 cm. Armorial bookplate in each volume. Bookplate in volume 1 has ms. note: A.J. Dearden, 1925. Half red morocco/marbled boards, marbled endpapers.
Anker 321. BM III p.1226. Nissen VBI 587. Yale p.182. Zimmer p.415.

* 1138

MANNING, THOMAS HENRY (1911-). *The birds of Banks Island.* By T.H. Manning, E.O. Höhn and A.H. Macpherson. Ottawa, 1956.

iv, 144 p. illus., map. 25 cm. (Canada. National Museum, Ottawa. Bulletin, no.143. Biological series, no.48). Paperbound.

* 1139

MARSHALL, CHARLES HENRY TILSON (1841-). *A monograph of the Capitonidae, or scansorial barbets.* By C.H.T. Marshall and G.F.L. Marshall. London: Published by the authors, 1871.

5 p.l., xli, [190]p. 73 col. plates. 31 cm. Body of text numbered in ms. (182 p.). Plates drawn and lithographed by J.G. Keulemans. Each plate has number and scientific name in ms. In ms. on fly leaf: Ernest Shoemaker, 6916-17th Ave., Brooklyn, N.Y., Nov. 16, 1934. Full brown mo-

rocco, marbled endpapers, gilt ornamentation, gilt edges.
Anker 324. BM III p.1245. Nissen VBI 591. Yale p.185. Zimmer p.416.

* 1140

MC ILHENNY, EDWARD AVERY (1872-1949). *The autobiography of an egret.* New York: Hastings House, c1939.

vi, 57, [1]p. incl. front., illus. 23.5 cm. Illustrated lining-papers. Author's autograph presentation copy. Blue cloth binding, dustjacket.
Yale p.180.
Gift of E.A. McIlhenny to LSU Library.

* 1141

MC ILHENNY, EDWARD AVERY (1872-1949). *Bird city.* Illustrated with photographs taken by the author...with introduction by Harris Dickson. Boston: The Christopher publishing House, [c1934].

3 p.l., 3-203 p. incl. front., illus., plates. fold. plate. 18 x 20.5 cm. Author's autographed presentation copy to the LSU Library. Original red cloth binding.
Yale p.180.

* 1142

MCLEAN, DONALD D. *The quail of California.* Sacramento: California State Printing Office, 1930.

47 p. illus., maps (4 fold.) 23 cm. (State of California. Division of Fish and Game. Game bulletin, no.2). Printed paper wrappers.

1143

MEINERTZHAGEN, RICHARD (1878-1967). *Birds of Arabia.* Edinburgh: Oliver and Boyd, [1954].

xii, 624 p. illus. maps (1 fold.) 28 cm. Original gold cloth binding.
Yale p.190.

1144

MEINERTZHAGEN, RICHARD (1878-1967). *Pirates and predators: the piratical and predatory habits of birds.* Edinburgh: Oliver & Boyd, [c1959].

ix, 230 p. illus. front. 28 cm. 41 color plates, 44 monochrome. Original orange cloth binding.

1145

MENABONI, ATHOS. *Birds.* Text by Sara Menaboni, paintings by Athos Menaboni. New York: Rinehart, [c1950].

xi, 132 p. illus., col. plates. 32 cm. Red library binding.
Yale p.190.
Copy 2 gift of the Herman Deutsch Estate.

* 1146

MERSHON, WILLIAM BUTTS (1856-), ed. *The passenger pigeon.* New York: The Outing publishing company, 1907.

xii p., 1 l., 225 p. col. front., 7 plates (2 col.), facsim. 24.5 cm. Dark red cloth binding.
Yale p.191.

1147

MEYER, BERNHARD (1767-1836). *Kurze Beschreibung der Vögel Livund Esthlands...* Nürnberg: J.L. Schrag, 1815.

282 p. hand col. illus., front. 21 cm. Red library binding.
BM III p.1298. Zimmer p.432.

1148

MEYER, BERNHARD (1767-1836). *Taschenbuch der deutschen vögelkunde, oder Kurze beschreibung aller vögel Deutschlands.* Von Hofrath Dr. Meyer und Professor Dr. Wolf. Mit illuminirten Kupfern. Frankfurt am Main: F. Wilmans, 1810.

2 v. col. fronts., col. plates. 22.5 cm. Paged continuously: v.1, xvii, [1], 310 p.; v.2, xii, [311]-614 p. Blue cloth binding.
Anker 336. BM III p.1299. Nissen VBI 1008. Yale p.192. Zimmer p.432.

* 1149

MICHELET, JULES (1798-1874). *The bird.* With 210 illustrations by Giacomelli. London, New York: T. Nelson, 1868.

x, 340 p. illus. 25 cm. Translated by "A.E." from the 8th edition of *L'oiseau.* In ms. on fly leaf: "T.W. Thacker, Esq., with the best regards of T. Nelson & Sons, London, Nov. 27, 1868." Full red morocco, gilt decoration, gilt edges.
Nissen VBI 631 (French ed.) Yale p.192. Zimmer p.434.

1150

MILLAIS, JOHN GUILLE (1865-1931). *British diving ducks.* With thirty-two plates (twenty-two of which are coloured) by Archibald Thorburn, O. Murray Dixon, H. Grönvold and the author. London, New York [etc.]: Longmans, Green and co., 1913.

2 v. col. fronts., 72 plates (37 col.) 41.5 cm. "Only four hundred and fifty copies of this book have been printed.

This copy is no.428." Original red cloth binding.
Anker 342. BM VII p.842. Nissen VBI 633. Yale p.193.

*1151

MILLAIS, JOHN GUILLE (1865-1931). *The natural history of British game birds.* With eighteen coloured plates, seventeen photogravures and two other illustrations by Archibald Thorburn and J.G. Millais. London, New York: Longmans, Green, 1909.

xi, 142 p. 36 plates (part col.) 42 cm. "Only five hundred and fifty copies of this book have been printed. This copy is no.107." Ex libris Rudolph Valentino. Red cloth binding.
Nissen VBI 636. Yale p.193.

1152

MONTAGU, GEORGE (1751-1815). *Ornithological dictionary; or Alphabetical synopsis of British birds.* London: Printed for J. White, 1802.

2 v. in 1. illus. front. 22 cm.

______ *Supplement.* London: Printed by S. Woolmer and sold by S. Bagster and Thomas Underwood, 1813. (1 v., unpaged. illus. 22 cm.) Half brown calf/marbled boards.
Anker 344. BM III p.1339. Nissen VBI 642. Zimmer pp.441-42.

1153

MONTES DE OCA, RAFAEL. *Hummingbirds and orchids of Mexico.* Editing and foreword by Carolina Amor de Fournier. Introduction and texts on the hummingbirds, by Rafael Martín del Campo. English version by Norman Pelham Wright. Mexico: Editorial Fournier, [c1963].

34 p. illus., 60 col. plates (incl. facsim.) 41 cm. "Reproduced from original watercolors [in the Galeria de Arte Mexicano, México, D.F.] painted by Rafael Montes de Oca from 1874-1878." "The facsimile is a reproduction of the title page designed by the author in 1878 for his MS., "Monografia de los colibries y apuntes sobre las principales orquideas de Mexico," written to accompany the watercolors. LSU owns copy no.0511. Beige cloth binding.

*1154

MORRIS, BEVERLEY ROBINSON. *British game birds and wildfowl.* With sixty plates, coloured by hand. 3d edition. London: J.C. Nimmo, 1891.

vi, 272 p. 69 col. plates. 33 cm. Half tan morocco/green cloth.
BM III p.1354 (1855 ed.) Nissen VBI 644. Yale p.198. Zimmer p.442.

*1155

MORRIS, FRANCIS ORPEN (1810-1893). *A history of British birds.* London: Groombridge and sons, 1868.

6 v. col. fronts., 352 col. plates. 26 cm. Vol.2 and 4 published 1866. First published in 1851-57. Original green cloth, gilt ornamentation.
Anker 346. BM III p.1354. Nissen VBI 645. Yale p.198. Zimmer pp.443-44.

*1156

MORRIS, FRANCIS ORPEN (1810-1893). *A history of British birds.* 4th edition, revised, corrected, and enlarged. With three hundred and ninety-four plates coloured by hand. London: J.C. Nimmo, 1895-97.

6 v. col. plates. 26 cm. Slip mounted on lining paper of v.2-3: Ernest F. Cockbain. Half blue morocco/cloth, marbled endpapers, marbled edges.
Anker 346. Nissen VBI 645. Yale p.198. Zimmer p.443 (2nd ed.)

1157

MORRIS, FRANCIS ORPEN (1810-1893). *A history of British birds.* 5th edition, revised and brought up to date, with an appendix of recently added species and with 400 plates specially corrected for this edition, and all coloured by hand. London: J.C. Nimmo, 1903.

6 v. illus. 27 cm. Stamped on fly leaf of all volumes: Roland Kiep. In ms. Xmas 1903. Bookplate of C. Roland Kiep. Original green cloth binding, gilt ornamentation.
Anker 346. Nissen VBI 645. Yale p.198.

1158

MORRIS, FRANCIS ORPEN (1810-1893). *A natural history of the nests and eggs of British birds.* 3rd edition. London: John C. Nimmo, 1892.

3 v. 248 col. plates. 27 cm. Green cloth binding, gilt decoration
BM III p.1354 (other eds.) Nissen VBI 646. Yale p.198 (earlier eds.) Zimmer p.443.

*1159

MORRIS, FRANK T. *Birds of prey of Australia.* Forewords by Allan McEvey [and] Joseph Burke. Melbourne: Lansdowne, [c1973].

171 p. 49 plates (24 col.) 25 cm. "A limited edition of 500 copies numbered and signed by the artist." LSU owns copy no.250. Label mounted on p.[3] of cover: Bound by P.L. Marsh, Melbourne. Full brown calf binding.

* 1160

MORRIS, FRANK T. *Finches of Australia: a folio.* Melbourne: Lansdowne Editions, 1976.

68 p. chiefly illus. 52 cm. This edition is limited to 350 copies numbered and signed by the author, this copy being number 287." Quarter calf/cloth binding.

* 1161

MOUTON-FONTENILLE, MARIE JACQUES PHILIPPE (1769-1837). *Traité élémentaire d'ornithologie, contenant: 1. Les principes et les généralités de cette science; 2. L'analyse du système de Linné sur les oiseaux; 3. La synonymie de Buffon; 4. Les caractères des genres; 5. La description et l'histoire des espéces Européennes; suivi de l'art d'empailler les oiseaux.* Avec dix planches en taille-douce. Lyon: De l'imprimerie de J.B. Kindelem, chez Yvernault et Cabin, Libraires de l'Académie, 1811.

3 v. 10 fold. plates. 22 cm. In ms. on title page of each vol — Ed: de Pury. Label mounted on lining paper of v.1: J.J. Blaise, Libraire à Paris, Quai des Augustins, No.61, à la descente du pont neuf. Salmon cloth binding.
BM III p.1362.

* 1162

MUDIE, ROBERT (1777-1842). *The feathered tribes of the British Islands.* 2d edition. London: Whittaker, 1835.

2 v. illus. 18 col. plates. 21 cm. Green cloth binding.
BM III p.1364. Nissen VBI 654. Yale p.200. Zimmer p.446 (4th ed.)

1163

MUNRO, GEORGE CAMPBELL (1866-). *Birds of Hawaii.* Colored illustrations by Y. Oda. Honolulu, Hawaii: Tongg publishing co., [c1944].

3 p.l., 189 p. illus., 20 col. plates. 23.5 cm. Errata slip inserted before preface. Author's autograph presentation copy. Orange cloth binding.
Nissen VBI 660.

* 1164

MURRAY, JAMES A. *The avifauna of British India and its dependencies; a systematic account, with descriptions of all the known species of birds inhabiting British India, observations on their habits, nidification, &c., tables of their geographical distribution in Persia, Beloochistan, Afghanistan, Sind, Punjab, N.W. Provinces, and the peninsula of India generally, with woodcuts, lithographs, and coloured illustrations.* London: Richardson; Bombay: Education Society's Press, 1887-90.

2 v. illus., plates (part col.) 25 cm. Issued in 7 parts. Vol.2 has imprint: London: Trübner & Co.; Bombay: Education Society's Press. Green cloth binding, gilt ornamentation.
BM III p.1382. Nissen VBI 147. Yale p.201. Zimmer p.450.

1165

MUSCHAMP, EDWARD A. *Audacious Audubon, the story of a great pioneer, artist, naturalist & man.* New York: Brentano's, [c1929].

5 p.l., 312 p. front., ports., facsim. 21 cm. Original green cloth binding.
Yale p.14.

* 1166

The Natural history of Birds; containing a variety of facts selected from several writers, and intended for the amusement and instruction of children. With copper plates. London: Printed for J. Johnson, 1791.

6 pts. in 3 v. 115 (i.e. 116) col. plates. 19 cm. Attributed to Samuel John Galton by the Blacker-Wood Library of Zoology and Ornithology. Dictionary catalogue, v.3, p.486, McGill University. Plate opposite p.26 in pt.2 is not listed in the "Directions for placing the plates in part II." Full calf binding.
Yale p.203.

1167

NEHRLING, HENRY (1853-1929). *Our native birds of song and beauty, being a complete history of all the songbirds, flycatchers, hummingbirds, swifts, goatsuckers, woodpeckers, kingfishers, trogons, cuckoos, and parrots of North America.* With thirty-six colored plates after water-color paintings by Prof. Robert Ridgway, Prof. A. Goering and Gustav Muetzel. Milwaukee: G. Brumder, 1893- .

v. col. plates. 30.5 cm. Also issued in 16 parts. LSU holdings in 1 vol. (xlix, 371 p.) In ms. on fly leaf: Dr. Leonhard Stejneger, Smithsonian Institution, Washington, D.C. With compliments of the author. Milwaukee, Wis., Aug. 23rd 1893. Red leather binding, gilt decoration.
BM III p.1409. Nissen VBI 672. Yale p.206. Zimmer pp.462-63.

1168

NEWMAN, KENNETH B. *Six eagles of Southern Africa.* [Cape Town?]: Purnell, [c1968].

6 col. plates in portfolio. 71 cm. Title from spine. *Lithography: Rufus & Joubert and beith process, South Africa.* Author's autograph on each plate. Edition limited to 1,200 copies. LSU copy no.765. Illustrated paper boards.

* 1169

NICOLL, MICHAEL JOHN (1880-1925). *Nicoll's Birds of Egypt.* By R. Meinertzhagen. Published under the authority of the Egyptian Government. London: H. Rees, 1930.

2 v. (xvi, 700 p.) illus., 3 fold. maps, plates (part col.), port. 33 cm. Green cloth binding.

Anker 363. BM VII p.924. Nissen VBI 149. Yale p.209.

1170

Notes on the birds of Kent. By R.J. Balston, C.W. Shepherd and E. Bartlett. London: R.H. Porter, 1907.

xix, 465 p. illus. (part col.) fold. col. map. 23 cm. Green cloth binding.

BM VI p.54. Yale p.19. Zimmer p.38.

* 1171

NOZEMAN, CORNELIUS (1721-1785?). *Nederlandsche vogelen; volgens hunne huishouding, aert, en eigenschappen beschreeven door Cornelius Nozeman. Alle naer 't leeven geheel nieuw en naeuwkeurig getekend, in 't koper gebragt, en natuurlyk gekoleurd door, en onder opzicht van Christiaan Sepp en zoon.* Amsterdam: J.C. Sepp, 1770-1829.

5 v. (500 p.) 250 col. plates (part fold.) 56 cm. Added title page engraved in colors. Tail-pieces. Bookplate in each vol.: "Je ne change qu'en mourant." Half calf/paper over boards.

Anker 369. BM III p.1455. Nissen VBI 684. Yale p.212. Zimmer pp.469-70.

* 1172

NUTTALL, THOMAS (1786-1859). *A manual of the ornithology of the United States and of Canada... The land birds.* Cambridge, [Mass.]: Hilliard and Brown, 1832.

viii, 683 p. illus. 19 cm. Stamped on verso of title page: American Museum of Natural History, Central Park, New York. Label on p. vi: For the People, For Education, For Science; Library of The American Museum of Natural History. LSU copy imperfect: p.[i-ii] lacking. Red cloth binding.

BM III p.1456. Nissen VBI 685. Zimmer p.470.

1173

NUTTALL, THOMAS (1786-1859). *A manual of the ornithology of the United States and of Canada. The land birds.* 2d edition, with additions. Boston: Hilliard, Gray and company, 1840.

viii, 832 p. illus. 20 cm. Black cloth binding.

Nissen VBI 685. Yale p.212. Zimmer p.470.

* 1174

NUTTALL, THOMAS (1786-1859). *A manual of the ornithology of the United States and of Canada. The water birds.* Boston: Hilliard, Gray and company, 1834.

vii, 627 p. illus. 20 cm. Bookplate on p.[2] of c.1: Library of Joseph Jones M.D., New Orleans, La., 1870. Copy 1, half calf/marbled boards, marbled endpapers, spine taped; copy 2, red cloth binding.

BM III p.1456. Nissen VBI 685. Yale p.212. Zimmer p.470.

1175

NUTTALL, THOMAS (1786-1859). *A popular handbook of the birds of the United States and Canada.* New revised and annotated edition, by Montague Chamberlain, with additions, and one hundred and ten illustrations in color. Boston: Little, Brown, and company, 1903.

xliv, 473, ix, 431 p. col. front., illus., XX col. plates. 22 cm. In ms. on fly leaf: Greta Bohmansson, Xmas 1905 from farbror Rujojo Karl. Original green cloth binding.

Yale p.212. Zimmer p.471.

1176

OATES, EUGENE WILLIAM (1845-1911). *A handbook to the birds of British Burmah, including those found in the adjoining state of Karennee.* London: R.H. Porter, 1883.

2 v. fold. col. map. 27 cm. Armorial bookplate of William Lutley Sclater. Half red morocco/cloth binding.

BM III p.1459. Zimmer p.471.

* 1177

OGILVIE-GRANT, WILLIAM ROBERT (1863-1924). *Game birds.* London: J.F. Shaw, [pref., v.2, 1896].

2 v. illus., map, 39 plates (38 col.) 20 cm. (Allen's Naturalist's library). Original blue cloth, gilt edges.

Anker 371. BM I p.30 (Allen's Naturalists Library). Nissen VBI 688. Yale p.213. Zimmer p.473.

* 1178

Les Oiseaux de Chine, de Mongolie et de Corée non passereaux. [Par] R.D. Étchécopar et François Hüe. Illustré par Paul Barruel et Francis Berille. Papeete: Éditions du Pacifique, 1978.

585 p. [12] leaves of plates, illus. (some col.). Map on endpapers. 24 cm. Tan cloth binding, dustjacket.

1179

Les Oiseaux de Chine de Mongolie et de Corée passereaux. [By] R.D. Étchécopar and François Hüe. Illustré par Patrick Suiro et C.G. Armani. Ouvrage publié avec le concours du Centre National de la Recherche Scientifique. Paris: Societe Nouvelle des éditions Boubée, 1983.

705 p., [24]p. of plates, illus. (some col.) 25 cm. Gold cloth binding, maps on endpapers, illustrated dustjacket.

1180

THE ORNITHOLOGIST AND OÖLOGIST. *Birds: their nests and eggs.* Hyde Park, Mass.: F.B. Webster Company, [1875]-1893.

18 v. illus. 27 cm. LSU owns Jan., 1884-July, 1893, vols.9-18. In original printed paper covers.

1181

The Osprey. An illustrated monthly magazine of popular ornithology. v.1-5, Sept. 1896-Dec. 1901; v.6 (new ser., v.1) Jan.-July 1902. Galesburg, Ill.: The Osprey company, 1897-1902.

6 v. in 4. illus., plates, ports., facsim., maps. 26 cm. LSU owns v.1, Sept. 1896-June 1897; v.2, Sept. 1897-June 1898; v.5, Sept. 1900-Dec. 1901; unbound; v.1, new ser., Jan. 1902-June 1902, unbound.
Yale p.217.

* 1182

PARKIN, THOMAS. *The Great auk, or garefowl (Alca impennis, Linn.)* St. Leonards-on-Sea, [Eng.]: J.E. Budd, printer, 1894.

20, [3]p., 3 leaves of plates. illus. 23 cm. Author's presentation copy to H.A. Hubberty, Jr. (?). Paper read before the Hastings and St. Leonards Natural History Society on June 28th, 1894. Three pages at end include additional comments and observations. Unbound.
BM VI p.983.

1183

Parrots and cockatoos. Paintings by Elizabeth Butterworth. Text by Rosemary Low. London: Fischer Fine Art, [1978].

2 v. illus. 54 cm. Text, 51 p. 25.5 cm. Edition A consists of ten bound copies numbered 1/60 to 10/60. LSU owns copy no.5/60. "Each set consists of twenty etchings, with aquatint. All the etchings are signed and numbered by the artist." Black cloth binding, issued in black cloth slipcase. Binding by Royal College of Art, London.

1184

PEÑA, MARTIN RODOLFO DE LA (1941-). *Enciclopedia de las aves Argentinas.* Santa Fe, República Argentina: n.p., 1978- .

v. illus. 29 cm. LSU owns Fasciculo 1. Red library binding, original covers bound in.

* 1185

PERROTT, CHARLOTTE LOUISA EMILY (d. 1836). *A selection of British birds.* With a new introduction and commentary by Philip J.K. Burton on the birds illustrated and described by Mrs. Perrott. London: Publishing Partnership, 1979.

[8] sheets, 5 leaves of col. plates. 55 cm. "Reproduced from the 1835 edition, of which there are only three known copies; the original was engraved and published by Robt. Havell, London and included Havell's advertisements inserted at end. This facsimile edition is limited to 250 copies. LSU owns copy no.5. This facsimile edition is half-bound in brown goatskin with hand-marbled paper and protected, together with its matching quarter-bound prospectus by a solander box quarter-bound in goatskin and brown cloth."

1186

PETERS, HAROLD SEYMOUR (1902-). *The birds of Newfoundland.* By Harold S. Peters and Thomas D. Burleigh. Illustrated by Roger Tory Peterson. Boston: Published in association with the Dept. of Natural Resources, Province of Newfoundland by Houghton Mifflin, 1951.

xix, 431 p. illus., 32 col. plates, map (on lining papers). 24 cm. Autograph of Roger Tory Peterson. Tan cloth binding. Nissen VBI 718. Yale p.225.

* 1187

PHILIP, DUKE OF EDINBURGH. *Seabirds in southern waters.* [1st ed.]. New York: Harper & Row, [c1962].

62 p. illus. 25 cm. Quarter blue cloth/paper over boards, dustjacket.

1188

PHILLIPS, JOHN CHARLES (1876-1938). *American waterfowl: their present situation and the outlook for their future.* By John C. Phillips and Frederick C. Lincoln, with illustrations by Allan Brooks and A.L. Ripley. Boston, New York: Houghton Mifflin company, 1930.

xiv, [1], 312 p. front., plates, maps. 25 cm. Frontispiece and plates accompanied by guard sheets with descriptive letterpress. Blue cloth binding.
Yale p.227.

* 1189

PHILLIPS, JOHN CHARLES (1876-1938). *A natural history of the ducks.* With plates in color and in black and white from drawings by Frank W. Benson, Allan Brooks and Louis Agassiz Fuertes. Boston: Houghton Mifflin co., 1922-26.

4 v. 102 plates (part col.), 118 maps. 32 cm. Vol.2 contains drawings by Henrik Grönvold, v.4 by Henrik Grönvold and S. Koboyashi. Quarter linen/paper boards.
BM VIII p.1008. Nissen VBI 728. Yale p.227.

1190

PINTO, OLIVERIO MARIO DE OLIVEIRA (1896-). *Ornitologia brasiliense; catálogo descritivo e ilustrado das aves do Brasil.* São Paulo: Departamento de Zoologia da Secretaria da Agricultura do Estado de São Paulo, 1964- .

v. illus. (part col.) 37 cm. LSU owns only v.1. Ivory cloth binding.

* 1191

POLLARD, JOHN RICHARD THORNHILL (1914-). *Birds in Greek life and myth.* [London]: Thames and Hudson, c1977.

224 p. [8] leaves of plates, illus. 23 cm. (Aspects of Greek and Roman life). Green cloth binding, dustjacket.

* 1192

PORTER, ELIOT. *Moments of discovery; adventures with American birds.* Text by Michael Harwood. 1st edition. New York: Dutton, c1977.

120 p. 117 col. plates, col. illus. 27 x 39 cm. Blue cloth binding, dustjacket.

* 1193

A Portfolio of Australian birds. Plates: William T. Cooper. Text: Keith Hindwood. Sydney: A.H. & A.W. Reed, c1968.

60 p., 25 col. plates. col. illus. 37 cm. Beige cloth binding, marbled endpapers.

* 1194

PROVANCHER, LÉON ABEL (1820-1892). *Les oiseaux insectivores: extrait d'une brochure publiée par l'abbé Provancher en 1874.* Quebec: Dussault & Proulx, 1905.

30 p. 18 cm. Stamped on cover, title page, p.[5] and p.30: Bibliotheca F.F. Minorum Quebec. Stamped on cover, title page, and p.[5]: O.F.M. Quebec. Stamped on cover: 905. Printed wrappers.
Zimmer p.497.

1195

PYCRAFT, WILLIAM PLANE (1868-). *Birds in flight.* Illustrated by Roland Green. London: Gay & Hancock limited, [1922].

x, 133, [1]p., incl. plates. col. front., plates (part col.) 25.5 cm. Green cloth binding.
BM VIII p.1038. Nissen VBI 747. Yale p.234. Zimmer pp.497-98.

1196

REICHENBACH, HEINRICH GOTTLIEB LUDWIG (1793-1879). *Aufzählung der Colibris oder Trochilideen in ihrer wahren natürlichen Verwandtschaft: nebst Schlüssel ihrer Synonymik.* Dresden: Carl Ramming, 1853?

24 p. 25 cm. "Cabanis Journ. Extraheft 1853." Unbound.
BM IV p.1670.

1197

REICHENBACH, HEINRICH GOTTLIEB LUDWIG (1793-1879). *Icones ad synopsis avium.* [Dresden und Leipzig: Expedition der vollständigsten Naturgeschichte, 1851-1854]

5 pts. plates. 25 cm. [1. Lfg.] Continuatio no. VIII. Alcedineae — [2. Lfg.] Continuatio no. IX. Meropinae — [3. Lfg.] Continuatio no. IX-[X]. Scansoriae. A. Sittinae — [4. Lfg.] Continuatio no. XI. Scansoriae. B. Tenuirostres. a. Dacninae. b. Certhunae. c. Trochilinae [not issued with this continuatio] d. Upupinae — [5. Lfg.] Continuatio no. XII. Scansoriae. C. Picinae — Plates CCCXCII-CCCCLXXI, CCCCLXXVII-DVI, DX-DCLXXXI. Originally issued in 5 Lfg. under the cover title, *Handbuch der speciellen Ornithologie,* as part of the author's *Die vollständigste Naturgeschichte.* LSU copy imperfect: [2. Lfg.] dedication wanting; [3. Lfg.] dedication, pp.145-149 wanting; [4. Lfg.] half title, dedication wanting; plates CCCXCII-CCCCLXXI, CCCCLXXVII-DVI, DX-DXLIX, DCXV-DCLXXXI wanting. Unbound.
BM IV p.1669. Yale p.238.

1198

REICHENBACH, HEINRICH GOTTLIEB LUDWIG (1793-1879). *Les oiseaux chanteurs — The song-birds —Die*

Singvögel als Fortsetzung der vollständigsten Naturgeschichte und zugleich als Central-Atlas für zoologische Gärten und für Thierfreunde. Ein durch zahlreiche illuminirte Abbildungen illustrirtes Handbuch zur richtigen Bestimmung und Pflege der Thiere aller Classen, herausgegeben von H.G. Ludwig Reichenbach. Dresden und Leipzig: Expedition vollständigsten Naturgeschichte, [1862].

2 p.l. 43, [1]p. 1 l., [45]-70, [2]p. 1 l., [73]-90 x p. XX, XXI-L plates (some col.) 25 cm. LSU copy bound as: 2 p.l., 1 l., x, 43, [1] [45]-70, [2]p., 1 l., [73]-90 p. LSU copy imperfect: pp.33-40 duplicated, plates wanting. Unbound.
BM IV p.1670. Nissen IVB 765. Yale p.238.

1199

REICHENBACH, HEINRICH GOTTLIEB LUDWIG (1793-1879). *Trochilinarum enumeratio; ex affinitate naturali reciproca primum ducta provisoria.* Editio...secunda, emendata et aucta. Leipzig: Friedericum Hofmeister, c1855.

12 p. 25 cm. Unbound.
BM IV p.1669. Nissen IVB 765.

1200

REICHENBACH, HEINRICH GOTTLIEB LUDWIG (1793-1879). *Die vollständigste Naturgeschichte der Tauben und taubenartigen Vögel: Wallnister, Erdtauben, Baumtauben, Hocco's, Columbariae, Megapodinae, Peristerinae, Columbinae, Alectorinae.* Mit 461 Abbildungen auf 65 Kupfertafeln, dazu folgen noch 72 Abbildungen Novitiae. Dresden: Expedition der vollständigsten Naturgeschichte, [1861-62].

206 p. 26 cm. LSU copy imperfect: illus. wanting? "Synopsis avium, iconibus coloratis hucusque rite cognitarum specierum illustrata...Columbariae," by Ludovico Reichenbach: [4]p. laid in. "Neu entdeckte Taubevögel und Nachträge zu den schon beschriebenen": pp.[161]-206. "This supplement was issued as Lief. 9 & 10 of the 'Central Atlas.'"
BM IV p.1669.

——— *Die vollständigste Naturgeschichte der tauben und taubenartigen Vögel: Columbariae, les Pigeons, les Pénélopes et les Hoccos.* Dresden, [1847]. ([1], l., 75 col. plates. 32 cm.) Includes plates 220-277, 230 b, 236 b, 240 b, 245 b, 253 b, 257 b, 272 b, 273 b, Novit. I-IX. Coral cloth binding, decorated endpapers.
BM IV p.1669. Nissen VBI 765. Yale p.238. Zimmer p.513.
Gift of George H. Lowery.

——— *Index zu L. Reichenbach's ornithologischen Werken.* Zusammengestellt von Adolf Bernhard Meyer. Berlin: R. Friedländer, 1879. [vii, 150 p. 28 cm.] Brown cloth binding.
BM III p.1297. Zimmer p.430.

* 1201

The Return of the brown pelican. Photographs by Dan Guravich. Text by Joseph E. Brown. Foreword by Roger Tory Peterson. Baton Rouge and London: Louisiana State University Press, 1983.

vii, 118 p. col. illus. 23 x 29 cm. Brown cloth binding with pelican silhouette, illustrated dustjacket.

* 1202

ROBERTS, AUSTIN (1883-1948). *The birds of South Africa.* [By Austin] Roberts; plates in colour by Norman C.K. Lighton; black and white illustrations by J. Perry and K. Hooper. Revised edition, 1st impression, by G.R. McLachlan & R. Liversidge. Cape Town: Published for the trustees of the South African Bird Book Fund; distributed by Central News Agency, 1957.

xxxviii, 504 p., 56 leaves of plates. illus. (some col.), maps. 23 cm. Beige cloth binding.
Yale p.243 (1948 ed.)

* 1203

ROBERTS, AUSTIN (1883-1948). *Birds of South Africa.* Plates in colour by Norman C.K. Lighton; black and white illustrations by J. Perry and K. Hooper. Revised edition, 2d impression by G.R. McLachlan & R. Liversidge. [Cape Town?]: Published for the Trustees of the South African Birds Book Fund; distributed by the Central News Agency [Johannesburg, 1958].

xxxviii, 504 p. illus., maps. 56 col. plates. 23 cm. Beige cloth binding.
Yale p.243 (1948 ed.)

* 1204

ROBINSON, HERBERT CHRISTOPHER (1874-1929). *The birds of the Malay Peninsula; a general account of the birds inhabiting the region from the isthmus of Kra to Singapore with the adjacent islands.* Issued by authority of the Federated Malay states government. London: H.F. & G. Witherby, 1927-76.

5 v. maps, plates (part col.) 28 cm. Vol.3 by H.C. Robinson and F.N. Chasen; v.4 by F.N. Chasen; v.5 by Lord Medway and D.R. Wells. Armorial bookplate of Kenneth Tlderton in v.1-4. Red cloth binding.
BM IV p.1083. Nissen VBI 791. Yale p.243.

* 1205

ROLES, D. GRENVILLE. *Rare pheasants of the world: a study of birds in captivity.* With drawings by the au-

thor. Hampshire: Spur Publications, c1976.
106 p., 2 leaves of plates. illus. (some col.) 31 cm. Orange cloth binding, dustjacket.

* 1206

ROOSEVELT, THEODORE, PRES. U.S. (1858-1919). *Revealing and concealing coloration in birds and mammals.* Author's edition. New York: American Museum of Natural History, 1911.

119-231 p. 25 cm. (American Museum of Natural History, Bulletin, v.30, article 8). Brown cloth binding, original covers retained.

* 1207

ROTHSCHILD, LIONEL WALTER ROTHSCHILD, BARON (1868-1937). *The avifauna of Laysan and the neighbouring islands: with a complete history to date of the birds of the Hawaiian possessions.* Illustrated with coloured and black plates by Messrs. Keulemans and Frohawk; and plates from photographs, showing bird-life and scenery. London: R.H. Porter, 1893-1900.

3 pts. in 2 v. (xiv, 126, xx, 127-320, 21 p.) illus., 83 plates (part col.) 39 cm. Title page bound in v.2. Description of plates 47 and 53 bound in v.1; plates bound in v.2 following p. xx. Original covers for pts.1-3 bound at end of vols. "With the author's compliments": in ms. on cover for pt.1. Bookplate in each vol.: Ex-libris Ernest Shoemaker, Brooklyn N.Y. Errata slip inserted at end of v.1. Half red morocco/cloth, marbled endpapers.

Anker 429. BM IV p.1739. Nissen VBI 794. Yale p.246.

* 1208

ROTHSCHILD, LIONEL WALTER ROTHSCHILD, BARON (1868-1937). *Extinct birds. An attempt to unite in one volume a short account of those birds which have become extinct in historical times — that is, within the last six or seven hundred years. To which are added a few which still exist, but are on the verge of extinction.* London: Hutchinson & Co., 1907.

xxix, 244 p. 1 l. 42 (i.e.49) plates (45 col.) 38 cm. Half red morocco/cloth, marbled boards.

Anker 430. BM IV p.1740. Nissen VBI 795. Yale p.246.

1209

ROURKE, CONSTANCE MAYFIELD (1885-1941). *Audubon.* With 12 colored plates from original Audubon prints; black and white illustrations by James MacDonald. New York: Harcourt, Brace and company, [c1936].

5 p.l., 342 p. col. front., illus., col. plates. 23 cm. Original blue cloth binding.

Yale p.14.

* 1210

ROWLEY, GEORGE DAWSON. *Ornithological miscellany.* [London: Trübner, 1875-78].

3 v. in 2. illus., 72 col. plates (1 fold.) 32 cm. Title page wanting; title from spine. LSU copy imperfect: v.1-3 have been bound in 2 v. without regard for vols. or pagination. Vols.1-3 incomplete. Stamped on fly leaf of v.1: "Bound by Walker & Son, Princess Place, Plymouth." "The naturalist. The wheatear on the south downs," by J.E. Harting: clipping and 4 illus. mounted in v.1, pp.404-[406]. Half green morocco/marbled boards, marbled endpapers.

Anker 432. BM IV p.1744. Nissen VBI 798. Yale p.246. Zimmer p.533.

* 1211

RUDBECK, OLOF (1630-1702). *Olof Rudbecks Fågelbok, 1693-1710. The book of birds, by Olof Rudbeck.* Planscher och text i urval av Bertil Gullander. Stockholm: E. Bokförlag, c1971.

144 p. col. illus. 30 cm. Title and text in Swedish, English, German, and French. Green cloth binding.

1212

RUSS, KARL (1833-1899). *The speaking parrots: a scientific manual.* Translated by Leonora Schultze and revised by Karl Russ. London: L. Upcott Gill, 1884.

viii, 296 p. col. plates. 20 cm. Translation of *Die sprechenden Papageien.* Original pictorial cloth binding.

BM IV p.1768. Nissen VBI 804. Zimmer p.536.

1213

RUSS, KARL (1833-1899). *Vögel der Heimat; unsre Vogelwelt in Lebensbildern.* Mit 40 Farbendrucktafeln nach original-Aquarellen von Emil Schmidt. Wien: F. Tempsky, 1887.

viii, 410 p. 32 col. plates. 25 cm. Ex Libris von Lengerken. In ms. on fly leaf: Hanns von Lengerken, Langfiches, 1905. Half calf/cloth binding.

Nissen VBI 805. Yale p.248.

1214

ST. JOHN, HORACE STEBBING ROSCOE, MRS. *Audubon, the naturalist of the New world. His adventures and discoveries.* By Mrs. Horace St. John. Revised and corrected, with additions, and illustrated with engravings by J.W. Orr, from original designs. Boston: Crosby and Nichols, 1864.

1 l., xxiv, 311 p. front. (port.), plates. 18 cm. Published also under title: *Life of Audubon, the naturalist of the New world.* Original red cloth binding, gilt decorated spine.
Yale p.14.

* 1215

SALMON, JULIEN. *Les oiseaux.* Avec planches en couleurs et nombreuses photogravures d'après les aquarelles et les dessins originaux de W. Kuhnert. Paris: J.B. Baillière, [1904-06].

2 v. illus., 63 plates (44 col.) 28 cm. (La Vie des animaux illustrée, sous la direction de Edmond Perrier, 3-4). Imprint date, 1890, in ms. on title page of v.1. Ex libris avium R.A.W. Reynolds. Label mounted on lining paper of v.1: Frederick R. Jones, "Adwell," Torre, Torquay and at 20 Victoria Parade, Beacon Quay. Half red cloth/marbled boards.
BM IV p.1550.

1216

SALOMONSEN, FINN (1909-). *Grønlands fugle. The birds of Greenland.* Text [af] Finn Salomonsen, planches [af] Gitz-Johansen. Med forard af. With a preface by Hans Hedtoft. København: E. Munksgaard, 1950-51.

3 v. illus., col. plates, map. 34 cm. Text in Danish and English. Orange library binding, original paper covers bound in.
Nissen VBI 808. Yale p.251.

1217

SAMUELS, EDWARD AUGUSTUS (1836-1908). *The birds of New England and adjacent states: containing descriptions of the birds of New England, and adjoining states and provinces arranged by a long approved classification and nomenclature; together with a history of their habits, times of arrival and departure, their distribution, food, song, time of breeding, and a careful and accurate description of their nests and eggs; with illustrations of many species of the birds, and accurate figures of their eggs; with an appendix containing supplementary notes.* 6th edition, revised and enlarged. Boston: Noyes, Holmes, and company, 1872.

vii, 591 p. col. front., illus., plates. 24.5 cm. In ms. on fly leaf: L.E. Geer. Original brown cloth binding.
Nissen VBI 813. Yale p.251. Zimmer p.544.

1218

SAMUELS, EDWARD AUGUSTUS (1836-1908). *Ornithology and oölogy of New England: containing full descriptions of the birds of New England, and adjoining states and provinces, arranged by a long-approved classification and nomenclature; together with a complete history of their habits, times of arrival and departure, their distribution, food, song, time of breeding, and a careful and accurate description of their nests and eggs; with illustrations of many species of the birds, and accurate figures of their eggs.* Boston: Nichols and Noyes, 1869.

vii, 587 p. front. (col.), illus., plates. 24 cm. Original red cloth binding.
Nissen VBI 813. Yale p.251. Zimmer p.543.

1219

SAMUELS, EDWARD AUGUSTUS (1836-1908). *Our northern and eastern birds. Containing descriptions of the birds of the northern and eastern states and British provinces; together with a history of their habits, times of arrival and departure, their distribution, food, song, time of breeding, and a careful and accurate description of their nests and eggs. With a supplement from Holder's "American fauna."* New York: R. Worthington, 1883.

vii, 600 p. col. front., illus., 40 plates (5 col.) 26 cm. First edition issued in 1867 under title, *Ornithology and oölogy of New England.* In ms. on fly leaf: "For Alexis Brooke Esq. from Miss Conins, J.E.S.W. Jan. 21, 1884. Original brown pictorial cloth binding.
BM IV p.1797. Nissen VBI 814. Yale p.251. Zimmer p.544.

* 1220

SAUER, GORDON C. *John Gould bird print reproductions.* 1st edition. Kansas City, Mo.: Distributed by Richland Enterprises, 1977.

v, 76 p. illus. 24 cm. "Two hundred numbered copies have been signed by the author. This is number 18." Paperbound.

* 1221

SAUER, GORDON C. *John Gould, the bird man; a chronology and bibliography.* Melbourne, New York: Lansdowne Editions, 1982.

xxiv, 416 p., [32]p. of plates, illus. 28 cm. Green cloth binding, illustrated dustjacket.

* 1222

SCHEITHAUER, WALTER. *Hummingbirds.* With 76 colour photos by the author. Translated by Gwynne Vevers. New York: T.Y. Crowell Co., [c1967].

176 p. illus. (part col.), map. 26 cm. Translation of *Kolibris*. Blue cloth binding.

1223

SCHLEGEL, HERMANN (1804-1884). *Revue critique des oiseaux d'Europe*. Leide: A. Arnz, 1844.

5 p.l., cxxxv, 116 p. illus. 25 cm. In French and German. Quarter red morocco/marbled boards, marbled endpapers.
BM V p.1838. Yale p.254. Zimmer p.554. All entered under German title.

* 1224

SCHLEGEL, HERMANN (1804-1884). *De toerako's afgebeeld en beschreven door H. Schlegel onder medewerking van G.F. Westerman*. Opgedragen aan B.M. den Koning. Uitgegeven door het Koninklijk Zoölogisch Genootschap Natura Artis Magistra.

24 p., 1 l. 34 plates (17 col.) 73 cm. Duplicate plates facing each other, one in color and one in black and white. Half tan calf/green morocco.
Anker 444. BM IV p.1839. Nissen VBI 833. Yale p.254. Zimmer pp.555-56.

1225

SCLATER, PHILIP LUTLEY (1829-1913). *Argentine ornithology. A descriptive catalogue of the birds of the Argentine Republic*. By P.L. Sclater...with notes on their habits by W.H. Hudson. London: R.H. Porter, 1888-89.

2 v. illus., XX col. plates. 26.5 cm. "The edition of this work being strictly limited to 200 copies for Subscribers, each copy is numbered and signed by the Authors." LSU owns copy no.52. Half red morocco/marbled boards.
BM VI p.73.

* 1226

SCLATER, PHILIP LUTLEY (1829-1913). *A monograph of the birds forming the Tanagrine genus Calliste*. Illustrated by coloured plates of all the known species. London: J. Van Voorst, 1857.

xvii, 104 p. 45 col. plates, col. map. 23 cm. Armorial bookplate, E. Bibl. Radcl. Half red morocco/cloth binding.
Anker 448. BM IV p.1882. Nissen VBI 839. Yale p.257. Zimmer p.559.

1227

SCOTT, PETER (1909-   ). *Wild chorus*. Written and illustrated by Peter Scott. New York: C. Scribner's sons, 1939.

ix, 118, [2]p. illus. plates. 29.5 cm. "Signed limited edition, November 1938: first ordinary edition, September 1939." Green cloth binding.
Nissen VBI 848. Yale p.258.

1228

SEEBOHM, HENRY (1832-1895). *The birds of Siberia; a record of a naturalist's visits to the valleys of the Petchora and Yenesei*... London: J. Murray, 1901.

xix, 512 p. incl. illus., port., fold. map. 23.5 cm. "Narrative of Mr. Seebohm's two Siberian expeditions in 1875 and 1877: first published under the respective titles of *Siberia in Europe*, 1880, and *Siberia in Asia*, 1882... Limitation of space has necessitated the omission of the lengthy foot-notes." — Preface. Tan pictorial cloth binding.
BM IV p.1891 (1st ed.) Yale. p.259. Zimmer p.567 (1st ed.)

1229

SEEBOHM, HENRY (1832-1895). *The geographical distribution of the family Charadriidae, or the plovers, sandpipers, snipes and their allies*. London [etc]: Henry Sotheran & co., 1888.

xxix, 524 p. illus., 21 col. plates (incl. front.) 33 x 27 cm. Armorial bookplate: Molineux. Original green cloth binding.
Anker 455. BM IV p.1891 (1887 ed.) Nissen VBI 165. Zimmer p.568.

1230

SEEBOHM, HENRY (1832-1895). *A history of British birds, with coloured illustrations of their eggs*. London: Published for the author by R.H. Porter, 1883-85.

4 v. illus. 25 cm. Plates bound in v.4. In ms. on fly leaf of v.1: R.D. Archer-Hind, Trinity College, Cambridge, 1886. Full calf, marbled endpapers, spines mended.
Anker 456. BM IV p.1891. Nissen VBI 851. Yale p.259. Zimmer p.568.

1231

SELBY, PRIDEAUX JOHN (1788-1867). *Illustrations of British ornithology*. Edinburgh: Printed for the proprietor, and published by W.H. Lizars, 1833.

2 v. 22 cm. Armorial bookplate of John W. Perry Watlington in each volume.
——— *Plates to Selby's Illustrations of British ornithology*. London: H.G. Bohn, 1841. (2 v. 228 plates (224 col.) 71 x 56 cm.) Half green morocco/marbled boards, gilt edges, gilt spine.
BM IV p.238. Nissen VBI 853. Yale p.260. Zimmer p.573.

* 1232

SELVA ANDRADE, CARLOS. *Love life of the birds*. Translated from the Spanish by Herbert M. Clark.

Illustrated by Axel Amuchastegui. Buenos Aires: Codex Editors, c1952.

194 p., [4]p. col. illus. 32 cm. "Corrigenda": [2]p. inserted. Green cloth binding, pictorial endpapers.
Yale p.260.

* 1233

SHARPE, RICHARD BOWDLER (1847-1909). *An analytical index to the works of the late John Gould.* With a biographical memoir and portrait. London: H. Sotheran, 1893.

xlviii, 375 p. port. 39 cm. Limited edition of 100 large paper copies. LSU Library copy imperfect: port. wanting. Green cloth binding.
BM IV p.1910. Yale p.114. Zimmer p.582.

1234

SHARPE, RICHARD BOWDLER (1847-1909). *A handbook to the birds of Great Britain.* London: W.H. Allen, 1894-97.

4 v. illus., CXXIV (i.e.128), col. plates (incl. fronts.) 20 cm. (Allen's naturalist's library). Quarter turquoise cloth/marbled boards.
Anker 466. BM IV p.1910. Nissen VBI 863. Yale p.262 (dif.ed.) Zimmer pp.583-84.

* 1235

SHARPE, RICHARD BOWDLER (1847-1909). *Monograph of the Paradiseidae, or birds of paradise and Ptilonorhynchidae, or bower-birds.* London: H. Sotheran, 1891-98.

2 v. illus., 79 hand col. plates. 56 cm. Bound by Bayntun Irivierei, Bath, England. Full green morocco, gilt decoration, gilt edges, marbled endpapers.
BM IV p.1910. Nissen VBI 865. Yale p.262. Zimmer p.581.

* 1236

SHELDON, WILLIAM G. *The book of the American woodcock.* Amherst: University of Massachusetts Press, 1971[c1967].

xvii, 227 p. illus. 26 cm. "Second printing, with corrections, 1971." Gold cloth binding, dustjacket.

1237

SHELLEY, GEORGE ERNEST (1840-1910). *A handbook to the birds of Egypt.* London: John Van Voorst, 1872.

viii, 342 p. 14 col. plates. 26 cm. Bound for Bernard Quaritch, ltd., by Sangorski and Sutcliffe. Half red morocco/cloth binding.
Anker 469. BM IV p.1913. Nissen VBI 872. Yale p.264. Zimmer p.588.
Gift of Dr. and Mrs. A. Brooks Cronan.

* 1238

SHELLEY, GEORGE ERNEST (1840-1910). *A monograph of the Nectariniidae, or family of sun-birds.* London: Published by the author, 1876-80.

cviii, 393, [1]p. 121 plates. 34.5 cm. Originally published in XII parts, July 28, 1876-February, 1880. Original covers of the 12 parts bound at end. Each plate accompanied by descriptive letterpress. Full red morocco/marbled boards, gilt ornamentation.
BM IV p.1913. Nissen VBI 167. Yale p.264. Zimmer p.588.

* 1239

SHUFELDT, ROBERT WILSON (1850-1934). *Contributions to the anatomy of birds.* Washington: 1882.

1 p.l.,pp.593-806. illus., plates. 23.5 cm. At head of title: Author's edition. Department of the Interior. United States Geological and Geographical Survey. F.V. Bayden, U.S. geologist in charge. Extracted from the Twelfth annual report of the Survey. Black library binding.
BM IV p.1919. Zimmer p.590.

1240

Singing birds; how to catch, cage, feed and teach them, by an old fancier. London: J. Turner, [1800?].

24 p. 16 cm. (Penny series of useful handbooks, 8). Unbound.

* 1241

SMART, GREGORY. *Birds on the British list: their title to enrolment considered, especially with reference to the British Ornithological Union's list of British birds: with a few remarks upon 'evolution' and notes upon the rarer eggs.* London: R.H. Porter, 1886.

xxiv, 148 p. 23 cm. Armorial bookplate of Richard Heywood Thompson. Author's holograph letter to Richard Heywood Thompson laid in. Note to the editor (leaf) laid in. Errata slip inserted. In ms. on fly leaf: With the author's kind regards, June 24 '86. Original green cloth binding.
BM IV p.1940.

1242

SMYTHIES, BERTRAM EVELYN. *The birds of Burma.* With 31 colour plates by A.M. Hughes. 2nd revised edition. Edinburgh: Oliver and Boyd, [1953].

xliii, 668 p. 31 col. plates, fold. col. map. 25 cm. Green cloth binding.
Nissen VBI 882. Yale pp.269-70.

1243

South African birds. First series. [By] Dick Findlay and Allan Bird. Johannesburg: South African Natural History Publication Co., 1959.

1 portfolio ([4]p., 12 col. plates). 36 cm. Title from portfolio.

* 1244

SPIX, JOHANN BAPTIST VON (1781-1826). *Avium species novae, quas in itinere per Brasiliam annis MDCCCXVII-MDCCCXX jussu et auspiciis Maximiliani Josephi I. Bavariae regis suscepto collegit et descripsit J.B. de Spix.* Tabulae...a M. Schmidt Monacensi depictae. Monachii: Typis F.S. Hübschmanni, 1824-39.

2 v. 222 col. plates. 39 cm. Signed in ms. on title page of each vol.: Geo. A. McCall. Index for each vol. in ms., laid in. Half green morocco/marbled boards, marbled endpapers.

Anker 483. BM V p.1992. Nissen VBI 891. Yale p.275. Zimmer p.600.

* 1245

STANLEY, EDWARD, BP. OF NORWICH (1779-1849). *A familiar history of birds; their nature, habits, and instincts.* The 5th edition. London: John W. Parker, 1851.

xi, 480 p. incl. front., illus. 17.5 cm. Green cloth binding, gilt decoration.

BM V p.2002. Zimmer p.601.

1246

STEARNS, WINFRID ALDEN (1852-). *New England bird life: being a manual of New England ornithology.* Revised and edited from the manuscript of Winfrid A. Stearns by Dr. Elliott Coues. Boston: Lee and Shepard; New York: C.T. Dillingham, 1881-83.

2 v. illus. 20.5 cm. Vol.1: editor's presentation copy. Original brown cloth binding.

BM V p.2006. Yale p.277. Zimmer p.602.

1247

STEVENS, LYLA. *Birds of Australia in colour.* Illustrated by Anne Lissenden. Melbourne: Whitcombe & Tombs, 1955.

60 p. illus. 25 cm. Pictorial boards.

1248

STOUT, GARDNER D. *The shorebirds of North America.* Editor and sponsor: Gardner D. Stout. Text by Peter Matthiessen. Paintings by Robert Verity Clem. Species accounts by Ralph S. Palmer. New York: Viking Press, [c1967].

270 p. illus., 32 col. plates. 37 cm. "This specially bound first edition...is limited to 350 numbered copies... This copy is number 290." Signed by authors and illustrator. Full leather binding, marbled endpapers, cloth slipcase.

1249

STRICKLAND, HUGH EDWIN (1811-1853). *Ornithological synonyms.* By the late Hugh Edwin Strickland. Edited by Mrs. Hugh E. Strickland and Sir W. Jardine. Vol. I. *Accipitres.* London: J. Van Voorst, 1855.

xlvi, 222 p. 23 cm. No more published. In ms. on fly leaf: Chas. W. Richmond. Original brown cloth binding, spine missing.

BM p.2036. Zimmer p.606.

1250

SUNDEVALL, CARL JACOB (1801-1875). *Försök till Fogelklassens naturenliga uppställning.* Stockholm: Samson & Wallin, 1872.

187 p. 22 cm. Issued in 2 parts, 1872-73. This copy in 2 parts, as originally issued, in wrappers.

BM V p.2049. Yale p.280. Zimmer p.611.

Gift of George Lowery.

* 1251

SUTTER, ERNST. *Oiseaux de paradis et colibris; images de la vie des oiseaux sous les tropiques.* [Par] Ernst Sutter [et] Walter Linsenmaier. Version française par Paul Géroudet. Zurich: Service d'images Silva, c1953.

127 p. 60 mounted col. plates. 30 cm. Translation of *Paradiesvögel und Kolibris.* Yellow cloth binding.

1252

SUTTON, GEORGE MIKSCH (1898-). *Portraits of Mexican birds: fifty selected paintings by George Miksch Sutton; foreword by Enrique Beltrán.* 1st edition. Norman: University of Oklahoma Press, c1975.

xi, 106 p. illus. (some col.) 37 cm. Blue cloth binding, dust-jacket.

1253

SWAINSON, WILLIAM (1789-1855). *On the natural history and classification of birds.* London: Printed for Longman, Rees, Orme, Brown, Green, & Longman [etc.], 1836-37.

2 v. illus. 17 cm. (The Cabinet Cyclopaedia. Conducted by...D. Lardner... Natural History. [v.117-118]). Marbled boards, taped spine.

BM V p.2054. Nissen VBI 913. Zimmer p.371.

* 1254

SWAINSON, WILLIAM (1789-1855). *A selection of the birds of Brazil and Mexico.* London: H.G. Bohn, 1841.

4 p. 78 col. plates. 25 cm. Half red morocco/green cloth binding.

Anker 494. BM V p.2055. Nissen VBI 912. Yale p.281. Zimmer p.615.

1255

SWANN, HARRY KIRKE (1871-). *A monograph of the birds of prey (Order Accipitres).* Edited by Alexander Wetmore. Illustrated by plates reproduced in colour from drawings by H. Grönvold, also by coloured plates of eggs and plates in photogravure. London: Wheldon & Wesley, ltd., 1930-1936.

2 v. plates (part col.) 31 cm. "The edition of this work is limited to 412 copies; it will not be reprinted." Issued in parts each with a special title page. Vol.[2] lacks general title page. Dark blue library binding.

BM VIII p.1276. Nissen VBI 917 (dif. ed.) Yale p.282. Zimmer p.619.

1256

SWANN, HARRY KIRKE (1871-). *A synopsis of the Accipitres (diurnal birds of prey) comprising species and subspecies described up to 1920, with their characters and distribution.* 2d edition, revised and corrected throughout. London: Wheldon & Wesley, 1922.

233 p. port. 22 cm. Green library binding.

Anker 497. BM VIII p.1276. Nissen VBI 916. Yale p.282. Zimmer pp.618/19.

1257

SWAYSLAND, W. *Familiar wild birds.* 1st[-4th] series. London, New York: Cassell, 1883.

4 v. col. illus. 20 cm. Original green pictorial cloth binding.

BM V p.2056. Nissen VBI 918. Yale p.282. Zimmer p.620.

1258

SYMONS, R.D. (1898-). *Hours and the birds; a Saskatchewan record.* [Toronto]: University of Toronto Press, [c1967].

xi, 224 p. illus. (part col.), maps. 26 cm. Blue cloth binding.

1259

TAKATSUKASA, NOBUSUKE (1889-1959). *The birds of Nippon.* [Tokyo]: Maruzen Co., 1967.

701 p. illus., maps, 3 col. plates, ports. 31 cm. Some chapters issued originally in serial form, 1932 to 1938, forming v.1 of a projected 2 volume work, v.2 of which was never published. Red cloth binding.

* 1260

TAVERNER, PERCY ALGERNON (1875-). *Birds of western Canada.* 2d edition, revised. Ottawa: F.A. Acland, 1928.

379 p. 84 leaves of plates, illus. (some col.) 26 cm. (National Museum of Canada, Bulletin, no.41; Biological series, no.10). Blue cloth binding.

Anker 499. BM VIII p.1289. Nissen VBI 925.

* 1261

TEGETMEIER, WILLIAM BERNHARD (1816-1912). *Pheasants; their natural history and practical management.* Illustrated from life by J.G. Millais, T.W. Wood, P. Smit and F.W. Frohawk. 4th edition, enlarged. London: H. Cox, 1904.

xii, 255 p. illus., 22 plates (6 col.) 23 cm. Original dark brown cloth binding.

BM V p.2081 (1873, 1881 eds.) Nissen VBI 928. Yale p.255 (1911 ed.)

* 1262

THOMSON, ARTHUR LANDSBOROUGH (1890-). *Britain's birds and their nests: described by A. Landsborough Thomson.* With introduction by J. Arthur Thomson. Illustrated with 132 drawings in colour by George Rankin. London: W. & R. Chambers, 1910.

xxviii, 340 p. 132 col. plates. 26 cm. Red cloth binding, gilt decoration.

Nissen VBI 937. Zimmer p.633.

* 1263

THORBURN, ARCHIBALD (1860-1935). *British birds, written and illustrated by Archibald Thorburn.* With one hundred and ninety-two plates in colour. New edition. London, New York: Longmans, Green, 1931-[35].

4 v. 192 col. plates. 23 cm. Vol.4 without edition statement. Vol.4 reprinted by novographic process, June 1935. Red cloth binding.
Anker 508. Nissen VBI 938. Yale p.289. Zimmer p.634.

* 1264

TODD, FRANK S. *Waterfowl: ducks, geese & swans of the world.* San Diego, Calif.: Sea World Press, c1979.

x, 399 p. col. illus. 30 cm. In ms. on title page: J.S. McIlhenny. Blue cloth binding, silver gilt decoration, illustrated dustjacket.

* 1265

TUNNICLIFFE, CHARLES FREDERICK (1901-). *Tunnicliffe's birds.* Measured drawings by C.F. Tunnicliffe, RA. With an introduction, commentary and memoir of the artist by Noel Cusa. London: Victor Gollancz Ltd., 1984.

160 p. col. illus. 33.5 cm. Brown cloth binding.

1266

TUNSTALL, MARMADUKE (1743-1790). *Ornithologia Britannica: seu Avium omnium Britannicarum tam terrestrium, quam aquaticarum catalogus, sermone latino, anglico & gallico redditus: cui subjicitur appendix, aves alienigenas, in Angliam raro advenientes, complectens.* London: Printed for the author by J. Dixwell, 1771. 1 p.l., 4 p. illus. 51.5 cm. Unbound, marbled paper wrappers.
BM II p.715. Zimmer p.641.

* 1267

TYSON, CARROLL SARGENT (1878-). *20 birds of Mt. Desert Island.* [Milan, 1921].

20 col. plates in portfolio. 77 cm. Limited edition of 250 copies. On verso of each plate: "Made in Italy, Roberto Hoesch." Red cloth portfolio.
Yale p.294.

* 1268

VAN SOMEREN, VICTOR GURNER LOGAN. *The birds of East Africa: a collection of lithographs prepared from the originals drawn from nature by V.G.L. van Someren between the years 1909 and 1937.* A limited edition. Sarasota, Fla.: Published by A.C. Allyn for the Allyn Museum of Entomology, 1973-

v. chiefly col. illus. 29 cm. Each plate accompanied by guard sheet with descriptive letterpress. Edition limited to 100 sets. LSU owns copy no.15. With author's autograph. Full dark blue morocco binding.

* 1269

VIEILLOT, LOUIS JEAN PIERRE (1748-1831?). *Analyse d'une nouvelle ornithologie élémentaire.* Paris: Deterville, 1816.

1 p.l., 70 p. 21 cm. On label mounted on p.[2] of cover: "Graham Alnwick, bookbinder." Manuscript notes on fly leaf. Armorial bookplate of Francis Hubert Barclay. With this is bound: Temminck, C.J. *Observations sur la classification méthodique des oiseaux, et remarques sur l'analyse d'une nouvelle ornithologie élémentaire par L.P. Vieillot.* Amsterdam: Gabriel Dufour, 1817 (1 p.l., 60 p.) Half calf/marbled boards.
BM V pp.2082, 2216. Yale p.285. Zimmer pp.626, 655.

1270

VIEILLOT, LOUIS JEAN PIERRE (1748-1831?). *Histoire naturelle des oiseaux de l'Amérique Septentrionale, contenant un grand nombre d'espèces décrites ou figurèes pour la première fois.* Paris: Desray, 1807.

2 v. 124 (i.e. 131) col. plates., fold. map. 52 cm. Quarter red calf/marbled boards, marbled endpapers.
Anker 515. BM V p.2216. Nissen VBI 957. Yale p.300. Zimmer p.655.

1271

WAIT, WALTER ERNEST. *Manual of the birds of Ceylon.* [Colombo], Ceylon: The Director, Colombo Museum; London: Dulau & Co., ltd., [c1925].

3 p.l., 496 p., 1 l. 1 illus., XX plates, fold. map. 25.5 cm. Each plate accompanied by guard sheet with descriptive letterpress. Tan cloth binding.
Nissen VBI 964. Zimmer p.660.

1272

WALPOLE-BOND, JOHN. *A history of Sussex birds.* With fifty- three coloured plates by Philip Rickman. London: H.F. & G. Witherby ltd., [1938].

3 v. col. plates. 25.5 cm. Brown cloth binding.
Yale p.304.

1273

The Warbler. v.1-2, Jan. 1903-Nov. 1904; 2d. ser., v.1-7, 1st quarter 1905-1913. Floral Park, N.Y.: The Mayflower publishing company, 1903-13.

9 v. illus. col. plates. 25.5 cm. LSU owns v.1, nos. 1-4, 8; 1905. Published bimonthly, 1903-1904; quarterly, 1905-1906; annually, 1907-1910, 1913. Publication suspended, 1911-1912. Official organ of the Long Island Natural History Club, 1904; bulletin of the John Lewis Childs Museum and Library of Natural History, 1907-1913. Unbound.
Yale p.305.

1274

WEIGOLD, HUGO (1886-). *Der Vogelzug auf Helgoland graphisch dargestellt.* Berlin: R. Friedländer, 1930.

24 p. 91 diagrs. 32 cm. (Abhandlungen aus dem Gebiete der Vogelzugsforschung. Hrsg. von der Vogelwarte der Staatl. Biologischen Anstalt auf Helgoland, nr.1). In green cloth portfolio.

* 1275

WETMORE, ALEXANDER (1886-). *The birds of Haiti and the Dominican Republic.* By Alexander Wetmore and Bradshaw H. Swales. Washington: U.S. Government Printing Office, 1931.

iv, 483 p. 26 plates (incl. fold. map). 24 cm. (Smithsonian Institution. United States National Museum. Bulletin, 155). Clipping relating to Alexander Wetmore mounted on p.[3] of cover. Paperbound.
BM VIII p.1421. Nissen VBI 981.

1276

WETMORE, ALEXANDER (1886-). *Song and garden birds of North America, by Alexander Wetmore and other eminent ornithologists.* Foreword by Melville Bell Grosvenor. Washington: National Geographic Society, [c1964].

400 p. illus. (part col.), col. map (on lining paper). 27 cm. *Bird songs of garden, woodland, and meadow,* by Arthur A. Allen and Peter Paul Kellogg (12 p. and phonodiscs: 12 s., 7 in., 33 1/3 rpm.) in pocket. Orange cloth binding.
Gift of Ella V. Aldrich Schwing.

1277

WETMORE, ALEXANDER (1886-). *Water, prey, and game birds of North America, prepared by National Geographic Book Service.* By Alexander Wetmore and other eminent ornithologists. Foreword by Melville Bell Grosvenor. Washington: National Geographic Society, [c1965].

464 p. illus. (part col.), col. map (on lining paper). 27 cm. *Bird sounds of marsh, upland, and shore,* by Peter Paul Kellogg (12 p. and phonodiscs: 12 s., 7 in., 33 1/3 rpm.) in pocket. Green cloth binding.
Gift of Ella V. Aldrich Schwing.

1278

WHISTLER, HUGH (1889-). *Popular handbook of Indian birds. Illustrated with twenty-one full-page plates (ninety-nine figures) of which six are coloured, and one hundred and five figures in the text, from drawings by H. Grönvold.* 3d edition, revised and enlarged. London: Gurney and Jackson, 1941.

xxviii, 549 p. illus. (6 col.) 24 cm. Original green cloth binding.
BM VIII p.1423 (earlier eds.) Nissen VBI 982.

1279

WHYMPER, CHARLES (1853-). *Egyptian birds for the most part seen in the Nile Valley.* London: A. and C. Black, 1909.

x, 221 p. illus., 51 col. plates (incl. front.) 23 cm. In ms. on fly leaf: "M.A. Murray, With all good wishes for 1921 from E.B." Tan cloth binding, pictorial spine.
Anker 529. BM V p.2313. Nissen VBI 986. Yale p.310. Zimmer p.673.

* 1280

WILLUGHBY, FRANCIS (1635-1672). *The ornithology of Francis Willughby... In three books. Wherein all the birds hitherto known...are accurately described. The descriptions illustrated by...LXXVIII copper plates. Tr. into English, and enl. with many additions throughout the whole work. To which are added, three considerable discourses, I. Of the art of fowling: with a description of several nets in two large copper plates. II. Of the ordering of singing birds. III. Of Falconry.* By John Ray... London: Printed by A.C. for J. Martyn, 1678.

6 p.l., 441, [6]p. 80 plates. 36.5 cm. LSU copy imperfect: pp.159-160 wanting, p.330 incorrectly numbered 332, p.335 incorrectly numbered 333, plates 79-80 wanting. Half calf/marbled boards.
Anker 532. BM V p.2331. Nissen VBI 991. Yale p.312. Zimmer p.677.

* 1281

WILSON, ALEXANDER (1766-1813). *American ornithology; or, the natural history of the birds of the United States: illustrated with plates, engraved and colored from original drawings taken from nature.* First edition, first issue. Philadelphia: Bradford and Inskeep, 1808-14.

9 v. hand col. plates. 36 cm. Clipping relating to the author and portrait of the author laid in v.9. Half red calf, original binding, each volume in red cloth portfolio.
Anker 533. BM V p.2332. Nissen VBI 992. Yale p.312. Zimmer p.679.

* 1282

WILSON, ALEXANDER (1766-1813). *American ornithology; or, The natural history of the birds of the United States. With a continuation by Charles Lucien Bonaparte. The illustrative notes, and life of Wilson, by Sir William Jardine.* London: Whittaker, Treacher, & Arnot; Edinburgh: Stirling & Kenney, 1832.

3 v. 97 plates. 23 cm. Half green calf/cloth, marbled endpapers, gilt spine.

Anker 534. BM V p.2332. Nissen VBI 992-997 (dif. eds.)Yale p.312 (dif. eds.) Zimmer p.683.

1283

WILSON, ALEXANDER (1766-1813). *American ornithology; or, The natural history of the birds of the United States.* By Alexander Wilson and Prince Charles Lucien Bonaparte. The illustrative notes and life of Wilson, by Sir William Jardine. London, New York: Cassell, Petter & Galpin, [1878].

3 v. 103 col. plates, port. 23 cm. Quarter green morocco/red cloth, gilt spine.

Anker 534. BM V p.2332 (dif. eds.) Nissen VBI 992-997 (dif.eds.) Yale p.312 (dif. eds.) Zimmer p.683.

* 1284

YARRELL, WILLIAM (1784-1856). *A history of British birds. Illustrated with woodcuts of each species.* London: J. Van Voorst, 1839-43.

3 v. illus. 22 cm. Vols.1-2, c.1 have temporary title page which was issued only with v.1-2; v.1-2, c.2 have permanent title page with imprint: London, J. Van Voorst, 1843. Originally published in thirty-seven parts, of three sheets each, at intervals of two months, July 1837 to May 1843. Signed in ms. on title page of v.1, c.1: John T. (?) Hervey, 11th Novr 1842. Stamped on verso of 1st fly leaf of v.1-2, c.2 and on last fly leaf of v.3: Clarke & Bedford. Armorial bookplate of Earl Cornwallis in v.1-2, c.2 and v.3. Full green morocco, gilt ornamentation, marbled endpapers, gilt edges.

BM V p.2372. Nissen VBI 1029. Yale p.318. Zimmer pp.697-98.

1285

Zeitschrift für die Gesammte Ornithologie. Herausgeben von Dr. Julius von Madarasz. 1-4 Jahrgang, 1884-1888. Budapest: Franklin-Verein, 1884-1888.

4 v. illus. 24 cm. Half black morocco/marbled boards, marbled endpapers.

Anker 545. BM V p.2384. Zimmer p.706.

* 1286

ZIMMER, JOHN TODD (1889-). *Birds of the Marshall Field Peruvian expedition,* 1922-1923. Chicago, 1930.

pp.233-480. map. 24 cm. (Field Museum of Natural History. Publication, 282; Zoological series, v.17, no.7). Ex libris Nils Gyldenstolpe. Half brown morocco/marbled boards.

* 1287

ZIMMERMAN, WILLIAM (1937-). *Waterfowl of North America.* Introduction and editorial supervision by Olin Sewall Pettingill. Louisville, Ky.: Published by The Frame House Gallery, c1974.

93 p. illus., 42 maps, 42 col. plates. 57 x 70 cm. Quarter green morocco/cloth, medallion on front cover.

PART VIII

Horticulture

Illustration from *Nouvelle Iconographie des Camellias* published by Ambroise Alexandre Verschaffelt, 1848-60.

1288

American Camellia Catalog; a loose-leaf compendium of reference information on all camellias, with handpainted lithographs of selected varieties, published in yearly installments on a subscription basis. Production by Robert Park Erdman. Research by Albert Fendig; art by Athos Menaboni. 1st edition. Savannah: American Camellia Catalog, [c1949-53].

v. (loose-leaf), col. illus. 30 cm. LSU owns 1950-53. In loose leaf binders.

1289

APPEL, OTTO (1867-1952). *Atlas der krankheiten der landwirtschaftlichen kulturpflanzen.* Farbige tafeln im format von 31: 45 cm. In der Biologischen reichsanstalt für land- und forstwirtschaft nach der natur gemalt von August Dressel. Mit beschreibendem text von prof. dr. O. Appel und der E. Riehm. Berlin: P. Parey, 1924-33.

3 v. 23 cm. and atlas of col. plates in 2 v. 45 cm. LSU owns atlas and v.2 of text. Vol.1, atlas, original boards with printed label; v.2, atlas, black library binding; v.2, text, in pamphlet binder.

1290

BARRY, PATRICK (1816-1890). *The fruit garden; a treatise intended to explain and illustrate the physiology of fruit trees, the theory and practice of all operations connected with the propagation, transplanting, pruning and training of orchard and garden trees, as standards, dwarfs, pyramids, espaliers, etc., the laying out and arranging different kinds of orchards and gardens, the selection of suitable varieties for different purposes and localities, gathering and preserving fruits, treatment of diseases, destruction of insects, descriptions and uses of implements, etc. Illustrated with upwards of 150 figures, representing different parts of trees, all practical operations, forms of trees, designs for plantations, implements, etc.* New York: Scribner, 1851.

xiv, 398 p. illus. front. 20 cm. Original green cloth binding.

* 1291

BARRY PATRICK (1816-1890). *The fruit garden; a treatise intended to explain and illustrate the physiology of fruit trees, the theory and practice of all operations connected with the propagation, transplanting, pruning and training of orchard and garden trees...the laying out and arranging different kinds of orchards and gardens...* Auburn, Rochester: Alden and Beardsley, 1855, c1851.

xiv, 398 p. illus. front. 21 cm. In ms. on fly leaf: W.A. Knowlton. Original green cloth binding.

1292

BEACH, SPENCER AMBROSE (1860-1922). *The apples of New York.* By S.A. Beach, assisted by N.O. Booth and O.M. Taylor. Albany, N.Y.: J.B. Lyon printer, 1905. (New York Agricultural Experiment Station report for the year 1903, II).

2 v. col. front. (v.1), illus., plates (part col.) 23.5 cm. Original green cloth binding, gilt ornamentation.

1293

La Belgique Horticole, t.1-35, 1851-1885. Liége: A la Direction Générale, 1851-1885.

35 v. illus. 26 cm. Vol.1-4 subtitle, *Journal des Jardins, des Serres et des Vergers* par Charles Morren. Vol.5-14 subtitle, *La Belgique Horticole annales de botanique et d'horticulture*, par Edouard Morren. Quarter red morocco/marbled boards, marbled endpapers.

BM I p.128. Nissen BBI 228. Pritzel 10699.

1294

Bonsai-saikei; the art of Japanese miniature trees, gardens, and landscapes. By Toshio Kawamoto and Joseph Y. Kurihara. [1st limited edition. Tokyo: Nippon Saikei Co., 1963].

362 p. illus., plates (part col.) 29 cm. "Limited first editions 500...number 136." Original dark green cloth binding.

1295

BOOTH, WILLIAM BEATTIE (1804?-1874). *Illustrations and descriptions of the plants which compose the natural order Camellieae, and of the varieties of Camellia japonica, cultivated in the gardens of Great Britain.* The drawings by Alfred Chandler. London: Published by J. and A. Arch, 1831; New York: American Engraving Society, [c1944].

2 pts. in 1 v. 12 col. plates. 41 cm. Original covers for each pt. bound in. Green library binding.

BM I p.199. Nissen BBI 209.

* 1296

BRIDGEMAN, THOMAS. *The young gardener's assistant; containing a catalogue of garden & flower seeds, with practical directions under each head, for the cultivation of culinary vegetables and flowers. Also directions for cultivating fruit trees, the grape vine, &c. To*

which is added a calendar showing the work necessary to be done in the various departments of gardening in every month of the year. 7th edition, improved. New York: Mitchell & Turner, printers, for sale by T. Bridgeman, 1837.
vi, 360 p. 21 cm. Original beige cloth binding.

1297

BROWNE, DANIEL JAY (b. 1804). *The sylva Americana; or, A description of the forest trees indigenous to the United States, practically and botanically considered.* Illustrated by more than one hundred engravings. Boston: W. Hyde & co., 1832.
vii, 408 p. incl. illus., plates, front. 23 cm. In ms. on title page: D.T. Brainard's. Brown library binding.
BM I p.262. Pritzel 1252.

1298

BURBANK, LUTHER (1849-1926). *Luther Burbank, his methods and discoveries and their practical application; prepared from his original field notes covering more than 100,000 experiments made during forty years devoted to plant improvement, with the assistance of the Luther Burbank society and its entire membership under the editorial direction of John Whitson and Robert John and Henry Smith Williams.* Illustrated with 105 direct color photograph prints produced by a new process devised and perfected for use in these volumes. New York, London: Luther Burbank press, 1914-15.
12 v. col. fronts., col. illus., col. ports., facsims. 24.5 cm. Original stamped leather binding.

1299

CANDOLLE, ALPHONSE LOUIS PIERRE PYRAMUS DE (1806-1893). *Origin of cultivated plants.* New York: D. Appleton and company, 1886.
viii p., 1 l., 468 p. 19.5 cm. Original French edition, 1883; 1st American edition, 1885. Original red cloth binding.
BM I p.309. Plesch p.171 (1896 French ed.)

* 1300

CHAPAIS, JEAN CHARLES (1850-). *Guide illustré du sylviculteur Canadien.* 3. éd. Québec: J.A. Langlais, 1891.
205 p. illus., 22 cm. Stamped on cover, half title, front., title page, p.[3], 93-99: "O.F.M. Quebec." Stamped on cover: "891" and "Avec les compliments du Ministre de la Colonisation et des Travaux Publics de la Province de Quebec." Original printed wrappers.

* 1301

COBB, JONATHAN HOLMES (1799-1882). *A manual containing information respecting the growth of the mulberry tree, with suitable directions for the culture of silk.* In three parts. New edition. Boston: Carter, Hendee, 1833.
xii, 98 p. 98 plates (part col.) 18 cm. "Published by direction of His Excellency Gov. Lincoln, agreeably to a resolve of the Commonwealth." In ms. on fly leaf: Geo. C. Wells. Original printed boards.

* 1302

La Culture des fleurs: ou il est traitte' generalemente de la maniere de semer, planter, transplanter & conserver toutes sortes de fleurs & d'arbres, ou arbrisseaux a fleurs, connus en France; et de douze maximes generales desquelles il est necessaire d'etre instruit pour pratiquer utilement cette sorte d'agriculture. Bourg en Bresse: J. Ravoux, 1692.
209, [22], 45 p. 16 cm. Ex libris A.P.M. de Kluijs. In ms. on p.[2] of cover: 1642. In ms. on title page: Dubois de Mayne (?) no.25. "Almanach jardinier perpetuel," by Emanuelis: 45 p. (2d group). Contemporary calf binding, unrestored.

* 1303

DEWOLF, GORDON P. *Flora exotica; a collection of flowering plants.* Text by Gordon DeWolf. Woodcuts by Jacques Hnizdovsky. Boston: D.R. Godine, 1972.
60 p. illus. (part col.) 32 cm. "A deluxe edition of 50 includes an extra suite of hand-colored prints each individually signed by the artist." Text with illustrator's autograph. Green cloth binding; prints in tan cloth portfolio; both boxed.

1304

DUCHESNE, ÉDOUARD ADOLPHE (1804-1869). *Répertoire des plantes utiles et des plantes vénéneuses du globe contenant la synonymie latine et française des plantes, leurs noms vulgaires français et l'indication de leurs usages...précédé d'un traité indispensable aux personnes qui veulent herboriser, et composer des herbiers.* Paris: J. Renouard, 1836.
1 p.l., xlviii, 572 p. 21.5 cm. Ex libris David L. Blondheim. Stamped on fly leaf: Fr. Louis Planchon, Montpellier, No.C 6. Quarter black morocco/marbled boards, marbled endpapers.
BM I p.484. Jackson p.206. Nissen BBI 537. Plesch p.208. Pritzel 2444.

1305

DUHAMEL DU MONCEAU, HENRI LOUIS (1700-1781?). *Du transport, de la conservation et de la force des bois; ou l'on trouvera des moyens d'attendrir les bois, de leur donner diverses courbures, surtout pour la construction des vaisseaux; et de former des pièces d'assemblage pour suppléer au défaut des pieces simples, faisant la conclusion du Traité complet des bois et des forets,...* Paris: L.F. Delatour, 1767.

xxxii, 556 p. 27 fold. plates. 27 cm. In ms. on half title page: G. Cusachs. Newspaper clipping pertaining to saws and sawmills mounted on fly leaf. Ex libris G. Cusachs. Full brown morocco binding, gilt decorated spine. Binding by William A. Payne.

1306

DUVERNAY, J.M. *Fleurs des jardins et des serres.* [Par] J.M. Duvernay [et] H. Romagnesi. Aquarelles de Madeleine Rollinat. [Paris]: Bordas, [c1962-].

v. col. plates. 38 cm. In portfolio.

1307

EARLE, ALICE (MORSE) (1851-1911). *Old time gardens, newly set forth by Alice Morse Earle; a book of the sweet o' the year.* New York: Macmillan, 1901.

xix, 491 p. illus., plates. 25 cm. "Of this large paper edition...three hundred and fifty (350) copies have been printed on a specially made paper." LSU owns copy no.93. Label mounted on lining paper: Brentano's Booksellers & Stationers, Union Square, New York. Original green cloth binding, gilt ornamentation, illustrated endpapers.

1308

EVELYN, JOHN (1620-1706). *Silva; or, A discourse of forest-trees, and the propagation of timber in His Majesty's dominions; as it was delivered in the Royal Society, on October XV, MDCLXII, upon occasion of certain queries propounded to that illustrious assembly, by the Hon. the principal officers and commissioners of the Navy. Together with an historical account of the sacredness and use of standing groves. To which is added the Terra: a philosophical discourse of earth.* With notes, by A. Hunter. The 5th edition, with the editor's last corrections. London: H. Colburn, 1825.

2 v. front. (port.), plates (part fold.) 30 x 24.5 cm. Manuscript notes on title page of v.1-2 and throughout text of each vol. Holograph note laid in v.1. Six newspaper clippings pertaining to trees laid in v.1. Green library binding.
BM II p.549. Nissen BBI 615. Plesch p.220. Pritzel 2766.

1309

Floricultural cabinet and Florist's Magazine, v.1-27, 1833-59. London: Whittaker & Co., 1833-59.

27 v. illus. 23 cm. LSU owns v.1, 1833; v.2, 1834; v.3, 1835; v.6, 1838; v.7, 1839. Vol.1 has second edition on title page. Superseded by *Gardener's Weekly Magazine*. Ms. notes in vol.1. Label in v.2, 3, 6, & 7: P.W. Humphries. Original black cloth binding.
BM II p.639. Jackson p.472. Nissen BBI 3 2229n.

1310

Gartenflora; blätter für garten- und blumenkunde. Berlin: v.1-87, no.3, Jan. 1852—Mar. 1938; n.s., v.1- Apr. 1938- . Berlin, 1852- .

v. illus. 26 cm. LSU owns v.1-85, 1936. Supersedes *Schweizerische zeitschrift für gartenbau*. Suspended publication Je.-Oct. 1920, Sept. 1922-Dec. 1923, 1940. Vol.71, no.9-v.72 never published. Some illustrations replaced with photostats. Binding varies.

——— *Beilage. Orchis.* Berlin, v.1- 1906-
Beilage, supplements 3-14 bound with *Gartenflora*.
BM II p.642. Pritzel 10743.

1311

GEERT, AUGUSTE VAN. *Camellias; contenant les figures et une courte description des douze plus rares et plus belles variétés de ce genre.* New York: Paris Etching Society; Camilla Lucas, distributeur, [1950].

[8]p., [12] col. plates. 33 cm. Green cloth binding, original printed covers bound in.

*1312

GOPALASWAMIENGAR, K. S. *Cultivation of bulbous plants in India.* Madras: Huxley Press, 1932.

108 p., [15] leaves of plates. illus. 18 cm. "Published under the auspices of the Mysore Horticultural Society." Stamped on title page: Taru, no. 3643/111, Mori Gate, Delhi-110006, India. Paper wrappers.

1313

GREY, CHARLES HERVEY (1875-). *Hardy bulbs, including half-hardy bulbs and tuberous and fibrous-rooted plants.* Illustrated by Cecily Grey. London: Williams & Norgate Ltd, [1937- .]

v. illus. col. plates. 25 cm. LSU owns vols.1-2. Original green cloth binding.
Plesch p.243.

*1314

HARRIS, THADDEUS WILLIAM (1795-1856). *A treatise on some of the insects injurious to vegetation.* 3d edition. Boston: William White, Printer to the State, 1862.

xi, 640 p., [8] leaves of col. plates. illus., front. 24 cm. First published, without illustrations, Cambridge, 1841, under title: *A report on the insects of Massachusetts, injurious to vegetation.* In ms. on fly leaf and title page: Columbus Tyler. Original brown cloth binding, gilt ornamentation.

1315

HEDRICK, ULYSSES PRENTISS (1870-). *The cherries of New York.* By U.P. Hedrick, assisted by G.H. Howe, O.M. Taylor, C.B. Tubergen, R. Wellington. Albany, N.Y.: J.B. Lyon company, state printers, 1915.

xii, 371 p. front. (port.) col. plates. 31 cm. (New York Agricultural Experiment Station report for the year 1914, v.2, pt. II). Original green cloth binding.
BM VI p.447.

1316

HEDRICK, ULYSSES PRENTISS (1870-). *The grapes of New York.* By U.P. Hedrick, assisted by N.O. Booth, O.M. Taylor, R. Wellington, M.J. Dorsey. Albany, N.Y.: J.B. Lyon company, state printers, 1908.

xv, 564 p. front. (port.) col. plates. 31 cm. (New York Agricultural Experiment Station report for the year 1907, v.3 pt. II). Original green cloth binding.
BM VI p.447.

1317

HEDRICK, ULYSSES PRENTISS (1870-). *The peaches of New York.* By U.P. Hedrick, assisted by G.H. Howe, O.M. Taylor [and] C.B. Tubergen. Albany, N.Y.: J.B. Lyon company, printers, 1917.

xiii, 541 p. front. (port.), col. plates, map, facsim. 31 cm. (New York Agricultural Experiment Station report for the year 1916, vol.2, pt. II). In ms. on fly leaf: Mary Q. Doty, Albany, N.Y., April 5, 1919. Original green cloth binding.
BM VI p.447.

1318

HEDRICK, ULYSSES PRENTISS (1870-). *The pears of New York.* By U.P. Hedrick, assisted by G.H. Howe, O.M. Taylor, E.H. Francis, H.B. Tukey. Albany, N.Y.: J.B. Lyon company, 1921.

xi, 636 p. front. (port.) col. plates. 31 cm. Original green cloth binding.

1319

HEDRICK, ULYSSES PRENTISS (1870-). *The plums of New York.* By U.P. Hedrick, assisted by R. Wellington, O.M. Taylor, W.H. Alderman, M.J. Dorsey. Albany, N.Y.: J.B. Lyon company, state printers, 1911.

xii, 616 p. front. (port.) col. plates. 31 cm. In ms. on p.[2] of cover: Experiment Station Louisiana, Baton Rouge, La. Original green cloth binding.
BM VI p.447.

1320

HEDRICK, ULYSSES PRENTISS (1870-). *The small fruits of New York.* By U.P. Hedrick assisted by G.H. Howe, O.M. Taylor, Alwin Berger, G.L. Slate [and] Olav Einset. Report of the New York State Agricultural Experiment Station for the year ending June 30, 1925 [pt.] II. Albany, N.Y.: J.B. Lyon company, printers, 1925.

xi, 614 p., incl. tables. front. (port.) col. plates. 31 cm. (New York State Dept. of Farms and Markets. 33rd annual report, 1924/25, pt. II). Original green cloth binding.

* 1321

[HEY, REBECCA]. *The spirit of the woods.* London: Longman, Rees, Orme, Brown, Green, & Longman, 1837.

xvi, 306 p. [26] leaves of plates, col. illus. 25 cm. Errata slip inserted. Illustrations by the author. In ms. on fly leaf: Mrs. N. Faxon, with the love & respect of W. A. W., Jan. 1, 1838. In ms. on title page: Mrs. Wm. Hey. Half green morocco, green cloth binding.
Plesch 260.

* 1322

HIBBERD, SHIRLEY (1825-1890). *The fern garden, how to make, keep, and enjoy it; or, Fern culture made easy.* 7th edition. London: Groombridge, 1877.

vi, 148 p., [8] leaves of plates. illus. (some col.) 19 cm. Embossed on fly leaf: W.H. Smith & Son Library, 186 Strand. Original green cloth binding stamped in black with gilt ornamentation.

* 1323

HOOKER, WILLIAM (1779-1832). *Pomona Londinensis: containing colored engravings of the most esteemed fruits cultivated in the British gardens, with a descriptive account of each variety.* By William Hooker. Assisted in the descriptive part by the president and members, and sanctioned by the patronage of the

Horticultural Society of London. Vol.1. London: The author, 1818.

3 p.l., 49 col. plates. 34 cm. No more published. Each plate accompanied by leaf of descriptive letterpress. Full calf, gilt ornamentation.
BM II p.874. Nissen BBI 913. Plesch p.268. Pritzel 4247.

1324

L'Horticulteur Universel, journal général des amateurs et jardiniers présentant l'analyse raisonnée des travaux horticoles français et étrangers... [Éditeurs Gerard, Charles Antoine Lemaire, Jacques Martin Victor]. Publié par une réunion de botanistes et d'horticulteurs français et étrangers...tome 1-6, 1839-44; 2. sér., [t.1], 1847; [nouv. sér., t.1], 1847. Paris: H. Cousin, 1839-[47].

8 v. illus. 25 cm. LSU owns v.1-5, 1839-44. Quarter green morocco/marbled boards, marbled endpapers.
Pritzel 10770.
Gift of Mr. & Mrs. Milton R. Underwood in memory of Mrs. Annie Lawrason Butler.

1325

HOUSSAYE, J. G. *Monographie du thé, description botanique, torréfaction, composition chimique, propriétés hygiéniques de cette feuille orné de 18 gravures par J.-G. Houssaye.* Paris: Chez l'auteur, 1843.

2 p.l., 160 p. front., plates. 24.5 cm. LSU copy imperfect: front. wanting. Original blue cloth binding, gilt illustration on front cover.
BM II p.880.

* 1326

HOWITT, SAMUEL (1765?-1822). *British preserve.* Drawn & etched by S. Howitt. [London]: T. Griffiths, 1829.

70 p. 36 plates. fronts. 30 cm. Engraved title page. Half red calf/cloth, marbled endpapers, gilt edges.
BM II p.882. Nissen ZBI 2014. Yale p.138.

1327

HUME, H. HAROLD (1875-). *Camellias in America.* Harrisburg, Pa.: J. Horace McFarland Company, 1946.

xvi, 350 p. col. front., illus. (incl. ports., facsims.) col. plates. 28 cm. Dark red cloth binding.

* 1328

HUME, H. HAROLD (1875-). *Camellias in America.* Revised edition. Harrisburg, Pa.: J.H. McFarland Co., 1955.

xvi, 422 p. illus. (part col.) ports. 29 cm. Dark red cloth binding.

1329

HYAMS, EDWARD S. (1912-1975). *The English garden.* Photographs by Edwin Smith. New York: H.N. Abrams: [c1964?].

287 p. illus. (part col.) 33 cm. Green cloth binding.

1330

Journal des roses (rosa inter flores) et revue d'arboriculture ornementale... [1]-38 année; jan. 1877-août 1914. Melun et Paris: A. Lebrun [etc.] and A. Goin [etc., 1877-1914].

38 v. in 13. illus., plates (part col., part fold.) ports. 28.5 cm. Monthly. Orange library bindings.

1331

KERNER, ANTON JOSEPH, RITTER VON MARILAUN (1831-1898). *Flowers and their unbidden guests.* Translation revised and edited by W. Ogle with a prefatory letter by Charles Darwin. London: C.K. Paul & co., 1878.

xvi, 164 p. III double plates. 20 cm. Translation of *Die Schutzmittel der Blüthen gegen unberufene Gäste.* In ms. on title page: Given to B.L. Ogle, Esq. by Dows, Lady Dashwood, Feb.-1889. Embossed on title page: Kirtlington Park, Oxford. Original green cloth binding.
BM II p.971. Jackson p.99.

1332

KUPPER, WALTER. *Orchidées.* Version Française par Jean Lupold. [Planches de Walter Linsenmaier]. Zurich: Service D'Images Silva, [c1953].

127 p. 60 col. plates. 31 cm. Translation of: *Orchideen.* Label mounted on title page: W.S. Heinman, imported books...New York 21, N.Y. Purple cloth binding.
Nissen BBI 3 1114n (Ger. & French eds.)

* 1333

LAFITAU, JOSEPH FRANÇOIS (1681-1746). *Mémoire présenté a son altesse royale Mgr. le duc d'Orleans, régent de France, concernant la precieuse plante du ginseng de Tartarie, découverte en Amérique par le Père Joseph-François Lafitau. Nouv. édition. Précédée d'une notice biographique par M. Hospice Verreau; et accompagne d'un portrait du Père Lafitau; d'un facsimile de son autographe et de la planche représentant le gin-*

seng. Montréal: Senecal, Daniel, 1858.
42, [2]p., [2] leaves of plates. illus., port. 20 cm. Embossed on title page, p.[3], 41, [43], and plate: Bibliotheca F.F. Minorum Quebec. Stamped on title page: 55 F. Original printed wrappers.
Pritzel 4980.

* 1334

LAW-SMITH, JOAN. *A gardener's diary*. Written and illustrated by Joan Law-Smith. Melbourne: Published by the Women's Committee of the National Trust of Australia (Victoria), 1976.
111 p. illus. (some col.) 36 cm. "This signed edition is limited to 275 copies of which 260 are for sale. This is copy number 171." Brown paper over boards, bee medallion in gilt on front cover, in slipcase.
Gift of Margaret Stones.

1335

LOUDON, JOHN CLAUDIUS (1783-1843). *Arboretum et fruticetum britannicum, or, The trees and shrubs of Britain, native and foreign, hardy and half-hardy, pictorially and botanically delineated, and scientifically and popularly described; with their propagation, culture, management and uses in the arts, in useful and ornamental plantations, and in landscape gardening; preceded by a historical and geographical outline of the trees and shrubs of temperate climates throughout the world*. London: Printed for the author by A. Spottiswode and sold by Longman, Orme, Brown, Green and Longmans, 1838.
8 v. illus., plates (part fold.) 22.5 cm. The partially coloured and coloured copies by James Ridgway and sons. Ex libris George J. Arnold. Half green morocco/marbled boards, gilt edges.
BM III p.1182. Jackson p.407. Nissen BBI 1238. Pritzel 5629.

* 1336

LOWE, EDWARD JOSEPH (1825-1900). *Beautiful leaved plants; being a description of the most beautiful leaved plants in cultivation in this country; to which is added an extended catalogue*. By E.J. Lowe assisted by W. Howard. With sixty coloured illustrations. London: Groombridge and sons, 1865.
viii, 144 p. illus., LX col. plates (incl. front.) 26 cm. Restored binding, original red cloth covers retained.
BM III p.1184 (1864 & 1868 eds.) Nissen 1247 (1859, 1867 eds.) Plesch p.318 (1st ed., 1861 ed.) Pritzel 5642 (1864 ed.)

1337

Magazine of horticulture, botany and all useful discoveries and improvements in rural affairs... v.1- , 1835. Boston: Printed for Russell, Shattuck and Williams..., 1835- .
v. illus. tables, diagrs. 22 cm. Title varies: v.1-2, *American gardener's magazine...* v.3-, *Magazine of horticulture...* LSU owns v.1-10. Bindings: v.1, brown cloth; v.2, black cloth; v.3-10, blue marbled paper over boards, brown calfskin spines.

1338

MAINICHI, SHIMBUN SHA. *Tsubaki*. Tokyo: Inoue Book Co., 1971.
180 p. illus. (mostly col.) 30 cm. In Japanese with English translations of varietal names. Dark green cloth, paperboard slipcase.

1339

MORETON, CHARLES OSCAR. *The auricula; its history and character*. With seventeen coloured plates reproduced from paintings by Rory McEwen. London: Ariel Press, [c1964].
51 p. 17 col. plates. 41 cm. Illustrated paper covers.
Nissen BBI 3 1406n.

1340

MORETON, CHARLES OSCAR. *Old carnations and pinks*. With an introduction by Sacheverell Sitwell and 8 coloured plates by Rory McEwen. [London]: George Rainbird, in association with Collins, 1955.
xi, 51 p. col. plates. 35 cm. Illustrated paper over boards, illustrated endpapers.

* 1341

MOULEN, FRED. *Orchids in Australia*. Lausanne: Edita S.A., 1958.
148 p. (chiefly col. plates). 28 cm. Beige cloth binding.
Plesch p.344.

1342

NICOL, WALTER. *The planter's kalendar; or, The nurseryman's & forester's guide, in the operations of the nursery, the forest, and the grove*. By the late Walter Nicol...edited and completed by Edward Sang. 2d edition, improved and enlarged. Edinburgh: A. Constable, 1820.
xxiv, 589 p. 3 plates. 21 cm. Armorial bookplate of Sir Matthew Barrington. In ms. on title page: Matt. Barrington, 21 Dec. 1827. Manuscript notes throughout text. Half brown calf/marbled boards.

* 1343

NOTRE DAME DU LAC DES DEUX MONTAGNES (TRAPPIST ABBEY). *Conseils pratiques sur la conduite des arbres fruitiers en verger.* Par les RR. PP. Trappistes de Notre-Dame du Lac, Oka. Baie St. Paul: Echo de Charlevoix, 1904.

36 p. 20 cm. (Québec, Département de l'Agriculture. Bulletin). Stamped on cover, title page, and p.36: Bibliotheca F.F. Minorum, Quebec. Embossed on cover, title page., p.[3], and 5: C.F.M. Quebec. Original printed paper wrappers.

* 1344

ORMEROD, ELEANOR ANNE (1828-1901). *A manual of injurious insects with methods of prevention and remedy for their attacks to food crops, forest trees, and fruit.* To which is appended a short introduction to entomology compiled by Eleanor A. Ormerod. 2d edition. London: Simpkin, Marshall, Hamilton, Kent, 1890.

xiv, 410 p., [1] leaf of plates. illus., port. 21 cm. Bookplate of the College of Agriculture, Downton, Salisbury, mounted on fly leaf. Brown calf, gilt ornamentation.
BM III p.1477 (1881 ed.)

* 1345

PENNSYLVANIA HORTICULTURAL SOCIETY, PHILADELPHIA. *From seed to flower: Philadelphia, 1681-1876: a horticultural point of view.* Philadelphia: Pennsylvania Horticultural Society, 1976.

119 p., [1] fold. leaf of plates, illus. 25 cm. In ms. on fly leaf: "For E.A. McIlhenny collection, compliments of Joseph Ewan, 3 October 1976." Green pictorial paper wrappers.

* 1346

PHILLIPS, CECIL ERNEST LUCAS (1898-). *The Rothschild rhododendrons; a record of the gardens at Exbury* [By] C.E. Lucas Phillips and Peter N. Barber. Photographs by Harry Smith. Drawings by Gillian Kenny. With a foreword by the Lord Aberconway. New York: Dodd, Mead, [c1967].

xviii, 138 p. illus., maps, 66 col. plates, port. 32 cm. Quarter green cloth/ivory cloth binding, silver gilt ornamentation.

* 1347

PORTER, GEORGE RICHARDSON (1792-1852). *The tropical agriculturist: a practical treatise on the cultivation and management of various productions suited to tropical climates.* London: Smith, Elder and Co., 1833.

xii, 429, [1]p. incl. col. front., illus., plates. 23 cm. Ex libris Arpad Plesch. LSU copy imperfect: front. in black and white. Original green cloth binding.
BM IV p.1598. Plesch p.368. Pritzel 7276.

1348

PRINS, JOHANNA M.C. *The rose to-day; a selection of modern varieties.* Painted by Johanna M.C. Prins. Introduced by Gerd Krüssmann. London: Ariel Press, [1966].

[65]p. illus., 18 col. plates. 41 cm. Paper over boards, dust-jacket.

1349

REDOUTÉ, PIERRE JOSEPH (1759-1840). *Fruits and flowers; comprising twenty-four plates selected from "Choix des plus belles fleurs et des plus beaux fruits," together with the original preface by P.J. Redouté.* Edited and introduced by Eva Mannering. New York: Crown Publishers, [c1956].

xvi, 24 p. illus., col. plates. 41 cm. Beige cloth binding.
Nissen BBI 3 1591 (Brit. ed.)

1350

REPTON, HUMPHRY (1752-1818). *The landscape gardening and landscape architecture of the late Humphry Repton, esq., being his entire works on these subjects.* By J.C. Loudon. A new edition: with an historical and scientific introduction, a systematic analysis, a biographical notice, notes, and a copious alphabetical index. Originally published in one folio and three quarto volumes, and now comprised in one volume octavo. Illustrated by upwards of two hundred and fifty engravings. London: Printed for the editor, and sold by Longman & co., 1840.

xxiv, 619 p. illus., plans, port. 23 cm. Orange library binding.

* 1351

Roses. [By] Eric Bois [and] Anne-Marie Trechslin. Translated by Jean W. Little. [Edinburgh]: Nelson, 1962.

128 p. 71 mounted col. illus. 30 cm. Beige cloth binding.
Plesch p.147.

* 1352

ROYAL HORTICULTURAL SOCIETY, LONDON. *Transactions of the Horticultural Society of London.* 3d edition. London: Printed by W. Bulmer & C.C. Row; sold by J.H. Piccadilly, 1820-30.

7 v. illus., 132 plates (70 col.) 29 cm. Vols.2-7 have imprint: Printed by W. Nicol, successor to W. Bulmer, C. Row. Vol.3: 2d edition; v.4-7 without edition note. Armorial bookplate of Samuel Ryland in each vol. Half brown calf/boards, gilt ornamentation, marbled endpapers, marbled edges.

*1353

Scriptores rei rusticae. Libri de re rvstica. M. Catonis lib. I. M. Terentii Varronis lib. III. L. Ivnii Moderati Colvmellae lib. XII. Eiusdem de arboribus liber separatus ab alijs, quare autem id factum fuerit: ostenditur in epistola ad lectorem. Palladii Lib. XIIII. De duobus dierum generibus: simulq; de umbris, & horis, quae apud Palladium, in alia epistola ad lectorem. Georgij Alexandrini enarrationes priscarum dictionum, quae in his libris Catonis: Varronis: Columellae. [Venetiis, in aedibvs Aldi, et Andreae soceri mense maio, 1514].

34 p.l., 308 leaves. 21 cm. First Aldine edition. Last preliminary leaf blank. LSU copy imperfect: leaves 117 and 123 incorrectly imposed; leaf 210 incorrectly numbered 220. Manuscript notes throughout the text. Armorial bookplate of E.M. Vellum binding.
BM I p.327 (dif. eds.) Hunt 15. Pritzel 1606.

1354

SITWELL, SACHEVERELL (1897-). *Old garden roses.* By Sacheverell Sitwell, and James Russell. With 8 reproductions from paintings by Charles Raymond, and a foreword by Graham Thomas. London: George Rainbird, in association with Collins, 1955- .

52 p. illus. front. 8 col. plates. 43 cm. Limited to 2160 copies. LSU owns copy no.821. Grey paper over boards, pink label on front cover, illustrated endpapers.

*1355

THOMPSON, ROBERT (1798-1869). *The gardener's assistant: practical and scientific: a guide to the formation and management of the kitchen, fruit, and flower garden, and the cultivation of conservatory, green-house, and hot-house plants; with a copious calendar of gardening operations.* London: Blackie, [1859].

xv, 774 p., 12 leaves of plates. illus. (some col.) 26 cm. LSU copy imperfect: plates 3-6, 8-12 wanting. Half red morocco/cloth binding.

1356

TRIGGS, HARRY INIGO (1876-). *The art of garden design in Italy. Illustrated by seventy-three photographic plates reproduced in collotype, twenty-seven plans and numerous sketches in the text taken from original surveys and plans specially made by the author and twenty-eight plates from photographs by Mrs. Aubrey Le Blond.* London, New York: Longmans, Green, 1906.

xii, 135 p. illus., 128 plates (incl. front., 27 plans). 45 x 34 cm. Original burnt orange cloth binding.

1357

URQUHART, BERYL LESLIE, ed. *The Camellia.* Sharpthorne, Sussex: The Leslie Urquhart Press, [c1956-].

v. col. plates. 47 cm. Reproductions from paintings by Raymond Booth and Paul Jones. LSU owns vols.1-2. Vol.1, half green cloth/paper over boards, v.2, grey cloth binding.
Plesch p.445.

1358

URQUHART, BERYL LESLIE, ed. *The rhododendron.* With 18 reproductions from paintings by Carlos Riefel. Sharpthorne, Sussex: Leslie Urquhart Press, [c1958].

40 p. illus., 18 col. plates. 40 cm. "The line drawings in the botanical text are by Susanne Kolasse." Red cloth binding, map on endpapers.
Nissen BBI 3 2029n.

1359

VASEY, GEORGE (1822-1893). *The agricultural grasses and forage plants of the United States; and such foreign kinds as have been introduced. With an appendix on the chemical composition of grasses, by Clifford Richardson, and a glossary of terms used in describing grasses.* A new, revised, and enlarged edition, with 114 plates, published by authority of the Secretary of Agriculture. Washington: Government Printing Office, 1889.

148 p. 114 plates. 23 cm. [U.S. Dept. of Agriculture. Report no.32, rev.] At head of title: U.S. Department of Agriculture. Botanical Division. Special bulletin. LSU copy imperfect: plates 105, 112-114 wanting. Stamped on p.[3]: J.B. Garrett, County agent, St. Francisville, La. In ms. on p.5: J.B. Garrett. Brown library binding.
BM V p.2195. Nissen BBI 2041 (1st ed.)

*1360

VERSCHAFFELT, AMBROISE ALEXANDRE (1825-1886). *New iconography of the camellias, containing the figures and the descriptions of the rarest, the newest and the most beautiful varieties of this species, 1848-1860.* Translated from the French by E.A. McIlhenny. Avery Island, La.: E.A. McIlhenny, 1945.

1 p.l., ix, 318 p. 26 cm. "This verbatim translation includes in one volume all of the descriptive material contained in the original thirteen." — Preface. Dark red binding.

Gift in memory of E.A. McIlhenny by Mrs. Rosemary McIlhenny Osborn, Mrs. Pauline McIlhenny Simmons, and Mrs. Leila McIlhenny Brown.

* 1361

VERSCHAFFELT, AMBROISE ALEXANDRE (1825-1886). *Nouvelle iconographie des camellias contenant les figures et la description des plus rares, des plus nouvelles et des plus belles variétés de ce genre...* Gand: A. Verschaffelt, 1848-60.

12 v. in 14. col. fronts., col. plates. 26 cm. Each volume contains 48 colored plates. Printer's error: title page of v.2 has date 1851, corrected to 1850 in mss. Errors in numbering: in v.8, text for liv. 10, plate 4 incorrectly numbered 9 e liv., plate 4; in v.9, text for liv. 2, plate 4 incorrectly numbered, 1 e liv., plate 4; and in v.10, text for liv.10, plate 3 incorrectly numbered 9 e liv., plate 3. LSU copy imperfect: v.1, pt.2, liv. 10 (4 plates and texts); v.3, liv. 6-7, 10-11 (16 plates and texts); v.9, liv. 4-5 (8 plates and texts), liv. 6, plates 2-3, and liv. 8-12 (20 plates and texts) wanting. Bound at end of v.9: 15 extra, unidentified plates of camellias. Vol.1, pt.1: photostatic copy with typescript of title page and introd. in English. Vol.1, pt.1 A: photostatic copy of original title page and introd. in French for set in pocket in pam binder. Original red cloth binding, v.1, rebound. BM V p.2208. Nissen BBI 2056. Pritzel 9743.

Gift in memory of E.A. McIlhenny by Mrs. Rosemary McIlhenny Osborn, Mrs. Pauline McIlhenny Simmons, and Mrs. Leila McIlhenny Brown.

1362

Vick's Flower and Vegetable Garden. Rochester, N.Y.: James Vick, [1878].

166 p. illus. 6 col. plates. 24 cm. LSU lacking plate 6. Original brown cloth binding.

1363

Vick's Magazine. v.1-35, no.4; 1878-Sept. 1909. Rochester, N.Y.: James Vick, 1878-1909.

35 v. illus., plates (part col.) 24 cm. Title varies: 1878- *Vick's monthly magazine.* LSU holdings: vols. 1, 3, 5, 7; scattered nos. of vols. 8, 9, 10, 14. Bindings vary.

* 1364

WARDER, JOHN ASTON (1812-1883). *American pomology. Apples.* 290 illustrations. New York: Orange Judd and company, [c1867].

vii, 744 p. illus. 20 cm. Original green cloth binding with gilt apple on front cover.

* 1365

[*Water colour drawings of the fruits and flowers of Southern India.* n.p., 1837?].

23 col. plates. 41 cm. Watermarks on paper: J. Whatman, 1837. Classification for each illustration in manuscript. LSU copy imperfect: 4 plates wanting. Full black morocco.

* 1366

WATERHOUSE, EBEN GOWRIE (1881-). *Camellia trail.* With 21 plates in colour by Paul Jones. [Sydney]: U. Smith, [c1952].

44 p. 21 col. plates. 34 cm. "This edition is limited to 1,000 copies, each one...signed by the author and artist." LSU owns copies no.645 and no.709. Green cloth binding, gilt ornamentation.

1367

WILDMAN, THOMAS. *A treatise on the management of bees; wherein is contained the natural history of those insects; with the various methods of cultivating them, both ancient and modern, and the improved treatment of them. To which are added, the natural history of wasps and hornets, and the means of destroying them.* The third edition. London: Printed for W. Strahan and T. Cadell, 1778.

xvii, [3], 318, [7], 16 p. III fold. plates. 20.5 cm. Full contemporary calf binding. Ex libris Stanton Smith.

Gift of Mrs. Stanton Smith in memory of her husband.

* 1368

The Woodland companion; or, A Brief description of British trees. With some account of their uses. By the author of *Evenings at home.* Illustrated with twenty-eight plates. 3d ed., corrected. London: Printed [By T.C. Hansard] for Baldwin, Cradock, and Joy, and N. Hailes, Juvenile Library, London Museum, 1820.

92 p. 28 fold plates. 18 cm. Signed: J.A.
Quarter black morocco/marbled boards.

* 1369

WRIGHT, RICHARDSON LITTLE (1887-1961). *The gardener's bed-book; short and long pieces to be read in bed by those who love husbandry and the green growing things of earth...* [1st edition]. Philadelphia, London: J.B. Lippincott Co., 1929.

341 p. 20 cm. Illustrated lining-papers. In ms. on fly leaf: M. Marie Harman. Original green cloth binding.

PART IX

Hunting and Sports

Title page from *The National Sports of Great Britian* by Henry Thomas Alken. London, 1821.

*1370

ALKEN, HENRY THOMAS (1784-1851). *The national sports of Great Britain.* With descriptions in English and French. London: T. McLean, [1821?].

[108]p. 50 col. plates. 48 cm. Plates colored by hand. Half red morocco, cloth binding.

1371

BAKER, SAMUEL WHITE, SIR (1821-1893). *Wild beasts and their ways; reminiscences of Europe, Asia, Africa, and America.* [2d edition] London, New York: Macmillan, 1898.

xiv, 455 p. illus. front. 21 cm. Original red cloth binding, gilt ornamentation.
BM I p.90 (1890 ed.)
Gift of Emile C. Freeland.

*1372

BOOSEY, EDWARD J. *Foreign bird keeping; forty years' experience in their breeding and management.* With photographs by Alec Brooksbank; foreword by the Rt. Hon. The Viscount Chaplin. London: Cage Birds, [1956?].

342 p. illus. 26 cm. In ms. on fly leaf: Leila McIlhenny Brown. Blue cloth binding.

1373

FRIEDRICH II, EMPEROR OF GERMANY (1194-1250). *De arte venandi cum avibus.* Ms. Pal. Lat. 1071, Biblioteca Apostolica Vaticana. [Erstaufl.] Graz: Akadem. Druck- u. Verlagsanst., [1969].

2 v. 37cm. (Codices e Vaticanis selecti quam simillime expressi, v.30[i.e. 31]). Books 1 and 2 of *De arte venandi cum avibus*, as edited by King Manfred. Vol.1, quarter brown suede; v.2, full brown suede; issued in a case.

1374

[GRANDJEAN, J.J.] *Secrets, anciens et modernes de la chasse aux oiseaux. Contenant la manière de fabriquer les filets, les divers piéges, appeaux, etc.; l'histoire naturelle des oiseaux qui se trouvent en France; l'art de les élever, de les soigner, de les guérir, et la meilleure méthode des les empailler. Ouvrage orné de huit planches, renfermant plus de 80 figures.* Par M. J.-J.G..., Amateur. Paris: À la librairie encyclopédique de Roret, 1850.

268 p. illus. front., 7 plates. 14 cm. Quarter green calf/ marbled boards, marbled endpapers.
Yale p.115.

* 1375

GREENWOOD, JAMES. *Wild sports of the world; a book of natural history and adventure.* With wood cuts from designs by Harden Melville and William Harvey; portraits of celebrated hunters from original photographs and 8 coloured illustrations by Harrison Weir, Zwecker and others. London: Ward, Lock, and Tyler, [188-?].

xxii, 426 p., [8] leaves of plates. illus. (some col.), ports. 21 cm. Original green cloth binding, gilt ornamentation.

*1376

HUNT, LYNN BOGUE. *Our American game birds.* From paintings by Lynn Bogue Hunt. Descriptive text by Edward Howe Forbush. Wilmington, Del.: E.I. Du Pont de Nemours, [c1917].

1 l, 18 col. plates. 38 cm. Descriptive text on verso of plates. Half brown cloth/faded marbled boards.

1377

LLOYD, LLEWELYN (1792-1876). *The game birds and wild fowl of Sweden and Norway. With an account of the seals and salt-water fishes of those countries.* 2d edition, with map woodcuts, and chromo illustrations. M. Korner, illustrator. Day and Son, lithographers. London: Frederick Warne, 1867.

xx, 599 p. illus., plates (part col.), fold. col. map (in pocket). 27 cm. Full green morocco with original green cloth covers preserved, gilt ornamentation.
Anker 312. BM III p.1159. Nissen VBI 569. Yale p.173.
Zimmer p.403.

1378

Management of migratory shore and upland game birds in North America. Edited by Glen C. Sanderson. Washington, D.C.: Published by the International Association of Fish and Wildlife Agencies in cooperation with the Fish and Wildlife Service, U.S. Department of the Interior, 1977.

358 p. illus., maps. 27 cm. Blue cloth binding, pictorial endpapers, dustjacket.
Gift of Glen C. Sanderson.

*1379

MAVROGORDATO, J.G. *A hawk for the bush; a treatise on the training of the sparrow-hawk and other short-winged hawks.* With illustrations by G.E. Lodge. [Revised edition]. London: Spearman, [c1973].

xviii, 206 p. illus. 26 cm. Beige cloth binding, dustjacket.

* 1380

MCILHENNY EDWARD AVERY (1872-1949). *The wild turkey and its hunting.* Illustrated from photographs. Garden City, N.Y.: Doubleday, Page & company, 1914.

xi, 245, [1]p. front., illus., plates. 21cm. "This work was begun by Chas. L. Jordan." "The turkey prehistoric" and "The turkey historic" (pp.26-103) by Dr. R.W. Shufeldt. Stamped on lining paper and front.: Larry Merovka, 533 Solano Drive, N.E., Albuquerque, N. Mex., 87108. Black cloth binding.
Yale p.180.

* 1381

MILLAIS, JOHN GUILLE (1865-1931). *The wildfowler in Scotland.* With a frontispiece in photogravure after a drawing by Sir J.E. Millais, bart., 8 photogravure plates, 2 coloured plates, and 50 illustrations from the author's drawings and from photographs. London, New York: Longmans, Green, 1901.

xv, 167 p. illus. 31 cm. Title page in red and black. Half vellum/paper over boards.

1382

O'NEIL, TED. *The muskrat in the Louisiana coastal marshes; a study of the ecological, geological, biological, tidal and climatic factors governing the production and management of the muskrat industry in Louisiana.* With photographs by the author and maps and diagrams by W.H. McBride. New Orleans: Federal Aid Section, Fish and Game Division, Louisiana Dept. of Wild Life and Fisheries, 1949.

xii, 152p. plates, maps (part fold., 1 fold. col. in pocket). 26 cm. Author's autograph presentation copy. Green cloth binding, dustjacket.
Gift of Ted O'Neil with dedication to the E.A. McIlhenny Natural History Collection, in memory of Mr. Ned McIlhenny.

1383

ROOSEVELT, ROBERT BARNWELL (1829-1906). *Florida and the game water-birds of the Atlantic coast and the lakes of the United States, with a full account of the sporting along our sea-shores and inland waters, and remarks on breech-loaders and hammerless guns.* New York: Orange Judd company, 1884.

443p. incl. front. (port.) illus., plates. 19 cm. Original green decorated cloth binding.
Yale p.245.

* 1384

SALVIN, FRANCIS HENRY. *Falconry in the British Isles.* By Francis Henry Salvin and William Brodrick. London: J. Van Voorst, 1855.

147 p., 24 leaves of plates. col. illus. 30 cm. Bookplate mounted on p.[2] of cover: F.F. Whitehead. Original green cloth binding with gilt falcon on front cover.
BM IV p.1795. Nissen VBI 147. Zimmer p.541.

* 1385

SCHLEGEL, HERMANN (1804-1884). *Traité de fauconnerie.* By Hermann Schlegel et A.H. Verster de Wulverhorst. Leiden: Chez Arnz, 1844-53.

90, vi, p., [16] leaves of plates, col. illus. 72 cm. Engraved illustrated title page. Issued in parts. Pencilled sketches laid in with plates "Le vol du Heron 1" and "Le vol du Heron 2." Half red morocco/marbled boards.
BM IV p.1839. Nissen VBI 832.

1386

SCHLEGEL, HERMANN (1804-1884). *Traité de fauconnerie.* Par H. Schlegel et A.H. Verster de Wulverhorst. London: Pion Ltd.; New York: Johnson Reprint Corp., 1979.

90, vi p., [16] leaves of plates. illus. 69 cm. Reprint of the original edition, 1844-53, Chez Arnz, Leiden. "This edition is limited to 270 copies of which 250 are for sale. This is number 58." "An introduction to H. Schlegel, and A.H. Verster de Wulverhorst *Traité de fauconnerie*," by Gavin Bridson: [8] p. laid in. *Catalogue raisonne des ouvrages de fauconnerie*: vi p. Half dark red morocco/cloth binding.
BM IV p.1839 (orig. ed.) Nissen VBI 832 (orig. ed.)

* 1387

[WILLIAMSON, THOMAS, CAPTAIN . *Oriental field sports.* London: E. Orme, 1807?].

xii, 455 p. col. front., 40 col. plates. 44 cm. Title page wanting; title from spine. Plates are signed: Williamson & Howitt; J. Clark etched. Ex libris Robert Moore. Presentation copy to L. Kemper Williams from Robert Moore, with his signed autograph inscription. Red morocco, restored, gilt oramentation, gilt edges.
BM V p.2328 (1807 ed.) Nissen ZBI 4416 (1808 ed.)
Gift of the L. Kemper Williams Foundation.

* 1388

WILLIAMSON, THOMAS, CAPTAIN . *Oriental field sports; being a complete, detailed, and accurate description of the wild sports of the East; and exhibiting, in a novel and interesting manner, the natural history of the elephant, the rhinoceros, the tiger, the leopard, the bear, the deer, the buffalo, the wolf, the wild hog, the jackall, the wild dog, the civet, and other undomesticated animals; as likewise the different species of feathered game, fishes, and serpents. The whole interspersed with a variety of original, authentic, and curious anecdotes, which render the work replete with information and amusement. The scenery gives a faithful representation of that picturesque country, together with the manners and customs of both the native and European inhabitants. The narrative is divided into forty heads, forming collectively a complete work, but so arranged that each part is a detail of one of the forty coloured engravings with which the publication is embellished. The whole taken from the manuscript and designs of Captain Thomas Williamson. The drawings by Samuel Howett, made uniform in size and engraved by the first artists under the direction of Edward Orme.* London: Printed by Thomas McLean, 1819.

3 p.l., ii, 146 p. 40 hand col. plates. 47 x 60 cm. Half black morocco/illustrated boards.
BM V p.2328 (1807 ed.) Nissen ZBI 4416 (1808 ed.)

Index to the Catalogue

All authors, titles, artists, engravers, lithographers, and selected printers and publishers are included in the index according to the *Catalogue* entry number.